Preparation for Aging

Preparation for Aging

Edited by

Eino Heikkinen
Jorma Kuusinen and
Isto Ruoppila
University of Jyväskylä
Jyväskylä, Finland

Springer Science+Business Media, LLC

Library of Congress Cataloging-in-Publication Data

On file

Proceedings of the 17th International Congress of the International Association of the Universities of the Third Age (IAUTA), held August 12–14, 1994, in Jyvaskyla, Finland

ISBN 978-1-4613-5815-2   ISBN 978-1-4615-1979-9 (eBook)
DOI 10.1007/978-1-4615-1979-9

© 1995 Springer Science+Business Media New York
Originally published by Plenum Press in 1995
Softcover reprint of the hardcover 1st edition 1995

10 9 8 7 6 5 4 3 2 1

PREFACE

The challenges presented to society as a result of the demographic transition currently taking place involve many concrete, multidimensional, and contextual issues that require focused attention from both the multidisciplinary scientific community and society at large. The social and educational movement "The University of the Third Age" (UTA), which started in Toulouse in 1973 and was spearheaded by Professor Pierre Velas, represents a phenomenon that indicates the need for new and creative educational and research activities in attempting to improve quality of life among older adults.

In two decades the idea of the University of the Third Age has spread to all continents and the number of UTAs has increased exponentially, amounting at present to several thousand units with varying structures and programs. The International Association of the Universities of the Third Age (IAUTA) connects the local universities under an umbrella that, among other activities, sponsors international congresses.

The 17th International Congress of IAUTA was held in Jyväskylä, Finland, on August 12-14, 1994. About 800 people from 31 countries took part in the event. The purpose of the meeting was to provide a forum for discussion on the broad theme of "Preparation for Aging." The underlying idea was that the future grows out of our present and past.

In order to be able to prepare ourselves for the future we have to be able to answer a number of key questions: To what extent is the quality of life in elderly people determined by their earlier experiences? Given the fast pace of sociocultural change in modern society, what sort of plans can people make for life in retirement? What chances do elderly people, as subjects of their own lives, have to lead satisfactory and independent lives?

The presentations given at the congress examined both the theoretical issues related to the meaning of life in old age and foundations of lifelong learning as well as practical questions focusing on the utilization of the existing body of knowledge in developing cultural, social and health services for older adults in order to help them grow old successfully.

This volume consists of selected contributions to the 17th IAUTA Congress. Given the broad theme of the meeting, the multidisciplinarity of the research interest, and the diversity of the educational activities within different UTAs, this publication necessarily covers the whole range of topics regarded as important for discussions focusing on the preparation for aging.

It is hoped that this volume will carry forward the main ideas of the UTA, provide stimulus for further research, and give new perspectives and practical advice for arranging educational activities, creating communication and networks within the UTA movement. It is understood that both aging and preparation for it are socioculturally constructed concepts that should be made visible in order to gain insights into the ways in which the parties involved understand them and to help us understand why programs for successful retirement succeed or fail.

We wish to acknowledge the support of the Executive Committee of the International Association of the Universities of the Third Age, who helped to arrange the 17th International Congress in Jyväskylä, Finland. We very much appreciate the collaboration with Dr. Hana Hermanova from the WHO Regional Office for Europe in the planning and implementing of the Congress.

The financial support received from the Ministry of Education, Finland, is also gratefully acknowledged. Without this support the Congress could not have taken place.

Given the limited local resources in terms of both manpower and funding, the only way of arranging big meetings is coordination and cooperation. The parties involved include the City of Jyväskylä, the Finnish Centre for Interdisciplinary Gerontology at the University of Jyväskylä, the Congress Office of the University of Jyväskylä, and the University of the Third Age, University of Jyväskylä. Their contribution is gratefully acknowledged.

More than one hundred individuals participated in the planning and implementing of the Congress, including the members of the organizing and program committees, the staff of the various scientific departments of the University of Jyväskylä, the personnel of the Leisure Office of the City of Jyväskylä, and the students of the University of the Third Age, University of Jyväskylä. We wish to direct our warmest thanks to all of them. Finally we would like to thank Ms. Pirjo Koikkalainen for her skillful technical assistance in the preparation of this volume.

Eino Heikkinen
Jorma Kuusinen
Isto Ruoppila
Jyväskylä

CONTENTS

Life Long Learning and Retirement

1. Meaning and Late Life Learning 1
   Harry R. Moody

2. The Third Age and the Disappearance of Old Age 9
   Peter Laslett

3. Life after Work: Lines, Boundaries, and Spaces 17
   Tom Schuller

4. Emergent Challenges for Universities of the Third Age 27
   Brian Groombridge

5. Retirement: A Truncated Rite of Passage 39
   Christian Lalive d'Epinay, in collaboration with Jean-François Bickel

6. The Role of a Preparation for Retirement in the Improvement of the Quality of
       Life for Elderly People 55
   René Jeanneret

7. In Search of the Meaning of Education: The Case of Educational Generations in
       Finland 63
   Ari Antikainen, Jarmo Houtsonen, Hannu Huotelin, and Juha Kauppila

8. Lifelong Learning Experiences from Norway 73
   Reidun Ingebretsen and Tor Endestad

9. If I Had My Life to Live over Again... 79
   T. I. Tikkanen and J. Kuusinen

Health and Quality of Life

10. The New Public Health Approach to Improving Physical Activity and Autonomy
       in Older Populations 87
   John B. McKinlay

11. Healthy Aging: Utopia or a Realistic Target? 105
    Eino Heikkinen

12. Adding Life to Years! Promoting Quality of Life in an Aging Europe 121
    Chris Todd

13. Health Related Quality of Life as an Outcome Measure for Health Care of
    Elderly People: The Emperor's New Clothes 129
    Chris Todd

14. Health-Related Quality of Life in Old Age: How to Define It, How to Study? 139
    Marja Jylhä

15. Health Related Quality of Life in Old Age: International Approach in Developing
    the LEIPAD Questionnaire 145
    Anna-Mari Aalto, Arja R. Aro, Anu Hämäläinen, and Jouko Lönnqvist for the
    steering group of the LEIPAD -project

16. Disability and Quality of Life in Old Age 151
    Luigi Ferrucci, Stefania Bandinelli, Francesca Cecchi, Bernardo Salani, and
    Alberto Baroni

17. Research and the Promotion of Quality of Life in Older Persons in the
    Netherlands 155
    Dorly J.H. Deeg

18. Does Work Stress Enhance the Rate of Aging? 165
    Willem J. A. Goedhard

Gender, Generation and Aging

19. Interrelations between Generations in Historical Perspective 175
    Birgitta Odén

20. Gender and Aging 181
    Christine Castelain Meunier

21. Gender, Aging, and Quality of Life 187
    Gerald E. McClearn, Pamela J. Maxson, and Debra A. Heller

22. Collectivism, Individualism and Grandchild-Grandparent Relations 191
    Helena Hurme

23. Mid-Life: Opening Remarks on Prevention 195
    Hélène Reboul

24. Life-Style and Its Determinants in Two Cohorts in the Elderly 199
    Pertti Pohjolainen

Pension Systems and Attitudes to Retirement

25. Psychological Issues of Aging and Work 205
    Pekka Huuhtanen

26. A New Concept for Productive Aging at Work 215
    Juhani Ilmarinen

27. What If the Disability Pension Application Is Denied? 223
    Raija Gould

28. How Will Finnish Pensions Be Working in the Future? 229
    Simo Forss

29. Life Situation as a Factor Explaining Retirement 233
    Jouko Lind

30. The Process of Early Retirement among the 55-Year Old Finnish Urban
    Population 239
    Aira A. Uusimäki, Ulla Rajala, Sirkka Keinänen-Kiukaanniemi, Hannu
    Virokannas, and Sirkka-Liisa Kivelä

Programs for Successful Retirement

31. Educating Health Professionals in Gerontology: A Canadian Perspective 245
    A. C. Beckingham

32. Preparation for Aging—the Role of the UTAs—Ten Reports 265

33. Preparing for Retirement—Good Housing in Old Age? A Review of the Provision
    of Housing and Care for Old People in 7 Systems 279
    John William Murray

34. Aging Well: European Health Programme for Older People 285
    Sally Greengross

35. Retirement Preparation in Subjects of Working Age 289
    Fiorella Marcellini and Norma Barbini

Index 293

MEANING AND LATE LIFE LEARNING

Harry R. Moody

Brookdale Center on Aging of Hunter College
New York City, New York

Let me begin with a semantic question: How do we describe the enterprise? What do we call the subject of this conference for which we are gathered here in Finland? Several labels suggest themselves.

First, there is "older adult education", a phrase that involves an euphemism "older adult" (older than what?). The phrase is an evasion of the truth about the last stage of life: not older, but final.

A second phrase suggests itself: "Third Age Education," the title of this conference. But the phrase is evocative only because the idea of the "third age" is itself ambiguous. For example, is there, as some have suggested, a "fourth age" (decrepitude)?

Finally, I come to "late life learning", a term I prefer because it designates at least where learning and education fit in to the life course as a whole: namely, late in the sequence, in the last act, in much the same way that "late style" refers to the distinctive style of artists in their old age (Münsterberg, 1983).

LEARNING IN LAST STAGE OF LIFE

In approaching this subject there is a need for honesty. We need to be candid in the matter of terminology in the same spirit as Confucius, who called for "the rectification of names:" that is, to speak of things as they really are. Indeed, if we were honest about describing this enterprise we would call it "end-of-life learning". But I will not use that phrase because "end-of-life" unavoidably evokes the aura of terminal illness. And what is of interest here is not so-called "death education"—yes, there really is a journal by that name—but rather education for living during the last stage of life.

To grasp the truth about late life learning we must keep in mind the aphorisms of two great philosophers, aphorisms which seemingly contradict each other. First, there is Plato: "Philosophy is the preparation for death" And, second, Spinoza: "The Freeman thinks of nothing less than of death". How can we reconcile these two perspectives on life and death? The wise have said that we repair the past and prepare for the future by living in the present. That spirit is surely what Spinoza was invoking in his statement. At the same time, Plato was right to speak of philosophy—more broadly, the search for meaning—as an existential enterprise, an activity of supreme seriousness, undertaken in the finality of death, as the example of Socrates taught Plato himself.

Preparation for Aging, Edited by E. Heikkinen *et al.*
Plenum Press, New York, 1995

Let me state again, late life learning is not a meditation about death but about life At the same time, it is learning undertaken when the clock is ticking, learning, so to speak, at the eleventh hour, in the last act of the play of life

The Denial of Finitude

I have stated the matter so badly because I believe that most of what we think of as "older adult education" today is, on the whole, an evasion and a conspiracy of denial a fruitless attempt to convince old people that they can remain young forever What better way to foster that illusion than to "keep up with the times" by being well-informed? What better way to relive youth than to return to a college campus? What better way to be constantly changing and growing than to learn new things?

Let us not underestimate the problem We are dealing here with a very powerful illusion, all the more mischievous because it is so widespread and because it clothes itself in the garment of "humanistic" self-fulfilment growth, change, openness to new ideas, and so on But at bottom, it remains an illusion It is the persistent illusion of our youth-intoxi-cated culture, which has convinced even the elderly that, to feel alive, they must feel young And late life learning, including the work that you and I and others in this room have devoted our lives to, contributes to that illusion

The psychologist Carl Jung understood very well the danger of this illusion In his classic essay, "The Stages of Life" Jung wrote

> "Aging people should know that their lives are not mounting and expanding, but that an inexorable inner process enforces the contraction of life For a young person it is almost a sin, or at least a danger, to be too preoccupied with himself, but for the aging person it is a duty and a necessity to devote serious attention to himself" (Jung, 1971, p 17)

It is this fundamental truth—finitude—that must remain the basis for any question about "meaning" and late life learning And it is this truth about finitude that is so difficult for the spirit of modernity to acknowledge Yet without remembering it, I believe, it becomes impossible to imagine the genuine personal growth, the real transformations that late life learning can make possible Jung speaks of such changes quite explicitly

> ' The worst of it all is that intelligent and cultivated people live their lives without even knowing of the possibility of such transformations Wholly unprepared, they embark upon the second half of life Or are there perhaps colleges for forty-year-olds which prepare them for their coming life and its demands as the ordinary colleges introduce our young people to a knowledge of the world? No thoroughly unprepared we take the step into the afternoon of life worse still, we take this step with the false assumption that our truths and ideals will serve us as hitherto But we cannot live the afternoon of life according to the programme of life's morning, for what was great in the morning will be little at evening and what in the morning was true will at evening have become a lie (Jung 1971 p 17)

The truths and ideals of the morning of life, alas, include our dominant conception of what education is all about If we take Jung seriously, we have to consider the possibility that education itself—including the much celebrated pursuit of "lifelong learning"—can itself become a lie The reason why the lie perpetuates itself is that we have recycled our existing institutions of higher learning—lectures, readings, discussions—to make them accessible to the elderly What we have not done is to think through the rationale—the meaning, if you will—of late life learning in the first place Thus, the educational enterprise still remains, not a college for seventy year olds or even forty year olds, but a college for twenty year olds

LEARNING FOR ITS OWN SAKE?

Now before I go further I must acknowledge that there is a view of "liberal education" which sees the matter in very different terms from what I have stated so far. There are well-informed, well-intentioned people who believe in something called "learning for its own sake". They believe that late life learning can, and should be, the crown of life, a fulfilment of a lifetime of intellectual growth and discovery. Their argument, if I understand it, is that there is no reason so seek any purpose or meaning in learning beyond the activity itself. Late life learning, in this view, would become analogous to art or to sport: it requires no teleology, no meaning beyond the activity to justify itself.

This outlook can be traced back to Aristotle, and it commands our respect. The functioning of the intellect, or of the body or the feelings, as the case may be, simply represents the fulfilment of a power (*dynamis*) distinctive to the human being. Happiness—*eudaimonia*, in Aristotle's phrase—self-actualization, in the modern jargon—means nothing less and requires nothing more. We are at our happiest at those moments when we score the goal, when we feel the climax of a concerto, when we understand the solution to an intellectual problem. These are moments of "learning for its own sake". What we have here, finally, is an aesthetic view of the purpose or meaning of late life learning, an activity justified at the end of life no less than at the beginning. (For a philosophical account of aging and the "stages of life" deriving from Aristotelian premises, see Norton, 1976).

This aesthetic view of late life learning will be congenial to the intellectual class, to professional academicians, because it puts their—shall I say our?—preferred way of life at the center of attention. It settles the philosophic debate without further ado. If "learning for its own sake" is a sufficient meaning for late life education, then to talk of "meaning" in any other sense is superfluous. Indeed, in a secular, pluralistic society, how can we speak intelligibly about the meaning of life in any case? (Moody, 1986).

The Idols of the Marketplace

But let us go further and ask why this outlook finds such resonance in our own society. Here we need to consider the political economy, the educational marketplace, and the role of "learning for its own sake" as an ideological rationale for adult education. The aesthetic view of late life learning—"learning for own sake"—will also appeal to another class, the educational entrepreneurs, the people who plan and promote programs. Why is that so? Because "learning for its own sake" relieves us of the responsibility for making decisions about what learning is worth while. Each individual must decide the question alone, as a consumer in the educational supermarket. One person will choose to explore opera, another to learn Latin, still another to explore computers, or to take up skiing. And what does the administrator, the educational entrepreneur say about those preferences? The answer is simple. Let a hundred flowers bloom, let the supermarket be stocked with every kind of good the older learner demands.

This view was familiar enough to Plato, who would have described it as the rule of unchecked appetites—preferences, in the modern jargon—arising in the lower part of the soul. But following Plato we can go further in our critique. The intellectual supermarket view of education is ultimately demeaning because it serves to reduce the activity of teaching and learning to a kind of *techne*: namely, a skill of marketing or appealing to whatever brings customers through the door. We see the consequences of that outlook around us all the time, in what Habermas describes as "the colonization of the life-world".

The danger is that we will not recognize the educational supermarket for what it is. The problem is that individual choice and learning for its own sake sound like lofty respect for individual differences. But the result is to avoid any discussion of meaning or purpose. To sidestep that discussion is fatal because it hastens the commodification of higher learning that will rob the university of its soul. The process, unfortunately, is already well advanced. The ethos of the modern university has long inclined to a view of learning that is instrumental: learning as a means to the end of economic productivity and human capital formation.

Critical Theory

We can appreciate the matter better if we look at the world of higher education from the standpoint of the Critical Theory of Jürgen Habermas (McCarthy, 1978). Habermas distinguishes three kinds of human interests: instrumental, hermeneutic, and emancipatory. An instrumental interest embodies the aim of domination over nature, an outlook familiar in modern science, and increasingly exercises hegemony over all intellectual life. But there is also a hermeneutic interest, which could include the goal of understanding or self-expression through the arts and humanities. This hermeneutic interest is addressed in some traditional views of liberal education as character formation: what the Germans call *Bildung*. Although it will sound nostalgic or out of date, there is a constituency for liberal education addressed to issues of character or virtue. Finally, there is what Habermas calls an emancipatory interest, which corresponds to a kind of learning that promotes empowerment of freedom. It is this last type of interest which is captured well by Rosenmayr in his book on old age, called provocatively, *Die Späte Freiheit*,"the late freedom" (Rosenmayr, 1983).

When we think of the "third age" or the last stage of life, we must put questions of meaning at the center of the enterprise of learning. "Meaning" in this respect entails a kind of transcendence or "going beyond" whatever we have taken ourselves to be in earlier stages of life. That is the stance implicit in what Jung speaks of as the transformations that are possible in later life.

MODELS OF TRANSCENDENCE

Transcendence takes different forms depending upon whether our focus is on past, present or future. For example, a past-oriented approach to late life learning would put life-review or autobiographical consciousness at the center of educational work. Here the remembered past becomes raw material to be shaped or recast into new forms. Learning builds upon the past but at the same time opens up new possibilities of what the past means. Moreover, learning focused on the past need not be subjective or preoccupied with the self. It can be a vehicle for self-transcendence. For instance, many older people become interested in genealogy and family history as part of an effort to locate an individual story in the larger narrative that binds generations.

There is equally a mode of transcendence focused toward the future. This mode of transcendence was given the name generativity by Erik Erikson in his famous account of the stages of human development (Erikson, 1982; Erikson et al., 1986). The psychologist John Kotre has suggestively transmuted Erikson's idea into the phrase "outliving the self" (Kotre, 1984). Kotre's phrase is felicitous because it evokes the dialectical truth of both mortality and hope. To outlive the self means to transcend the despair of transitory existence by projecting ourselves outward, by investing ourselves in ideal objects that survive us: children, works of art, or accomplishments in the public world. Transcendence through the future is the watchword of modernity and will therefore always have a special attractiveness today. We can for that reason appreciate the appeal of "productive aging", a phrase heard more and more in gerontology circles in the U.S.A.

Finally, there is a mode of transcendence focused on the present, a form of consciousness that we see in artistic contemplation, in meditation and prayer, in forms of life where "being" rather than "doing" are at the center of the existence. Although far removed from the spirit of modernity, it may be that transcendence-in-the-present constitutes one of those "gifts reserved for age" that have much to teach people of all generations.

I note that all three modes of transcendence—life-review, generativity, and contemplation—are very different from "learning for its own sake". The point is that not just any kind of learning will do. Acquiring more information or keeping up with current events, for example, can be vehicles for distraction, for evading the deeper questions of life. Is this the reason why television sets are kept on so often in nursing homes? To distract residents from the task of transcendence—the search for meaning—that remains the goal of the last stage of life?

The point of this question is not to suggest that every older learner ought to be a philosopher. Many forms of learning are vehicles of transcendence. Indeed, the ultimate questions can never be far from the surface in the last stage of life. What gnaws at the heart of the very old is a disturbing question, a question framed in the title of the famous book by American geriatrician Robert Butler's *Why Survive?* (Butler, 1975).

The Search for Meaning

Why, indeed? I heard that question asked many times in the last years of his life by a dear friend, Larry Morris, who lived in our house with my family and who died two years ago at the age of 97. Larry often wondered and asked me why God had kept him alive on this earth so long. I had no answer for him. But I remembered Jung's comments in the essay I quoted earlier:

> "A human being would certainly not grow to be seventy or eighty years old if this longevity had no meaning for the species The afternoon of human life must also have a significance of its own and cannot be merely a pitiful appendage to life's morning" (Jung, 1971, p 17).

In his last years Larry Morris, bedbound, always continued to learn. His curiosity about the world and about people was endless. He regularly read the *New Republic*, a prominent American news magazine on which he had worked when a young man. Alongside his bed he kept a copy of St. Augustine's *Confessions*, which he was rereading. In a nearby room stood an IBM-386 computer, which he bought as a gift for my children and which he urged me to explain to him. It is this spirit of inquiry, vigilant but never vain, that reminded me of the aged Goya, who inscribed at the bottom of one of his late paintings *"Aun aprendo"* (I am still learning). In Larry Morris's world, now shrunk to a single room and a few books, we can see, not acts of casual curiosity, but stages on life's way, icons of life-review, generativity, and contemplation: the magazine of past and present, the gift to successor generations, the text of classic autobiography that opens out into prayer and meditation.

Old Age: The Undiscovered Country

It is at this "teachable moment" when the spirit of late life learning takes flight. Florida Scott-Maxwell understood it and expressed it in her marvelous journal of old age, *The Measure of My Days*. There she writes:

> "Another secret we [old people] carry is that though drab outside—wreckage to the eye, mirrors a mortification—inside we flame with a wild life that is almost incommunicable In silent, hot rebellion we cry silently—'I have lived my life haven't I? What more is expected of me?'. This we do with a certain haughtiness, realising now that we have reached a place beyond resignation, a place I had no idea existed until I had arrived here" (Scott-Maxwell, 1979)

This is a recurrent theme for Florida Scott-Maxwell: old age is an undiscovered country, a mysterious stage of life where something great is demanded of us: a task, a challenge—that is to say, something quite different from distraction or entertainment, from the "leisure time activities" that are described as the proper agenda for later life.

"What more is expected of me?" Can we have the courage to hear, and to ask, that question? Ladies and gentlemen, the occasion for this conference is a celebration of late life learning, of education for the third age. But we are gathered here in the presence of a mystery, an undiscovered country whose boundaries we have barely begun to explore. As the historian Peter Laslett reminds us, the "third age" is the stage of life whose meaning has yet to be determined, at least on a collective level, by society as a whole (Laslett, 1991). Laslett notes that the existence of large numbers of older people is an event without precedent in history. The meaning of the third age is a matter to be shaped collectively by those in this room, by all who are alive today. That is our historical task and challenge.

CONCLUSION: THE HISTORICAL TASK

I have spoken today about the dangers we face and also about the opportunities. We must resist, I believe, the temptation to "commodify" late life learning: the danger of turning education for the third age into a kind of consumer commodity to be dispensed in the marketplace. Equally, we must resist the "post-modern" impulse to abandon the traditional idea of "stages of life" in favor an "age-irrelevant society". That impulse would convert old age into a prolongation of middle-age and thereby deny any possibility of arriving at that undiscovered country evoked for us by Florida Scott-Maxwell.

What we must do instead is to fashion an educational agenda which respects the strengths and weaknesses of old age itself; an agenda which is ruthless in its realism but lofty in its imagination of possibilities, a vision of transformations undreamed of in our conventional ideas of human development. We need institutional practices that embody what Habermas calls the hermeneutic and emancipatory category of human interest. These are what I have called the dimension of transcendence—life-review, generativity, and contemplation. The poets have said it better than I could hope to say. Reviewing her own troubled life, the poet Louise Bogan put it this way: "At first we want life to be romantic, Later to be bearable. Finally to be understandable". And in his magnum opus, the Four Quartets, T.S. Eliot wrote of the journey of life in these words:

"We shall not cease exploring
And the end of all our exploring
Will be to arrive where we started
And know the place for the first time."

The knowledge is within us already; it remains for us to discover it and to make it our own. The task of education for the third age is to empower us all to recognize what we have forgotten and to be worthy of what the last stage of life gives us a precious chance to remember.

REFERENCES

Butler, R , 1975, "Why Survive? Being Old in America", Harper & Row, New York
Erikson, E , 1982, "The Life Cycle Completed A Review", Norton, New York
Erikson, E , Erikson, J , and Kivnick, H , 1986, "Vital Involvement in Old Age", W W Norton, New York
Jung, C., 1971, The Stages of Life, in. "The Portable Jung", J Campbell, ed., Viking, New York, p 16-17

Kotre, J , 1984, *"Outliving the Self Generativity and the Interpretation of Lives"*, Johns Hopkins Press, Baltimore

Laslett, P, 1991, *"A Fresh Map of Life, The Emergence of the Third Age"*, Harvard University Press, Cambridge, MA

McCarthy, T , 1978, *"The Critical Theory of Jurgen Habermas"*, MIT Press, Cambridge, MA

Moody, H R , 1986, The Meaning of Life and the Meaning of Old Age, in *"What Does It Mean to Grow Old? Views from the Humanities"*, T Cole, and S Gadow, eds , Duke University Press, Durbam NC

Munsterberg, H , 1983, *"The Crown of Life, Artistic Creativity in Old Age"*, Harcourt Brace Javonovich, San Diego

Norton, D , 1976, *"A Philosophy of Ethical Individualism'* , Princeton University Press, Princeton

Rosenmayr, L , 1983 *"Die Spate Freiheit Das Alter - ein Stuck bewusst gelebten Lebens"*, Severin und Siedler, Berlin

Scott-Maxwell, F , 1979, *' The Measure of My Days"*, Penguin, New York, p 32

2

THE THIRD AGE AND THE DISAPPEARANCE OF OLD AGE

Peter Laslett

Aging Unit
Cambridge Group for the History of Population and Social Structure
27 Trumpington Street
Cambridge CB2 1QA
United Kingdom

THE TITLE UNIVERSITY OF THE THIRD AGE

I begin this essay by referring to the tendency in some countries, especially in France I believe, but probably also elsewhere, to avoid the title University of the Third Age for the institutions making up the association to which we all belong. They have adopted alternative titles, such as L'Université de Tous les Ages (in English, University of All Ages), and I believe there may be instances of French societies which began by calling themselves Universities of the Third Age, but which have changed to the title I have quoted, or to some other.

Now of course I applaud any intention to open university life to persons of all ages. We older people have been excluded for so long from established universities, at least in my own country, that it is not for us to exclude our juniors from our own now that we have them. We should do so I think on their application rather than making the overtures ourselves, and certainly not at the behest of Second Age professors, teachers or administrators.

Something of the same hesitation about the titles of our many associations exists in Britain, for they are similarly unwilling to be exclusive. But with us it is the elitist association with the word university which seems to be the important reason for avoiding the phrase. In general the English have tended to side-step the difficulty by using the initials, U3A, for all purposes and for all of our societies. My conviction is that any weakening of the tie between U3A's, even if it is only wanting a common name, reduces our national and international effectiveness. To me, and I hope to all my British colleagues, U3A's form the intellectual vanguard of the Third Age as a whole. U3A's are, or should be, leaders in meeting the challenge of an age-transformed society everywhere. We need each other, individually, nationally and internationally, and nothing, not even slight variations in our self-descriptions, should come in the way of that. What I assert of individual U3A's is of course intended also for national and international associations of them.

Preparation for Aging, Edited by E Heikkinen *et al*
Plenum Press, New York, 1995

LOSS OF STATUS THROUGH USE OF THE DESCRIPTION THIRD AGE

I am further informed, however, that older French persons are reluctant to use Third Age as a personal description. If they do, it would seem, they lose dignity and self-respect. In short they become subject to that process of stigmatization which has always and everywhere attached to older people as superannuated and unimportant, without influence and with little or nothing to offer, at least in the public sphere. They are most decidedly not sexy. The adoption of the description Université de Tous les Ages may indicate that le Troisième Age, the Third Age, is becoming identified in France with *la Vieillesse*, with Old Age, in fact. The Troisième Age is losing its novelty, freshness, distinctiveness.

If all this is true, then we have also to accept the following. Even in that country which gave birth to the University of the Third Age and to the concept of the Third Age itself, Old Age (Vieillesse), the social construct of Old Age which I shall be discussing here, lives on. It has simply acquired one more title by cannibalizing the phrase The Third Age.

In considering the subject at the head of this essay, we should I believe insist that all blanket expressions for the condition of being in late and later life must be critically handled, as well as the phrase Old Age itself. These blanket phrases, which include "The Elderly", "Senior Citizens", "The Retired", and so on, have ceased to be appropriate now that the vital necessity of recognizing differentiation during that lengthy phase of life has become apparent. "The Third Age" with the complementary title "The Fourth Age" should be adopted as the proper replacement of the obsolete expression Old Age, and soubriquets of an euphemistic or apologetic kind must be avoided too. "Golden Oldies" and similar phrases simply will not do.

For the Third Age is not an apologetic euphemism for those elements implied in the term Old Age which people find themselves unable to contemplate. On the contrary, it is an assertion of just those features of later life which Old Age in its ordinary usage so conspiciously fails to convey or indeed seems intended to conceal. This term of ours is emphatically not a piece of self-deception, devised and maintained to direct our attention away from the acceleration of the aging process which affects us all, or even to conceal it from us. The Third Age stands for the dignity and creativity, the social importance and public significance, the self-respect and civic virtue of older people which certainly continue indefinitely into later life, unless or until a Fourth Age of decrepitude intervenes, and often even after that.

EUPHEMISMS AND THE STIGMATIZATION OF OLDER PEOPLE

In championing the usage of the Third Age, then, we have a battle on our hands which is much more than verbal, and a battle which we may well be losing, if the reluctance to use the description Third Age is really spreading. If we do lose that battle, the Third Age will indeed go the way of all other words for unmentionable things, the most conspicuous of which are the terms in use to indicate, but never to describe, what goes on behind those doors marked Ladies and Gentlemen with all their language equivalents in every public building all over the world. Ultimately the Third Age may come to be used as an opprobrious epithet, and indeed certain expressions associated with later life have already suffered such a fate. It is decidedly an insult to be called a geriatric, just as it would be an insult to be called a toilet.

Old Age, then, and its euphemistic or jocular equivalents, get in the way of under-standing. What is more, they are dangerous to us in our Third Age position in the life-course.

This should be obvious now that we recognize that the process of stigmatization, which is what we are facing, has begun to threaten our chosen title of self-description – the Third Age It was originally selected, I imagine, because it was an anonymous, simply a numerical expression, with no reference to decline or disability and omitting old This was certainly in my own mind when I put Le Troisieme Age into English as The Third Age, and I believe I was the first ever to do so *

I was not at all confident in adopting the phrase that it would survive for very long without becoming tainted in the way I have described But on the whole I have been encouraged by its persistence in England The reports from France and French Universites du Troisieme Age have been disappointing but not surprising Nevertheless it is disconcerting to have to assume that this vital ground may have been yielded there without perhaps realizing what was at issue An indication that this may have happened is to be found even in the programme of this meeting, where my original title for this piece, *The Disappearance of Old Age*, is rendered as *La Disparition du Troisieme Age*, as if Troisieme Age is now identical with Old Age, with *vieillesse*

REJECTION OF P.C. AND OF TOTAL PROSCRIPTION OF THE EXPRESSION OLD AGE

Let us now turn to the reasons why terms like Old Age have become obsolescent and try to assess their relationship with that tendency to stigmatize persons in the later decades of life which is such a liability to all of us I should like to make two intervening comments in the way of provisos One is to reject that facile solution to our problem which is represented amongst the Anglo-Saxons by the initials P C , standing originally for political correctness, and a mark of the most extreme form of nominalism which has ever appeared in Western thinking, if thinking is the right word to use Changing or retaining names and usages cannot of itself help very much, and tends very easily to ridicule and trivialization, as it is quite evidently doing for the feminist cause, the anti-racist cause, and that of the disabled None of us is going to benefit from a campaign to insist, for example, that those who continue in the Third Age after a late birthday be called "chronologically advantaged" and those who fall into an early Fourth Age as "chronologically disadvantaged", expressions which I have heard used

My second proviso may be thought to be in the way of a concession, though it is not intended as such It would be as silly to require the prohibition of Old Age from ever being used at all as it would be to accept the absurdities of P C It is evident that under the special conditions of the genuinely neutral study of aging and of older organisms, the words Old Age prove a useful, perhaps an essential term, especially for the scientists Such observers and analysts are unlikely nowadays to fail to recognize the limitations of a single, overall term for such a very various condition, and most of them probably by now appreciate that distinction between the Third Age and the Fourth Age which is so important to us

---

* See the references cited in Laslett, 1995 ("Necessary Knowledge") being the introduction to Kertzer, D I and Laslett eds , *Ageing in the Past*, University of California Press In what follows, much use is made of Laslett, *A Fresh Map of Life The Emergence of the Third Age*, (1st edition out of print 1989, American edition, Harvard. 1991 Italian edition, Una Nuova Mappa Il Mulino, Bologna, 1991 [translator Paolo Viazzo] A German edition is in course of publication, and a Spanish edition in contemplation) See also Laslett, 1993 ("What is Old Age? Variation over Time and between Cultures", paper given at a conference in Sendai, Japan, on Health and Mortality among Elderly Populations, *International Union for the Scientific Study of Population*, in course of publication) and Laslett, 1994(a) ("The Third Age, the Fourth Age and the Future", being a review article in *Ageing and Society* (Vol 13, No 3), of the Carnegie Enquiry into the Third Age

DANGERS TO OLDER PEOPLE OF THE SOCIAL CONSTRUCTION OF OLD AGE

It is doubtful, however, how far these objective witnesses are yet aware of the possible dangers to people like ourselves in allowing Old Age to stand for a complete social construct relating to those in late and later life. It is when the supposedly objective observer uses terms like Old Age outside the context of his observations or hers that the possibility arises of damage being done. Under these conditions, and they are of course the general conditions even for the scientists, the reference can only be to Old Age as a social construct, a social construct potentially disadvantageous to the people to whom it refers. The authority with which experts speak is fraught with unfavourable possible consequences. Over the years, unaware, insensitive, irresponsible doctors and medical experts have done enormous harm in this way to the body of older people.*

This second proviso of mine, then, leads to complications, complications which lie at the very centre of our subject. In explicating this theme further, I shall turn to a closely related topic, in fact one continuous with that which we are examining. In the Anglo-Saxon school of social gerontology, and particularly in the British where historical sociology plays a considerable part, the claim is being made that the use of ages last birthday in any context where an individual is being judged, selected, admitted or rejected is inequitable and a breach of individual rights. A conspicuous illustration is the illegality of requiring a university professor in the U.S.A. to leave his post and retire on the sole ground of age.

PROHIBITION OF THE USE OF BIRTHDAY AGES?

Some social scientists and political leaders go much further. They insist that all public uses of birthday age are impermissible, and should be made illegal. Hence the charging of higher insurance premiums to those past a certain birthday, their exclusion from service on juries in Britain, the policy of British grant-giving bodies in refusing to entertain, solely on the grounds of age, applications for research funding should all be abolished forthwith.† Such a prohibition in this view should by no means be confined to birthdays in later life. It should apply to girls and boys being excluded from school before a certain birthday, or from practising certain forms of sexual relationship, from examinations for driving licences, and so on. It is not supposed, of course, that exclusion of this kind should not exist, only that they must not be legitimated on grounds of age alone.‡

The reason for the demand that birthday ages should be prohibited from use, at least from public use, is once more a matter of social construction. In this construction the situation of all persons past relatively high birthday ages, over 65 for example, or even over 60, is

---

* For example the great Anglo-Saxon physician of the earlier 20th century, Sir William Osler See below Administrators, politicians, advertisers and above all media men and women have done their bit as well, most of them by virtue of their power and influence rather than through their expertise

† As an example, the project on *Maximum Length of Life* at the Cambridge Group for the History of Population and Social Structure, of which I am the Director, was refused support for this reason since the applicant, who had of course to be the responsible head of the project, was over the university retirement age The fiction that a younger person was directing the activity had to be resorted to This project is an association of researchers themselves in the Third Age, working with one younger researcher, to investigate a set of issues of great general significance, but of particular significance to themselves, i.e 3rd Agers.

‡ See the references set out in the first note, and especially those citing the publications of Michael Young The most substantial argument is in Michael Young and Tom Schuller, *Life after Work the Arrival of the Ageless Society*, Harper Collins, 1991

construed on the manifestly incorrect assumption that all of them are no longer fully capable of doing what persons with lower birthday counts can do. She or he is judged in fact to have reached the supposedly unitary and continuous condition "Old Age", with all its connotations of incapacity and decrepitude. The point of importance here is that some, a very small and we hope a decreasing proportion, but still some, persons in their sixties are indeed subject to a degree of incapacity and this tendency does increase as further birthdays pass.[*] It is this fact which makes it possible to persuade the uncritical and unaware that all persons of advanced maturity have a roughly equal degree of disability, and experience a roughly equal increase of that disability as birthday year succeeds to birthday year.

INTERESTED REASONS FOR MISDESCRIBING OLDER PEOPLE

This is the essence of the case for Old Age itself being a social construct which bears unfairly on older people, and the words to emphasize are uncritical and unaware. It has to be noted that persuasion is possible because a part is being taken for the whole, because some people do correspond to the stereotype, then all can be claimed to do so. There are more sinister factors and possibilities too. If there are established interests in a society which would be furthered by the acceptance of this distortion of the facts about aging, for example if insurance companies can increase their profitability by applying an inflexible age rule in charging for cover for travel by air, something which Third Agers may well have experienced in getting to this meeting, then the distortion will be intensified and the stereotypes of older persons as incapable will be reinforced. No effort will be made to judge each case on its merits, taking account both of the enormous variability between older individuals, and of the improvement in their general condition which has been taking place recently and which continues to proceed, as we all devoutly hope, especially reduction in the rate at which such disabilities as there are worsen as the years pass.

Weighty and important as I find these arguments about the use of birthday ages to be, I have to confess that I do not go as far as Michael Young in calling for the entire prohibition of the public or social use of calendar age. This is firstly because I think that it is too idealistic, and would therefore not be acted on. No advocate for a cause like that of objectively analyzing age and aging can benefit adopting unrealizable aims and policies.[†] The further reason is identical with the one already given about the use of the expression Old Age itself, but the argument is stronger. Objective, genuinely neutral, "scientific" investigation of aging must require the use of birthday age calibration. Administrators and organizers are in an analogous position.

Once more the difficulty is ensuring that the use of such calibration is confined to the specified purposes and occasions, and that those who observe and deal with aging persons should always be on their guard against stereotyped descriptions which add to the distortions within the social construction we have under scrutiny. It particularly applies when these specialists use the words "Old Age" when talking in public and in general about the subjects

---

[*] The extent of the increase of disability with age is discussed in several papers in the present symposium. See especially that of Eino Heikkinen (in this volume).

[†] Reflecting as I have to do on the positions argued in my own books, especially *A Fresh Map of Life*, now being prepared for translations and a second edition, I have to confess that some of its idealistic tendencies give cause for concern There is, however, a countervailing argument Idealism, even exaggerated and impractical idealism, can be realistic if the situation is extreme, when a statement of lofty objectives may be required

of their studies and activities. As I have said elsewhere, research on aging is dangerous stuff, and those engaged in it have to be aware of the fact.

HISTORICAL SOCIOLOGY OF AGING AND DESCRIPTIONS OF OLDER PEOPLE

This brings us back to Old Age itself as a social construction, and to a highly abbreviated argument about the concept having become anachronistic to a considerable degree, and about its character as a social construct.* The conception of age, aging, and Old Age which was in position when Western populations began to grow old a century ago, since then attaining the irreversible status of the oldest populations the world has ever known, did indeed already consist in a set of stereotyped propositions which were unfavourable to older persons. At that period such expert opinion as there was, and especially medical opinion, was given to making what I have named "Hostile and Demeaning Descriptions of the Elderly" (title of *A Fresh Map*, chapter VII). Sir William Osler, the most prominent name in Anglo-Saxon medicine in the earlier part of this century, could even suggest, if as he claimed only ironically, that the life of everyone should be ended at age 61. It was insisted that decrepitude and uselessness did in fact objectively describe everyone of an age to retire, and this was the period in Western history when retirement was institutionalized. It was indeed the earliest time at which Western nations could afford pensions for the mass of older persons.

Even before radical population aging began, therefore, the social construction of Old Age took the part for the whole, the broken-down, incapable elderly for all of the elderly. But there are three circumstances which made this already inequitable judgement more understandable at that time. One is that those in the older age groups were so few in relation to the whole population, and lived in later life for so short a time, in comparison of course with the present day. The second is that the physical condition of those older people was absolutely, if not quite so clearly relatively, worse than it has since become, and their social and economic position worse as well. Many, perhaps most, of them did display those characteristics which impelled even informed and supposedly responsible commentators to suggest that they were otiose, even disposable.

The third and by far the most important circumstance about the historical situation as it was when the elements of the social construction of Old Age as we now recognize it came into existence is of consequence to the young, as well. The Third Age did not then exist. It could not have been present for demographic reasons before the 1950's, and the other circumstances which are required for its emergence had not established themselves. The term Old Age, even though biassed, was arguably the only available description for most or all purposes. What is fundamentally at issue for the historical sociologist about the tendency of Old Age as a blanket term to persist in our day is that we are still at the mercy of assumptions as to aging which were part of the immemorial social structure and intellectual life of our forebears, all our forebears up to the time that the Third Age did finally emerge. The tendency has been and still is to refuse to differentiate when referring to older persons. In my own language and typology, this means being unwilling to recognize the division between a Third Age of self-realization and a Fourth Age of decrepitude and decline. The

---

* The argument which follows is set out at length in *A Fresh Map of Life*, especially as to the justification of the division of the life-course into stages (stadial analysis of the First Age, the Second Age, the Third Age and the Fourth Age), the character of the Secular Shift in Aging, its historical suddenness and consequent effect on perceptions or misperceptions of aging in our time in Western developed societies, all of them having very old populations

enormous and fundamental social structural changes, which rapid, unprecedented – completely unprecedented in all previous time – aging brought with it have not yet been recognised or absorbed. We as Western societies are, as I have insisted on many occasions, still in a state which could be called false consciousness as to aging.

THE SYMBOLIC IMPORTANCE OF WHITE HAIR AND THE SOCIAL SOVEREIGNTY OF YOUTH

Glancing back over this compressed exposition of the historical sociology of aging, I am left with the apprehension that I may be misunderstood, and shall have to rely for clarification on the reading which I have recommended. I had intended to develop a further and much more sombre set of reasons why the stereotypes so obstinately persist and why the social construction of Old Age continues to be as it still is.* This argument was to have taken the whiteness of our hair as being a symbol of death to our juniors. The imperial reign of the cult of youth in the oldest societies which have ever existed in the world is one of the great paradoxes of our time. It is nowhere more blatant, more outrageous shall we say, than in the enormous mass of publicity and persuasion which characterizes the western media, an area of action and ideology which has never been so potent before, or anywhere else.

Fundamental to this gigantic contemporary phenomenon is the excoriation of every iota of white hair, even of grey hair, of wrinkling ("wrinklies" is one further playful but insulting epithet for people like ourselves); of every hint of the process of aging or even of maturation. The shapes, colours, capacities and characteristics, physical and mental, of everyone accepted as appropriate in the media has to be located in the early Second Age, in birthday terms between the teens and the early forties. Everyone older has to be referred to euphemistically or apologetically. As for those in the Third Age, prompt action is taken to ensure that they appear as rarely as possible, those in the Fourth Age for practical purposes not at all.† How could we expect objectivity about Old Age to establish itself in a society like ours, so dominated by the media and where the media has such unrealistic, inappropriate, anachronistic attitudes?

THE THIRD AGE AND THE PRESENCE OF DEATH

It is evident that the association of white hair with death and dying, while understandable, is grossly exaggerated, intensified by deep-seated and unconscious fears. But it is also true, though perhaps not very widely known, certainly to the journalists in their sublime ignorance of demography, that death does in our time happen almost exclusively to older people and that it can be maintained that the Third Age does the dying for the whole of the

---

* The relation of the Third Age, and especially the Universities of the Third Age (U3A's) in England with death and decline, is most impressively expounded by the Israeli anthropologist Haim Hazan in *Old Age Constructions and Deconstructions*, Cambridge, 1993 Hazan, however, presents a picture of hopeless immobility in the condition of persons in late life and seems to think them to be incapable of independent action to change their social structural position and to bring about more realistic attitudes, analysis and nomenclature in the society as a whole He appears to suppose the current social construction of Old Age to be unchangeable and irremovable

† See the Research Report Number 1 of the Research Committee of the University of the Third Age in Cambridge, *The Image of the Elderly on TV*, 1983 This, as far as I know, is the only piece of research by the Third Age on its own image The unwillingness of British television to allow any news broadcaster or commentator to perform if she or he has white hair is notorious

population. This is again historically entirely novel, but it is the inevitable accompaniment of the remarkable prolongation of everybody's lives. All of the circumstances so briefly run over bring us face to face as members of the Third Age with the fundamentals of our personal and physiological as well as social position. I make no apology for having touched upon these fateful facts. We have to know where we are.* We have to recognise that we have more to overcome than destructive verbal distortion and a maiming social construction of our condition as a collective Third Age population spread all over the developed world.

A PLEA FOR THE INDEPENDENCE OF THE THIRD AGE IN RUNNING THEIR OWN UNIVERSITY

I have one last remark to make, as an Englishman talking on behalf of members (not students) of the University of the Third Age organized on the English model. Under the circumstances which I have here tried to survey so cursorily, in considering the disappearance or wished-for disappearance of Old Age, as it is still socially construed, is it not clear that the Third Age needs its own intellectual instrument, quite independently organized from existent universities, which are quintessentially of the Second Age? Can you, my Continental sisters and brothers, really wonder why we English (and Australian and New Zealand) people in the Universities of the Third Age, in the U3A as we call them, undertake to teach ourselves, to think together for ourselves and by ourselves, to take from the Second Age what it has to tell us of our condition, but to fashion our own intellectual autonomy?

REFERENCES

Hazan, H , 1993, *"Old Age, Constructions and Deconstructions"*, University Press, Cambridge

Kertzer, D I , and Laslett, P, eds , 1995, *"Ageing in the Past"*, University of California Press, Los Angeles

Laslett, P, 1989, *"A Fresh Map of Life The Emergence of the Third Age"*, Weidenfeld, London, 1991, Harvard, 1991, Bologna, Il Mulino, 1991, Second edition in press

Laslett, P, 1993, "What if Old Age, Variation over Time and between Cultures", paper given at a Conference in Sendai, Japan, on Health and Mortality among Elderly Populations, International Union for the Scientific Study of Population, in course of publication

Laslett, P, 1994, The third age, the fourth age and the future, a review article of the Carnegie Enquiry into the Third Age, *Ageing and Society,* 13(3)

Laslett, P, 1995, Necessary Knowledge, in *"Ageing in the Past"*, D I Kertzer, and P Laslett, eds , University of California, Los Angeles

University of the Third Age in Cambridge, 1983, *"The Image of the Elderly on TV"*, Cambridge U3A, Cambridge

Young, M , and Schuller, T , 1991, *"Life after Work The Arrival of the Ageless Society"*, Harper-Collins, London

---

* I here respond to the challenge made in the impressive keynote address by Henry Moody which he gave after the opening ceremony at this meeting Living in truth, as Vaclav Havel has proclaimed, is the first duty of those in the Third Age

3

LIFE AFTER WORK

Lines, Boundaries, and Spaces

Tom Schuller

University of Edinburgh
Centre for Continuing Education
11 Buccleuch Place
Edinburgh EH8 9LW
United Kingdom

INTRODUCTION

The paper has three disparate but linked sections. I begin by challenging the notion, implicit in the title 'Preparation for Aging', of implied support for a linear sequential model of the life course. I shall then immediately qualify that challenge, and suggest that one major task for those of us concerned with the relationship between education and age is to explore the ways in which appropriate balances can be struck between the unavoidability and the desirability of predictable changes accompanying the aging process on the one hand, and the goal of enabling people to defy, undermine or master that predictability on the other hand.

In the second part of the paper I turn to the pictorial or metaphorical, and suggest that we can benefit from exploring the way in which images are used to depict the life course. This can help us to construct alternatives to the stereotypes which constrain older people's development, both in their own perceptions and in the opportunities which are available to them. Finally, having focused primarily on the temporal dimensions of aging I shall turn to one spatial issue, and ask whether emerging notions of cities and towns as learning entities might offer particular opportunities for older people to participate in and contribute to learning.

The paper is not an analysis of existing learning opportunities, nor does it set out to be of direct practical application. By dealing with the wider issues of how we see the structure of the life course I hope only to open up some new angles on the debate. The argument can be summed up in terms of three interrelated questions: What kinds and degrees of predictability are helpful, in the life course as a whole and in daily life? How can we expand the range of trajectories which are sanctioned for individuals? What kinds of boundaries are needed to provide a supportive environment for aging?

Preparation for Aging, Edited by E. Heikkinen *et al.*
Plenum Press, New York, 1995

'PREPARATION FOR AGING': AGAINST LINEARITY

In the 1970s I worked at the Organisation for Economic Co-operation and Development, helping to prepare policy documents which made the case for recurrent education. The strategy was for a redistribution of learning opportunities over the lifetime, based on a critique of existing national educational systems for their essentially 'front-loaded' character, i.e. for concentrating resources on initial education at the expense of adults (CERI/OECD, 1973). Some of that work now looks rather crude—for example, the models we showed of working careers, which conspicuously ignored the complexities of female career patterns. But the message, although now familiar, still has validity and still demands political mobilisation.

I mention this because the title of the congress, 'Preparation for Aging', triggered off in my mind some concerns about the image which it generates, concerns which resemble those which we were addressing two decades ago in respect of the educational system overall. The image is one of a linear sequence, where education is seen as a preparatory phase, preceding and separate from the phase for which it is a preparation.

Separating education temporally from working life, with the implication that once prepared as a youngster the worker will no longer need further education, is now something no one would be heard to support. Everyone agrees that the two should be integrated, and that no one can expect to go through life without acquiring new skills and expertise—though how far reality matches this perception is another matter. In the same vein, the congress' title causes me to ask how far aging is something for which 'preparation' is appropriate, since preparing is by definition something that is done beforehand, and not concurrently with the event or state to which it is geared. After all, aging is a continuous, if uneven, process, and it is artificial to suggest that it is something which can be prepared for in the same way as one might, for example, prepare for a marathon or for an examination. So is the use of 'preparation' diverting attention, and perhaps resources, away from where they should be, on the integration of aging and development? The front-loaded education system concentrated on the First Age at the expense of the Second; is there a danger of repeating this mistake one stage further along the life course line?

To some extent this is, I acknowledge, a semantic point, made for provocation. No one would contest, in principle, the need for learning to continue as aging occurs, beyond retirement or other significant markers in the life course. Moreover preparation does not necessarily imply that it occurs all in one block. There is no necessary conflict between thinking creatively about what is needed to prepare for aging and what is needed to live the aging life. But I make the point because I am increasingly aware of the power which metaphors—verbal and pictorial—exercise at many levels, from practical policy issues to matters of personal psychology. I return to this in later sections of the paper. As a generalisation, it is hard to shift public and political perceptions which equate education with initial education; and at the personal level it is hard for most of us to accept how much learning there is to be done, recurrently. Anything which reinforces a simplistic model of preparing-then-doing needs challenging. I therefore make no apology for urging that preparation for aging must be explicitly seen in the wider context of continuing development.

In that wider context there will be considerable variations, since 'aging' of course encompasses a wide range of possible processes and events and the notion of preparation will vary accordingly. The transition from employment is one of these, and formed the subject and the title of 'Life after Work', a book which I co-authored with Michael Young (Young and Schuller, 1991). The book was based on a study of 149 men and women, living in South London, who had recently left full-time employment, voluntarily or otherwise. One of the problems we had in selecting a title for the book was that we precisely did not want to suggest that work, in the sense of purposeful and socially valued activity, ceases when employment in the formal and remunerated sense ceases. Nevertheless 'Life After Employment' did not have quite the same ring to it, so we compromised. I have subsequently argued, following Peter Laslett's case for Third Agers as cultural trustees (Laslett 1989, pp 199-200), for a 'social economy of the Third Age' which recognises the contribution made by unpaid older people, in the same way as domestic work is now at least partially acknowledged as work, following decades of feminist pressure (Schuller and Bostyn, 1992, p 86). However, for some of our Greenwich sample purposeful activity was wholly equated with paid employment. Moreover—and significantly for the purposes of this paper—some were wholly unprepared for the transition out of work, even when it was wholly predictable and indeed predicted. In other words, there were individuals who knew years beforehand the day on which they would be retiring and yet psychologically closed their minds to its consequences. These, it should be added, were all men.

The convulsions of the labour market over the last two decades are all too familiar, with unemployment now at the head of the issues confronting countries in the European Union and elsewhere. The drop in the economic activity rates of older men has been precipitous (see, for example, Kohli et al., 1991). The consequence has been a collapse in the predictability of careers and the dramatic erosion of the boundaries separating the statuses of economic activity and retirement, blurring the transition out of paid employment. The effects of this have been extremely varied. In some cases unexpectedly early ends to working lives came as a welcome release, in others as tragedy; conversely, predictable retirements could signal smooth progression to the next stage or an event systematically ignored until it was too late.

In *Life After Work* we categorised, not always comfortably, our sample into the unemployed, the medically retired, the early retired and the normally retired, and examined the differences in the ways adjustments were made by each category to life after work. To illustrate the point made above, I shall quote three examples from the category of the normally retired:

Mr Welsh had worked all his life at the Royal Arsenal as a clerical officer The standard retirement age was sixty for both men and women, but they were allowed to carry on till sixty-five, subject to an annual review of their competence He lasted out till sixty-five, and thereby completed his fifty years of service with the Arsenal His wife prepared him for it 'The wife got me used to it over the two years before She broke me in gently, helping me to buy spare clothes and so on for the time we would not be able to afford them '

Mrs Weedon had been works manager of a small factory where the retirement age was sixty-five but people could go at sixty if they wished and Mrs Weedon did She had planned for it over several years 'It was a general winding-down, or planned decline All my life I'd been told what to do You should have freedom from other people and from the clock, and there's no freedom till you retire '

Mr Wheale could not settle at all For twenty-seven years he had been a despatch clerk in British Petroleum and had known all that time he would have to retire at sixty The company could not be faulted, giving him a pre-retirement course and an extra lump sum of £2000 on top of his entitlement, but all to no avail He enjoyed his freedom for the first two months, it was like a fine holiday extended longer than he had ever had before But then he became restless for it to be over so he could get back into 'harness', if not BP's harness then someone else's He began to think he would not even live to sixty-five, he was so miserable without his harness If this was freedom it was so disagreeable he thought freedom was killing him

(Young and Schuller, 1991, pp 79-81)

Mr Wheale brings powerfully to mind Simone de Beauvoir's observation:

The adult behaves as though he will never grow old Working men are often amazed, stupefied when the day of their retirement comes Its date was fixed well beforehand, they knew it. they

ought to have been ready for it In fact, unless they have been thoroughly indoctrinated politically,
this knowledge remains entirely outside their ken (1972, p 4)

The examples illustrate the difference, thoroughly borne out in the rest of our sample,
between men and women's apparent capacities to anticipate and adapt They also raise the
issue of predictability How far does it help people to know beforehand what changes are
likely to occur when they make a transition in their life course, or at least to know when the
transition will take place?

There is no easy answer to these questions I want here only to make the point that
to be effective preparation must be broadly conceived In particular, it must include the direct
experience of managing and structuring time in new ways, as much as formal learning At
least in respect of those habituated to full-time employment I would argue that it is only
when men (especially) and women actually come to deal with a way of life which involves
a different temporal pattern that they will effectively prepare themselves for the removal of
the temporal scaffolding which formal employment supplies

It is therefore interesting to note some innovations which address themselves directly
to this issue I cite here only the examples of two French companies, drawn from helpful
work carried out by Genevieve Reday-Mulvey (1994) AXA allows employees aged 55 and
over who have been with the company for at least 15 years to leave work, on a 65% salary,
with all benefits safeguarded The company has the option of calling them back to work,
when they resume on full salary, the intention being to call them back for about eight months
in their first year, six months in their second year and so on The AXA pool currently numbers
about 500 Similarly Rhone Poulenc encourages part-time retirement, with some employees
working one week in two, some three days one week and two the next, and so on The aim
of both schemes is to allow a graduated retirement, if their patterns of work change,
individuals can no longer ignore the impending change of status

Naturally, so long as such schemes are voluntary the determinedly myopic will
continue to close their eyes Even if more large firms are introducing such schemes, the scale
is still small But my argument is that they pay attention to what is in essence the learning
environment, in a crucial way, they create, as it were, the temporal space within which
learning—or preparation—can take place

It is worth pausing for a moment to place this in some kind of theoretical context
For Anthony Giddens , one of the key features of modernity is the absence of predictability
and the problems this poses for personal identity "In the context of a post-traditional order,
the self becomes a *reflexive project* Transitions in individuals' lives have always demanded
psychic reorganisation, something which was often ritualised in traditional cultures in the
shape of *rites de passage* But in such cultures, where things stayed more or less the same
from generation to generation on the level of collectivity, the changed identity was clearly
staked out—as when an individual moved from adolescence into adulthood In the setting
of modernity, by contrast, the altered self has to be explored and constructed as part of a
reflexive process of connecting personal and social change " (1991, pp 32-33) In a whole
range of spheres of activity and knowledge, including scientific knowledge, certainty and
predictability have been undermined

The issue of predictability and choice are also taken up in Prado's book, *Rethinking
How We Age* (1986) He introduces the concept of 'interpretative parsimony' to describe the
way in which, as we grow older, we rely on fewer and fewer strategies (or 'plots') in life
This, he stresses, is a function of success, not the reverse, we re-use those strategies which
have worked Up to a point, therefore, this is highly functional But beyond that point such
interpretative parsimony may switch to being dysfunctional, inhibiting the exploration and
adoption of new strategies

Parsimony is a product of success or at least perceived success As reliance on standing plots
increases fewer new plots occur for what is encountered is dealt with more and more in familiar

terms....It is not that failure of creativity leads to fewer new plots and hence greater reliance on standing ones. Rather, greater reliance on standing plots inhibits the production of new ones. Older people are seen as somehow unable to generate new plots for various physical and psychological reasons. But in fact what occurs is that a lifetime of success or perceived success leads to overreliance on the familiar, to parsimony, and that inhibits plot creation. (1986, p 71.)

Arguably preparation for aging has a dual function. On the one hand it is to enable people to cope with, and benefit from, predictable challenges, whether these derive from biological or social sources, form inside or outside. Part of this must be to help people anticipate change, and must therefore be based on predictability. Secondly, however, the ability to cope with unpredictability must be enhanced, to encourage the creation of what Prado calls new plots, and to enable people to make positive use of events and opportunities which could not be foreseen. The unpredictability may be at the individual level—a sudden illness, the break up of a relationship, or the receipt of a large inheritance—or at the social level, in the shape of political upheaval, environmental change, or alterations in social security arrangements. Getting the right balance between drawing on past experience and regularities in order to give people confidence in their ability to anticipate their futures on the one hand, and on the other hand giving them the confidence precisely to abandon that kind of security and to be able to meet, even create, new challenges which cannot be foreseen seems to me to be a major task for educational programmes.

IMAGES OF THE LIFE COURSE

In this section I want to make two claims. The first is that it is important to pay attention to the images or metaphors of the life course which we hold in our collective minds, and to extend the range of these images. The second is that there is a particular need to provide alternatives to the linear sequential model of the life course, which calls for fresh imagination.

There is no space or time here for a discussion of the notion of metaphor or image as such. Morgan (1986) and Ortony (1979) provide these in stimulating fashion. As Morgan says:

"Metaphor is often regarded just as a device for embellishing discourse, but its significance is much greater than this. For the use of metaphor implies a *way of thinking* and *a way of seeing* that pervade how we understand our world generally." (1986, p 12.)

Petrie goes further:

"They [metaphors] may provide the most memorable ways of learning and thus be our most efficient and effective tools. But further, they are epistemologically necessary in that they seem to provide a basic way of passing from the well-known to the unknown." (1979, p 460.)

If one of the tasks facing those with a concern for expanding the horizons of the Third Age is to provide alternatives to current assumptions and practices, then the generation of fresh images and metaphors for the Third Age in its life course location forms part of that task.

In mediaeval times, the use of the circle to depict the life course was dominant, reflecting exactly the premodern condition which Giddens refers to. Just as childhood gave way to adulthood (allowing for the fact that the status of childhood itself was not clearly defined, as Aries has argued), so old age curved back to link up with birth in the definitive form of repetition. Obviously it was not the same person being reborn, but the circle epitomised the way in which new lives were expected to recapitulate the old. Then the ends of the circle were unhooked and it was, so to speak, bent back to form an arch. This is not to imply a simple chronological succession, but Thomas Cole has provided a marvellous iconography in his book *The Journey of Life* (Cole, 1992) which locates the process historically. At some point in the sixteenth century it appears that the life course began to be shown as an arch, rising to a peak somewhere in middle age—however that was defined—and falling again to old age and death, without this reconnecting to birth. Cole shows the emergence of a new schedule of life as a career, which "found its definitive iconographic representation in a rising and falling staircase (or a pyramid of stairs)—*a shape which became the standard bourgeois image of a lifetime for the next 350 years.*" (p 19, my italics) He links this solidly to material changes in the political culture: "Reformed ideas about time and its preciousness as a commodity gave clear priority to the characteristics of middle age—the age when men were most capable of participating in the market. The new iconographic shape of a man's life from cradle to grave implied a hierarchy of values consistent with the north-western and central European marriage pattern, where power over economic resources rested primarily with middle-aged men. The growth of urban markets intensified this hierarchy, encouraged efficiency in intergenerational transmission, and fostered a more economically 'rational' approach to aging." (ibid., pp 23-24).

If there is a dominant modern image, I suggest that it is the line. Modernity has meant the flattening of the arch into single dimensional linearity, chopped usually into decades or some other unit of chronological uniformity. The lines appear as vertical or horizontal, or stepped. Claims are made to varying degrees of universality; as usual, most of the models are drawn from Western industrialised countries, and mainly from male patterns. The linearity is imposed by the dominance of chronological age in our lives; it conveys an image of life as a sequence of stages, end on to each other and—since time in this sense cannot run backwards—without overlap or recursiveness.

I want now to return to *Life After Work* in order to present one alternative to this. In our discussion of the differences between men and women in the ways in which they managed their time following the transition from employment, we used two images.

> For men, time has traditionally been more like a single huge span, a central cable whose core has been his full-time continuous work; wound around, to be sure, with regular short periods of leisure chopped up to match the daily, weekly and annual rounds, wound around with the great events of family life, the births, the marriages and the deaths, wound around with unemployment, but nevertheless a continuous single strand running high above individual events all the way from adolescence to senescence Women have had to knit together different strands, especially their two sorts of work, whose rewards may be monetary or more fundamental than that, into a more complex whole The pattern can be likened not so much to a cable as a swaying cat's cradle of twists and turns and overlaps

We argued that this meant that women were better prepared for the third age than men are:

> Men have customarily had more of a commitment to paid work, even if they have not been able to get it, and because of what can be lost when they no longer have what has been central to their lives Women, even if in this respect they are moving closer to men, have not moved all the way by any means Their experience in weaving together different strands, their experience in coping with transitions between different statuses, their experience in shaping their identities from a variety of models rather than so much from one—all are conducive to making the best of life after retirement that does not differ so much from their lives before. Partly because of this, partly because they have been *the* carers in society, women's personalities may well be less brittle, more elastic, than men's If one strand in the cat's cradle is snipped, the rest can hold together without unravelling, the cradle subtly adjusting itself to take up the slack (Young and Schuller, 1991, pp 126-127) (I can confess here that we never put this last point to empirical test, to see whether a cat's cradle does hold together; but I make this confession also to make the point that it does not detract from the fruitfulness of the image.)

Inga Elqvist-Saltzmann has also explored gender differences in a life course perspective, and the role of images in helping us conceptualise and analyze these. She uses the term—and persuaded her publisher to use the physical image—of a 'green hill' to represent the household work done by women over the life course, and explains why:

Life-line curves and 'declines' have proved to be excellent tools for illustrating the different kinds of work affecting the educational and professional career In a society where the evaluation of educational effects has to a large extent been measured in terms of production, we have been anxious to find a way of showing why women's roads through education and the labour market have become so 'winding' paths My work in analysing the 'declines' has shown, however, that it is easy to adopt a traditional way of thinking, to upgrade professional activity and public like and to downgrade the reproductive duties It is easy to fall into the trap of regarding the lack of studies and professional activity as a flaw, a hole, and decline One way of encouraging a less traditional way of thinking was to turn the life-line curves upside-down and to make work in the home visible in the form of surface We painted the surfaces of these reversed pictures green Work in the home appeared as *green hill* taking up a considerable part of the life-span (1994, p 150)

There are many other metaphors available. Music is a particularly rich potential source, and provides the title for Mary Catherine Bateson's book *Composing a Life*, an account of the lives of five women, including the author herself. Weaving and oriental puzzles are other images she makes particular use of, but jazz is perhaps the most appropriate of all, exemplifying "artistic activity that is both repetitive and innovative, each participant providing background support and sometimes flying free." (1990, pp 2-3) For me the image of a tree has many strengths, with its rhythms of growth—in several directions—decay and regrowth. The point, however, is not to pick a 'winner' as the best image, but to enhance our awareness of alternative images, in order to have available to ourselves, and to make available to others, a plurality of possibilities.

SPACE AS WELL AS TIME: THE CITY AS A GEOGRAPHICAL LEARNING UNIT

One of the themes of the discussion so far has been the breaking down of established boundaries. The boundary between public and private time, always much sharper for men than for women, is being eroded by a number of factors, the most pertinent example of this being the boundary between employment and retirement. Boundaries, it should be clear, are not only confining constructions; they also provide enabling structures to people's lives. In this section I want simply to make some rather speculative points linking aging and spatial location.

I want to explore the notion of a learning city. I am well aware that many older people live in rural areas or in small towns or villages. There are particular issues associated with rural life, notably as far as access to social and educational facilities is concerned. But I use here the notion of 'city' very loosely, to refer to a cluster of inhabitants of almost any number, however small. It could as well be learning town or learning village.

The essence of the notion is that most of us live in geographical units which contain enormous reservoirs of learning potential which are largely neglected. These may be formal educational institutions, such as the university I come from in Edinburgh, or other visibly informational or cultural institutions such as libraries or theatres. But there are also myriads of informal learning sources, notably other people in their various capacities: as fellow-employees, as neighbours and so on. We need to find some way of crossing the boundaries which keep people apart in the sense that they are denied the opportunity to learn from each other. This has been the subject of a most interesting OECD study of seven cities scattered around the world each of which has some claim to have a strategy for lifelong learning (OECD, 1993), and is the focus of an initiative which I am involved in Edinburgh under the title *Edinburgh City of Lifelong Learning*.

Why do I bring this up here? For two reasons. The first is that older people tend to be less mobile. They change jobs and residence less frequently than the younger people. This means, at least in principle, that they have a greater commitment to and dependence on their locality, and often greater knowledge of it. Almost any extramural department in the UK will testify, for example, to the popularity of local history classes and to the depth of expertise which is often brought to it by local inhabitants, many of whom are third agers. A focus on the idea of a learning city as involving people collectively in changing the culture of the city towards one of inquiry and development should have a particular appeal to longer-term residents who have a greater stake in the place. The essentially social nature of learning—epitomised by practice in Universities of the Third Age—has a particular echo in strategies which focus on geographically bounded areas.

The second reason is that the idea of the learning city, at least as we have construed it, is based on the principle that everyone is both a learner and a teacher, that the boundaries between tutor and student are broken down. This is again of the essence of U3A practice, and what we are trying to do is to extend that principle across the board: into professional life, where people can learn from each other across occupational and sectoral divides, and into social policy, to counteract the social segregation which characterises too much of our societies today. This is not to argue for the abolition of boundaries. In many instances, people simply are repositories of knowledge and expertise, whether they are formally designated as teachers or not, and it would be perverse to propose some kind of wholesale educational egalitarianism. But it would be a huge step forward if we could convince more people that they had things to teach as well as to learn—and that these two activities can actually co-exist.

In short, the exploration of cities, towns or villages as learning units, combining the benefits of local knowledge and personal contact with the openness enabled by new technologies, could have particular implications for aging and the access people have to learning opportunities.

CONCLUSION

In this paper I have set out to challenge from one particular perspective the notion that aging can be conceived of as a linear sequential process, as is the implicit assumption of many psychological and even sociological models of the life course. With specific reference to the transition out of employment, especially full-time employment, I have argued as an adult educator that effective learning is more likely to take place if changes are made in the temporal environment, such that people—and men in particular—directly experience changes in their time structures before they actually leave formal employment. I have suggested that the use of images and metaphors may be fruitful, both analytically and as a way of opening up new opportunities for people to manage their own aging. And I have concluded with the image of the city as a learning unit within which citizens can prepare for and manage their life after work.

REFERENCES

Bateson, M.C., 1990, *"Composing A Life"*, Penguin, London.
Beauvoir, S. de, 1972, *"Old Age"*, André Deutsch, London..
CERI/OECD, 1973, *"Recurrent Education: A Strategy for Lifelong Learning"*, OECD, Paris.
Cole, T., 1992, *"The Journey of Life"*, Cambridge University Press, Cambridge.
Elgqvist-Saltzman, I., 1994, Declines and Green Hills: Competence Development in a Life Perspective, in: *"Gender and Education in a Life Perspective"*, G. Bjerén, I. Elqvist-Saltzman, eds., Avebury, Aldershot.
Giddens, A., 1991, *"Modernity and Self-Identity—Self and Society in the Late Modern Age"*, Polity Press, Cambridge.

Kohli, M, Rein, M, Guillemard, A -M, van Gunsteren, H, 1991, *"Time for Retirement"*, Cambridge University Press, Cambridge

Laslett, P, 1989, *"A Fresh Map of Life"*, Weidenfeld and Nicolson, London

Morgan, G, 1986, *"Images of Organisation"*, Sage, London

OECD, 1993, *"City Strategies for Lifelong Learning"*, Gothenburg

Ortony, A, ed, 1979, *"Metaphor and Thought"*, Cambridge University Press, Cambridge

Petrie, H G, 1979, Metaphor and Learning, in *"Metaphor and Thought"*, Ortony, A, ed, Cambridge University Press, Cambridge

Prado, C G ,1986, *"Rethinking How We Age A New View of the Aging"*, Greenwood Press, Westport

Reday-Mulvey, G, 1994, *"The Future of Work—Redefining Productive Work, with Case Studies in Germany and France"*, The Four Pillars, 16, International Association for the Study of Insurance Economics, Geneva

Schuller, T, Bostyn, A M, 1992, *"Learning Education, Training and Information in The Third Age"*, Carnegie U K Trust, Dunfermline

Young, M, Schuller, T, 1991, *"Life After Work"*, Harper Collins, London

4

EMERGENT CHALLENGES FOR UNIVERSITIES OF THE THIRD AGE

Brian Groombridge

University of London and
University of the Third Age
London
United Kingdom

In choosing this title—Emergent Challenges for Universities of the Third Age—the organisers have also presented me with a challenge. In accepting it, I shall take a bold and ambitious line, inspired by some remarks by the President of AIUTA, Professor Jacques Lefevre. He believes the University of the Third Age has made such progress in concept and practice that its mission and responsibilities should be redefined. The world increasingly and urgently needs our experience, our brain power and our participation as well informed citizens. In words echoing President Kennedy's famous phrase, he said:

> In 1994, the real problem is no longer wondering what society is going to do for its 'old folk', it is enquiring what older people can do for society To prepare older citizens for a social, cultural and ecological commitment is becoming a new challenge for the Universities of the Third Age (Lefevre, 1994) *

What might this new challenge this *'neuveau defi'* mean for the U3A movement in practice? Here for consideration is a personal response, including a major proposal and concluding with an outline six-point plan.

In view of Professor Lefevre's global perspective, and his use of the word, '*ecologique*', ecological, let us begin with the planet itself, and then consider some of the 'emergent challenges' of the title.

THE GLOBAL CHALLENGE

The international community examined the multiple crises facing the world at planetary level at the UN Conference on Environment and Development (UNCED) in June 1992 in Rio de Janeiro, the so-called Earth Summit. The results were summarised in five

---

* En 1994, le veritable probleme n'est plus de se demander ce que la societe va faire pour ses 'vieux', il est de s'interroger sur ce que les aines peuvent faire pour la societe Preparer les seniors a un engagement social, culturel et ecologique devient un nouveau defi pour les universites du troisieme age

Preparation for Aging, Edited by E Heikkinen *et al*
Plenum Press, New York, 1995

Agreements, and especially in the masterplan known as Agenda 21, outlining in its 500 pages action to be taken on just about any issue related to environment and development.

The chapter that caused least argument at Rio was Chapter 36, 'Promoting Education, Public Awareness and Training'. "Environment and development issues should", it said, be integrated into educational activities, formal and non-formal, and seen as a 'cross-cutting' issue from school to universities, employment training and adult education. (Grubb et al., 1993)

However reluctantly, governments are increasingly recognising that the problems cannot be solved by them, nor by politicians or experts, nor by the citizens, acting alone, but only by all these parties co-operating together, locally, nationally and internationally. As well as transcending national boundaries, the issues themselves are also interdisciplinary and intersectoral. (Groombridge, 1993)

There is no time to elaborate these thoughts, but everyone is familiar with the situations they are meant to evoke. Universities the world over are being mobilised to help solve the problems itemised in Agenda 21. Finland, along with the Nordic countries generally, is well known for its commitment. Earlier this year I was at the University of Joensuu, internationally famous for its knowledge of forestry. There I learned how the University, through its Education Faculty, is sharing its knowledge with the schoolchildren of the region and their teachers. In Lahti, 100 kilometres north of Helsinki, specialists from the University of Helsinki are working with the local people to make the lakes fit for fish again (Groombridge,1994). There are many universities in many countries, realising, like the Social Science University in Grenoble, France, that Agenda 21 concerns the human and social scientists as much as it is the business of hydrologists, glaciologists and chemists of the atmosphere. This University (Pierre Mendes France) has just set up a centre, l'Institut de l'environnement, with two objects: (1) to co-operate with the other university in Grenoble, which specialises in the physical sciences; and (2) to make effective contact with agencies in the region with which the universities need to collaborate in practice: "Grenoble wishes to become a turntable"—as it were a manoeuvrable point of contact—'for the environment'. (Université Pierre Mendes France, 1994) [*]. What has all this to do with us? What challenges emerge for us by measuring ourselves against that scale of need? In such terms, do Universities of the Third Age have anything to contribute?

There's a lot we can do. At the simplest level, as consumers, our personal choices are relevant, and as citizens, we all need to be as well informed as possible. U3As already study many subjects relevant to improving the environment.

Our next task, if we are not doing it already, is to examine our curricula, just as the regular universities themselves are having to do, to make sure that environmental matters are featured in a significant way. Having taken the trouble to study some of these questions, to make ourselves better informed than if we had just relied on what we read in the newspapers, or see on television, and then, having meditated on the moral and political implications of these issues, we need to make a contribution to the wider society.

I believe U3As could respond to this challenge, not just for their own benefit, but for the general good. There are three reasons for believing this, all stemming from the essential characteristics of Universities of the Third Age.

1. U3As are well placed in that they have one foot in the world of higher education, and the other in adult education. Adult education in this context is crucial because its raison d'etre is the sharing of knowledge, not so much its generation. I would like to see us linking up with others in the world of adult education, who are working out ways to share knowledge and values about the environment more effectively.

---

[*] 'Grenoble veut devenir une plaque tournante pour l'environnement'

To give a concrete example—I think of the people who met together from all over the world a year ago in Hadeland, Norway, to set up a network for developing and improving what they called 'A Pedagogy of Environmental Responsibility'. They were concerned with 'access to knowledge, the transfer of knowledge and the sharing of knowledge.' Knowledge needed to be shared through what they called 'a reflective community'. (Norwegian Association for Adult Education, 1993)

'A reflective community'. Is that not what we are?

Being at the interface between higher education and adult education, U3As are one of the bridges between the universities and the community. The academy desperately needs such connections. It cannot itself undertake all the necessary work of diffusion, of popularisation, requiring, as that does, interdisciplinary and practical consideration of the issues.

2. Not all old people are wise, but most old people do have maturity of judgement, especially perhaps studious old people—like U3A members and students. Such maturity needs to be at the disposal of our societies just now. In an important book Earth in Balance, which he wrote about the global environmental crisis just before he became Vice President of the United States, Al Gore points out that shortage of relevant information is not the problem. On the contrary :

> We are drowning in information We have generated more data, statistics, words, formulas, images, documents, and declarations than we can possibly absorb And rather than create new ways to understand and assimilate the information we already have, we simply create more Our current approach to information resembles our old agricultural policy. We used to store mountains of excess grain in silos throughout the Midwest and let it rot, while millions around the world died of starvation Now we have generated vast mountains of data that never enter a single human mind as thought Information while it may be valuable is no substitute for knowledge—much less wisdom (Gore, 1992)

The key concepts in that passage are 'understanding', 'assimilation', 'knowledge' and 'wisdom'. These concepts are in large part what study with the University of the Third Age is all about.

Aino Sallinen, Rector of this University of Jyväskylä, makes a similar point to Al Gore's and links it to her vision of a university's responsibilities. In a recent interview she says that people's confusion is increased by the speed at which information is propagated as well as its having become globalised:

> One of the university's important responsibilities is to help individuals organise reality The highest level of education must teach students not only how to acquire information but also how to relate to it critically Values   are important The ideal of the educated individual includes intellectual flexibility, creativity and the ability to comprehend wholes (Sallinen, 1994)

Universities of the Third Age have a headstart in some aspects of Rector Sallinen's ideal. Hence their aim to create communities in which everyone is a teacher as well as a learner. They too, just like the Institute at Grenoble University, may need to convert a number of their study circles into something more, into nodal points, into *'plaques tournantes"*, turntables, capable of 'organising reality' in Sallinen's phrase, connecting assimilated knowledge with the decision-making process in the locality, region, or further afield.

3. It has become more and more obvious in our time that government has become impossibly difficult. That is one reason why there have to be Non-Governmental Organisations representing thoughtful, well-informed citizens at major events such as the Rio Summit. The responsibility for policy-making has to be more widely shared. Democracy can no longer be equated with choosing a handful of people, mostly middle aged men, most of them no brighter than we are, to decide everything. That is the third reason why U3As must get involved in a more deliberate, overt way.

It is not just that complexity has overwhelmed governments' capacity to cope. At the daily level, politicians and governments as a class do not have time to study and think. They hurtle from crisis to crisis. Third Age students, in contrast, are not distracted (they make good students precisely because they are not devoting their best energies to raising a family or succeeding at work) and do have time to think. That capacity needs now to be harnessed.

The local policy level is as important as the global. The British community development specialist, Tony Gibson, has said: "Unless local people have ownership or control of projects they will not take responsibility. People have local knowledge. Their ideas are always superior to those of experts,[because they] know the social variables. Experts tend only to know technical variables" He continues: "The real economy is the proper use of resources, which include all the time wasted by people out of work and in retirement." (Gibson, 1994). The U3A is one medium through which some of that waste could be remedied, through which local wisdom could be combined with intellectual and technical expertise.

Some of this may seem fanciful. If it does, then it is worth noting that the process advocated here has already begun, in, among other places, this very city. The University of the Third Age in Jyväskylä is working with the local council by providing it with considered evaluations of health, education, housing and other social services.

For these reasons, I propose the following response to the main emergent challenge proclaimed by Professor Lefevre:

That U3As locally, nationally, and internationally, should inaugurate a service of Citizens' Commissions, aspiring in a scholarly way, not in a partisan or sectarian way, to make significant contributions to the process of policy formulation. They should do this in association with like-minded partners, through appropriate networks, and in a whole variety of fields, of which the environment itself is the most important and the most inclusive.

GROWTH AND ACHIEVEMENT

It would not be possible to envisage such an important social role for the U3A movement if it were not already a success. The Citizens' Commissions proposal is credible because the U3A movement is succeeding (1) at the level of individual students; (2) in terms of intellectual seriousness; (3) in demonstrating that it already accepts responsibilities to society as a whole; and (4) in showing that it is capable of expansion and adaptation in different countries and cultures.

Success at the Personal Level

Human beings are a curiosity-driven species, and in old age curiosity prompts a growing number of us to want to keep up to date with our interests, or develop new ones, to promote the work-suppressed aspects of ourselves, to make sense of our lives and the times in which we have lived, and in which we are living—of modernity (and, if you will, post-modernity as well!).

We're also more or less gregarious, we often do our best in association with others, we want to belong, we want to be recognised. We want to go on making a contribution to society. The demographic restructuring of society and related developments mean that increasing numbers of healthy, well educated, occupationally and professionally experienced people will be surfacing, often prematurely, from the labour market. Many will wish to satisfy their curiosity in a serious, studious way. Like many of us here, they will be glad that U3As were invented, catering to these intellectual, expressive and social needs as they do in a very special way.

Seriousness

Statutory universities are having to devise new forms of teaching, for a wider range of purposes, and in more milieux, than in the past. In many countries, they have until recently been reluctant to widen the scope of their teaching to include continuing education. French universities were encouraged to do so by progressive legislation in the 1960s, underpinning continuing education with a vocational emphasis. Instituts d'Education Permanente, or Centres de Formation Continue were established. The invention and creation of the Universite du Troisieme Age in Toulouse, with a general, academic emphasis, could be seen as a daring manifestation of the same historic trend. Not only were higher education opportunities opening out to mature men and women: henceforward they would even include many of the oldest. Consequently, U3As have a special mission, making them distinctive among bodies providing educational opportunities for older people. This mission mirrors the classic responsibilities of chartered or statutory universities. They both exist to undertake research, to teach, and to serve the wider society and community, albeit in very different proportions, and with quite different emphases.

This seriousness of purpose is the key membership criterion used by AIUTA. As Professor Lefevre said in Barcelona, AIUTA is not an exclusive club, not primarily concerned with institutional structure:* The aim is to maintain the level of instruction, and of course, also of the research and the service. It was on that basis that AIUTA made the bold decision to admit the British national umbrella body as a member. AIUTA now (with many variations) embraces at least two main ways of working: the original francophone model (the U3A as a major development of university extension work); and the anglophone model, adapting the concept enthusiastically imported from France, (the U3A embodying a culture of mutual aid and voluntarism in adult education). The strengths and weaknesses of these different approaches have yet to be analysed in depth across a range of issues, though a good start has been made by, among others, the journal TALIS, (Third Age Learning International Studies), published by the University of Toulouse.

Continuing success depends also on the standards being kept high, to justify everywhere the concept and rhetoric of 'University', as well as, in many countries, to justify a formal relationship with higher education. In practice, there is bound to be considerable variation between one U3A and another, but what seems constant throughout the movement is an awareness of the issue, a quest to improve the quality of what is done. These standards may show in the curriculum; in the deployment of well qualified teachers; in the range of formats (long specialised courses lasting several semesters, seminars, large scale lectures, study travel, and so on). They appear increasingly in the development of research by Third Age students, either in the form of project work as an active method of learning, or as original and useful contributions to knowledge.

Social Responsibility

There are many ways in which U3As are already being of direct or indirect benefit to society. Indirectly, U3As help society understand the implications of the demographic revolution. As Peter Laslett's colleague, the sociologist Michael Young has said: "In the long run society needs to be decisively tilted towards the interests of older people" (Young, 1993). U3As encourage that tilt by making older people visible, creating a public space for generations whose presence can be overlooked and whose worth is often underestimated.

---

* 'Le but d'AIUTA c'est de maintenir le niveau de l'enseignement donnes par les Universités du Troisième Age'

They facilitate the elaboration of roles for older people (the French have that expressive word *'epanouissement'* for this process). In so doing, they also make a tangible contribution to the question, urgent in many parts of the world, of how to redefine useful work so that it is not the same as paid employment.

They help society and governments understand the actual meaning of the words 'lifelong learning'. That concept recently received the accolade of recognition by the G7. The communique from the Naples economic summit urged "Increasing investment in people through better education, training and the development of a culture of lifelong learning". When UNESCO promulgated the doctrine of 'Education Permanente' over 20 years ago, it meant more than instrumental education for flexible production skills. The promotion of civil society and cultural life was also intended, and we here are emphatically part of that. Two leaders of the Third Age University in Ljubljana, Slovenia, Professor Dusa Findeisen and Professor Ana Kranjc, are among many to claim that U3As serve society by reminding the regular universities and the rest of the formal education system of their own fundamental, but often neglected, values: "The Third Age University... leads other educational programmes by example. School children and young people can see that it is possible to learn for enjoyment and interest as well as for vocational motives. the Third Age University contributes to humanising and indirectly counteracting the alienating influences of formal schooling"(Findeisen and Krajnc, 1991).

Nearly ten years ago, at the Thirteenth International Gerontological Congress in New York, I was with one of the founders of the movement in France, Pierre Brasseul, when he made a similar claim: "Education for older adults", he said, "represented a real challenge to the traditional model of education geared towards production.., profitability... and social advancement" (Brasseul, 1985).

Direct contributions to society are also of many different kinds, such as the provision of outreach educational opportunities for people who cannot move out of their residential homes; the foreign language translation service which the U3As in Britain offer without charge to voluntary organisations; or, as in many countries, the conversion of reminiscence into the riches of oral history (a transformation which could be celebrated in a volume to itself).

Expansion

Consequently and fortunately, the movement is buoyant. The statistics, still being collected, show there are now around 1500 U3As in over 20 countries, with more springing up all the time in more countries and all continents—290 in France, the country of origin, 400 (counting only AIUTA members) in China alone, 17 already in Poland, and about 250 in my own country (where, as noted, they are organised as voluntary study circle organisations). In Finland, study with the U3A goes on at 28 centres provided by the continuing education departments of eight universities. And so it grows.

Given the challenges, personal and global, how many are needed? How many can be brought into being? Looking towards the imminent millenium, would this Congress accept as a slogan: 3000 by the year 2000, or is that far too modest?

A SIX-POINT PROGRAMME

Challenges of different kinds evidently emerge to U3As from the increase in the numbers and proportions of alert elderly people; from the problems facing society, and even from the deteriorating condition of the planet. U3As are united in their fundamental aims, but vary in their missions, aspirations and programmes, as well as in their structures. They vary in size and resources and their development is inevitably uneven. Different U3As will therefore respond in different ways to these challenges. Despite such variations, the movement could, as it enters its third decade, debate the following six-point programme, confronting challenges, locally, nationally and internationally.

1. Continue to promote the growth and expansion of the movement;
2. Encourage an enlightened public policy for diversity in the sphere of third age education;
3. Continue to review the U3A's values and aims;
4. Further improve the quality of Third Age Learning and the relevance of curricula;
5. Make more use of media and communications;
6. Promote more international co-operation, especially through AIUTA.

The comments that follow on each of these suggestions are unavoidably brief and superficial, but they may help to open up thoughts and discussion.

Challenges of Growth and Expansion

As U3As expand, they face the challenges of successful growth itself, problems of size and resources, affecting the key challenge of the quality of what they do – especially important, as we have seen, in view of the distinctive claim made by calling themselves Universities.

In relation to the teaching, colleagues from Slovenia have identified securing enough able tutors to be one of the biggest problems, and one of the founders of the Finnish U3A, Professor Anna-Liisa Sysiharju, gave early warning of this as a likely difficulty (Sysiharju,1986). Experience suggests there are several ways of tackling it. The best news is that the word soon gets round that third age students are very satisfying to teach, but there are university systems which fail to take account of continuing education when considering staff for promotion. That needs attention and advocacy in the right quarters. Additional tutors are being recruited from staff who are themselves emeritus and emerita. Graduates from other professions are increasingly and valuably involved in the work. In this respect, staffing U3A programmes resembles the British university 'extra-mural' tradition, which could never have expanded—as it has done for well over a century—had it depended solely on internal academic staff to do the teaching. These people, almost 'adjunct faculty' in the American sense, often benefit from guidance about how older adults learn (as indeed can internal faculty, accustomed to lecturing young people with little experience of life). Effective ways of doing that are to be found in Slovenia, here in Finland, and elsewhere.

In Britain this particular problem takes a different form, because, as noted already, they are essentially self-programming study circle organisations. Hence they depend utterly on volunteers to play all the roles—not just tutorial, but administrative as well. It is essential for them to keep a balance between the recruitment of new members and the availability of voluntary staff to run the programme. London and other larger centres are fortunate. They can draw on a pool of highly qualified recently retired men and women, but some U3As have agreed to operate on a small scale in the interests of efficiency and intimacy.

Encourage Enlightened Public Policy

An enlightened public policy for diversity in the sphere of third age education is essential. It is a mistake to expect the U3A to be an all-purpose education body for all older people. Many different ways of providing educational opportunities are needed, simply because older men and women vary so much one from another. This must be recognised practically in terms of auspices, programmes, idioms, levels, curricula, and modes of delivery.

U3As themselves should therefore make a knowledgeable contribution to the formation of public policy about lifelong education, especially as in many countries the current emphasis on education for the labour market excludes many older people. U3As (and their students acting as individual citizens) should use their weight to press the authorities to take seriously not only the need, but, in the full legal sense, the Human Right to educational opportunities in the third age (Groombridge, 1989).

Continue to Review the U3A's Values and Aims

Statutory and chartered universities in the formal higher education system are in many countries re-appraising what they do in order to improve the quality of their research and teaching. U3As may need to do something similar, albeit without the elaborate and often expensive systems of quality assurance proliferating in the formal sector.

It may be time for some U3As to reconsider their purposes. For entirely laudable reasons, the stress has been strongly placed so far on the merits of study for its own sake, on the irrelevance of credits and degrees. This may however carry the risk of denying some older learners the chance to have their work publicly validated and assessed if they wish. One hears now of people being emboldened by their U3A experience to embark on a regular university course, and perhaps such access should be facilitated and not frowned upon. The British Open University's experience proves that third age students are among the very best, and that studying with a qualification in mind is not incompatible with studying for the joy of it. (Clennell et al, 1984).

In some quarters to raise this issue is so controversial that it constitutes what the English call a hot potato (for the Finns: "Kuuma Peruna")! I am even heretical enough to think that sometimes—rarely no doubt, but sometimes—older students may be wishing to study seriously with a late career move in mind—and why not?

The essence of the matter can be summed up in a memory of the Paris X campus of the University of Paris, which I visited in 1978. It was ten years almost to the day since *'les evenements'*. Young students were handing out political anniversary leaflets. They looked grim. The campus was bleak, the walls covered in graffiti. But I was with a group of buoyantly cheerful women, of all ages from about 55 to much older. They told me that being a student in middle and old age had brought them happiness, that they studied for the joy of it: "When you hear laughter on the campus", they said, "That's a Third Age student".

Further Improve the Quality of Learning and the Relevance of Curricula

It might be helpful to think in terms of improving the quality of our learning, as individuals, as individual U3As, and as a movement. At the level of the class or course, are the students sometimes too docile, too passive, as though listening alone were enough? The quality of the learning experience will also be affected by the skill and sensitivity of teaching, and by different modes of delivery. As to the curricula: I sometimes wonder whether they are as bold and original as they could be, given the U3As' freedom to innovate on the one hand and the nature of the external challenges on the other. Do they sometimes look too much like traditional lists of subjects, and not like modern interdisciplinary frameworks? The environmental challenge is not the only one to provoke an imaginative review of the curriculum. The curriculum needs also to be negotiated with or determined by the most vital interests of these exceptionally mature students. Learners in the Third Age know what matters most to them, so they know what, for them, is worth knowing.

As conventional universities are also discovering, tutorial capacity must literally be made to go further. Teaching at a distance, in some situations and for some purposes, often combined with face-to-face work, will be used more and more. In Britain, some U3A groups are studying with course materials supplied by the Open University. In Britain and Finland, telephone conferencing is used to involve people who live in residential homes and are no longer mobile. The University of Helsinki also sends audio tapes, related publications including study guides to old people's homes. The residents find this an efficient way to learn. They can play the tapes over and over. If they wish, they can listen to them in the middle of the night.

Make More Use of Media and Communications

Distance teaching depends on communications media, whether using old technologies or new. There are projects in many countries to show that U3A members are keen to make use of them, from Information Superhighways to modest audio-tapes and, of course, books. Those U3As associated with parent universities which are modernising their communications, for example to access libraries or to devise new modes of scholarly discourse, should be well placed to extend the benefits to older people. Rector Sallinen, in the interview already quoted, extolled the value of communications technology to the international academic enterprise: "Communications technology can be used as a means allowing 'universitas', the totality of the sciences, to be implemented as concrete activity on a world-wide basis, as a universal." (Sallinen, 1994). This 'electronic campus' vision is being implemented at the University of Jyväskylä, and was exemplified at this conference by the Video-Conferencing Workshop. This event, organised by the Helsinki University of the Third Age, brought together U3A students in Jyväskylä, Toronto and Ingria in Russia, and showed that such aspirations are not fanciful. Moreover research shows that older people have much to give as well as much to gain from the modern media. (Ruoppila, 1988).

There are even promising developments in the UK, without support from mainstream higher education. In any one U3A, the number of people interested in a particular aspect of, say, science or technology, may be small, but 'Computer literate' members have seen that critical mass can be achieved by linking up with others, not only by post and telephone, but also using electronic mail and computer conferencing. Such means are beginning to sustain networks of students belonging to U3As in different parts of the country and in different parts of the world. (Cloet, 1994). There are moves to go truly international and prepare to use the Information Superhighways now being set up, by joining a group, such as the one already on the Internet, called Elders. Al Gore, currently the leading political champion of Information Superhighway developments (and stressing their direct relevance to his pro-environment agenda) has urged that these means should be available to everyone, including people in remote and rural areas and from all social classes. To which should be added: and all generations.

For all the glamour and excitement of new technology, print remains important and accessible, not least because U3A students are succeeding as authors. The point is illustrated at this conference by the useful and attractively produced books published by the Italian Associazione Nazionale Universita della Terza Eta (including Irma Re's four volumes, Arrows of Memory), the Summer University of Jyväskylä, and the Lahti Centre for Research and Training (University of Helsinki).

Promote more International Co-operation, especially through AIUTA

The main strength of AIUTA is plain to see in this congress itself: this is an international gathering, with all that implies for the sharing and comparing of knowledge and experience; it brings together academics with third age students; the students have different interests; the academics come from different disciplines and backgrounds. Some know a great deal about how adults learn, about how to improve group work, about how to make better use of libraries, data bases and other information sources. Some are dedicated to science and theory; others care most about better policies and above all better practice. Students in particular know what works and what doesn't; what they like, and what they don't. Co-operation across boundaries is par excellence what AIUTA itself is for. As a result of these ingredients coming together, the Congress constitutes an opportunity to combine resources for improved quality and development.

In a modest way this process continues between congresses through AIUTA NEWS. But that, as its name makes clear, is a news bulletin. The challenges are so great, the opportunities so immense, and we have so much to learn from one another, that ways should be explored to increase the channels for the international exchange of knowledge, ideas and experience. These could take a variety of forms, according to practicality: a publication, a magazine perhaps, a manual, a series of leaflets or booklets, *'fiches'*, as the French might say, available (perhaps in several languages) in conventional forms or electronically.

These media would allow U3As from all over the world to exchange every kind of information, to swap ideas, to describe their responses to the challenges of growth, the needs of the time, and the burgeoning interests of students. News and analysis could be shared about various creative initiatives—new developments in the curriculum, for example, and research projects. Research ideas could be explored, research partners in other countries contacted, and findings reported.

Resources would be needed: political, intellectual and financial backing would be essential. I believe they could be found through the several relevant international organisations, in co-operation with other, often national publications, and with the support of the growing number of private and public enterprises seeking the goodwill of older people. The success in 1993 of the European Union's Year of Older People and Solidarity Between Generations showed the power of such alliances.

REFERENCES

Brasseul, P, 1985, "The Impact of Education for Older Adults on Traditional Education in France", paper read at the 13th Congress of the International Association of Gerontology, New York.

Clennell, S , et al , 1984, *"Older Students in the Open University"*, Open University, Milton Keynes

Cloet, R , ed , 1994, *"Science and Technology Network Newsletter"*, University of the Third Age (UK)

Findeisen, D , and Krajnc, A , 1991, The Third Age University in Slovenia, *Journal of Educational Gerontology*, 6(1) 31-45

Gibson, T , 1994, Quoted in *The Guardian*, 8 7 1994

Gore, A , 1992, *"Earth in the Balance Forging a New Common Purpose"*, Earthscan, London

Groombridge, B , 1989, Education and Later Life, in. *"Human Ageing and Later Life"*, A M Warnes, ed., Edward Arnold, London, pp. 178-191.

Groombridge, B , 1993, International Politics as an Educational Activity, *LEIF* (Life and Education in Finland) 4/93 Translated as Kansainvalinen Politiikka Kasvatustoimintana, *Aikuiskasvatus* 3/94.

Groombridge, B , 1994, Beyond the Campus: Adult Education and the Extended University, in· *"Educational Studies and Teacher Education in Finnish Universities 1994"*, J Vahasaari, ed , Ministry of Education, Helsinki, pp 22-28

Grubb, N , et al, 1993, *'The Earth Summit Agreements A Guide and Assessment'*, Royal Institute of International Affairs and Earthscan, London

Lefevre, J , 1994, Editorial in *AIUTA News* No 2

Norwegian Association for Adult Education, 1993, *"Towards a Pedagogy of Environmental Responsibility"*, NAAE, Oslo.

Ruoppila, I , 1988, Aging and Technological Development, in *"Livsstil, Livslopp och åldrande" ["Life-style, Life-Course and Ageing"]*, J -E Ruth, ed , Dokumentation från Pedagogiska Fakultet , Åbo Akademi, Åbo, pp 115-140

Sallinen, A , 1994, Vivat Academia, *Finland* (33rd edition)

Sysiharju, A-L , 1986, "The Beginning of the Third Age Universities in Finland A Link in the Tradition of a Dialogue with the Society", paper read at the AIUTA Conference, Tournai

Université Pierre Mendes France, 1994, *Intercours* (le journal de l'Université No 213)

Young, M (Lord Young of Dartington), 1993, in *"A Tenth Anniversary Greeting to the British U3A"*

5

RETIREMENT

A Truncated Rite of Passage

Christian Lalive d'Epinay, in collaboration with Jean-François Bickel

University of Geneva
Department of Sociology and
Centre for Interdisciplinary Gerontology (CIG)
59, Route de Mon-Idée
CH-1226 Thônex, Geneve
Switzerland

INTRODUCTION

The word "retirement" has different meanings but its most current usage today refers to stopping work and drawing a pension. From a sociological standpoint, three aspects of this phenomenon may be distinguished (cf. Atchley, 1982). Firstly, retirement is a modern *institution*, a "social fact" in the Durkheimian sense, displaying its three defining features: generality, externality and constraint. Secondly, this institution imposes upon the individual one of the two most important *transitions* (change in social status) in his life course, starting school being the other. Thirdly, retirement refers to the new status resulting from such transition, a status associated with a particular stage in an individual's life course. *Approaching retirement age, taking retirement, reaching retirement, leading a pensioner's life*: these phrases point to the various facets of the social phenomenon.

This paper focuses on the aspect of transition. Understanding this transition as an institutionalised process leads us to analyze retirement through the ethnological theory of rites of passage, as it was put forward by Van Gennep at the beginning of the century (1969; first edition: 1909).

Although as a general phenomenon, retirement is a modern institution, it has, nonetheless, a long history. The kings and nobles of the *Ancien Régime* used to endow their old and faithful servants, members of certain groups, primarily army officers and other State officials did enjoy pensions. However retirement depended on the goodwill of the Prince and the official's health. In Prussia and England, towards the end of the nineteenth century, and in France, at the beginning of the twentieth century, some social security systems were set up. But even then it was often ill-health and senescence, ie., unfitness for work, rather than any specific chronological age, that defined entitlement (cf. Köhler and Zacher, 1982; Bois, 1989).

Therefore, we would like to underline that, whatever the historical origins of retirement, the pace of implementation and modes of operation in Western European

Preparation for Aging, Edited by E. Heikkinen *et al.*
Plenum Press, New York, 1995

countries, it was only after the Second World War that it became an institution covering the whole of the working population as from a specifically defined chronological age. Generalised retirement did not arise in an economically prosperous period; it did not derive from affluence. Indeed, at the time, Europe was struggling to heal the wounds of war. Rather, retirement was an expression of the configuration of the sociopolitical field of market-economy democracies, the desire for national unity to tackle post-war challenges, the agreement on a new conception of solidarity, and the need to show that Marxist regimes did not have a monopoly on social concerns.

The post-war period ushered in a new historical phase, described by Fourastié (1979) as the "Thirty Glorious Years", which was to effect the break between industrial and post-industrial society (Touraine, 1969; Bell, 1976). This period of unprecedented and regular economic growth brought about a general revolution in values and lifestyles. It is hardly surprising, therefore, that the institutional functions of retirement should have evolved very quickly during the period and that the original meanings assigned to it by industrial society should have blended with the values of the newly emerging society.

For our purposes, this historical period had another important feature: a profound transformation of the social structure. Industrial society had been basically understood in terms of a class system, a system where individual and family social membership depended on their position in the economic production system. After the war (in the United States even earlier) the development of the middle strata and the generalisation of wage-labour sparked off the debate on the relevance of the concept of "class" (cf. Clark and Lipset, 1991).

But it was only during the last decade that researchers became aware of the theoretical impact of yet another transformation. A society exerts organizational and channelling pressures upon an individual's life course, adapting it to its own demands through the definition of social age groups. Both demographic changes and the extension of life expectancy resulting from the dynamics of industrial society entailed a revision of this programme which established new stages and transitions. Today, the programme defines four main positions, four clearly distinguished sets of individuals classified according to their life course situation, each having its own specific status, rights and duties. The four positions are: young people in training, economically active adults, retired adults, the very aged ("old" and "old-old", following B. Neugarten).

In view of the above, the social structure understood as the distribution of the main groups within society must be conceptualised today along two different lines. To the traditional "horizontal" axis (social classes or strata) which should be redefined, a new vertical axis is added: the different positions along a life course (Kohli, 1985; Lalive d'Epinay, 1994). The former remains, despite changes, a positioning axis within the economic organization; the latter is mainly, but not totally, a positioning axis within the labour market.

We are confronted here with a modern metamorphosis of the age classes of traditional societies (Bernardi, 1985; Tornay, 1988). Thus, the application of the ethnological concept of rites of passage to the study of life course transitions appears quite in order.

The empirical material underpinning this study is not the result of systematic data collection. The material comes from both Switzerland and France. Although our solidly drawn conclusions can be generalised to all Western European advanced industrial societies, they are, strictly speaking, not definitive results but rather "grounded hypotheses" (Glaser and Strauss, 1967).

1. The Notions of Rite and Rite of Passage

According to Centlivres (1981), a *rite* is a celebration unfolding in a space and time which differs from the profane space and time. During the celebration symbolic actions take place, expressed in visible, material and corporeal emblems which establish an exchange of meaning and emotions among participants and witnesses. These meanings refer to representations, or rather, to founding myths shared by the group.

As far as the actors involved are concerned, three main sets can be distinguished:

- the initiates, candidates or applicants, who are going "to cross the bridge";
- the initiators, priests or officiants, who define the ceremonial activities, prepare the initiates and accompany the candidate during the passage;
- the community or witnesses, divided into the witnesses from the group left behind, those from the group one enters into, and general witnesses.

The concept of *rites of passage* is associated with Arnold Van Gennep and his 1909 book. These rites accompany the unfolding of a life course, from the cradle to the grave, highlighting the transitions: changes of place, state, social status and age, just as other rites often accompany natural cycles (Van Gennep, 1969).

The rite of passage has a tripartite-sequence: separation, transition and reincorporation phases (preliminal, liminal and postliminal); the emphasis on one or other of these phases can vary from rite to rite. The passage is often symbolised by a material object (door, room, forest and so on). Belmont (1986) stresses the homology between ritual and physical passages: during the ceremony, the initiate must cross a frontier, a point of no return. For the initiate, the passage entails the mourning of his former self, but also the achieving of a new identity. Following Turner (1969), Belmont further underlines the importance of the second ("liminal") phase, the suspension or threshold that allows the separation between what was and what will be (the new integration).

A rite is the set of "ceremonial sequences accompanying the passage from one situation to another, and from one (cosmic or social) world to another" (Van Gennep, op. cit.). These sequences have a three-fold function: firstly, to express the separation of the novice from his former environment; secondly, to make a momentary pause; and thirdly, to reincorporate the applicant into a new environment and thus reintegrate him into the general order of things (cf. also, Van Gennep, 1908).

The Hungarian ethnologist and psychoanalyst G. Róheim (1942) argues that this form of rite is a response to the existential anguish of separation: by dramatizing it, it makes it bearable. It may be added that, by transforming anguish into a community act, rites make separation both inevitable and trivial.

Exploring the contemporary relevance of rites of passage, Centlivres pinpoints the three characteristics of their current evolution: the decline of emblems, the eclipsing of the local community by the family group, the lessening role of the specialist (the initiate sometimes becoming the officiant). He also notes a change in rites associated with the life cycle: some become obsolete (such as engagement ceremonies) whilst others arise, such as retirement (Centlivres, 1981, 165).

In the following sections, the three ritual phases will be considered in turn; then the rite's mythical referents will be analyzed.

2. Initiation

Classes and seminars preparing for retirement are a growing industry today. However, their history, still uncharted, is brief. The very idea of preparing for retirement would have been a contradiction in terms at the time the struggle for retirement rights first developed. The issue then was to release the exhausted worker from a job for which age had made him unfit, and no new competencies were thought to be required in retirement. The current pre-retirement programmes appeared and flourished in recent times in step with the growing dissociation between retirement from work and individual senescence (see Section 5 below).

Their availability, however, remains unequally distributed, geographically and socially speaking. Widely available in the cities, State administration and large corporations, these courses are virtually nonexistent in the rural areas and small enterprises.

The fact that actually only a minority of retired people have attended these programmes is explained by their recent origin: in Switzerland, in 1990, a mere 6%, 9% in the cities, 1% in the rural areas (Winterthur-Vie, 1991), but this percentage is increasing with successive cohorts. In a survey of students at Neuchâtel's University of the Third Age (ie., relatively young retired people, highly schooled though not necessarily university-educated, and interested in cultural and intellectual activities) it was found that four out of ten men, and four out of five women, had not followed any retirement preparatory course. Forty per cent claimed they had been unable to do so due to lack of programmes or information (Jeanneret, 1992).

Nowadays many programmes are on offer. The lowering of retirement age and the current deregulation of the institution contribute to the expansion of the market. The redundancy plans accompanying "restructuration" often include this type of training opportunities. The generalisation of the idea of a "second career" (Gaullier, 1988) paves the way for many programmes of continuous education. Today, preparation for retirement has become a growing market over which business and non-profit organizations compete.

The duration of the courses varies widely. From a few lectures a year or six months before retirement, mostly providing information on existing rights, to a company-based programme for all employees aged 50 (that is, fifteen years before the official pensionable age) with a first phase on general information concerning retirement issues and, ten years later, a complete programme over the last three or four years of employment. Classes take place partly during, partly after working hours. Participation is optional but encouraged.

Reviewing studies undertaken in the early 1980s, Schneider shows that the contents of the programmes offered were not uniform but varied according to different organizers and approaches. Whilst companies focused primarily on the material aspects of their employees' future retirement, religious and parish organizations tended to introduce the question of meaning and values (Schneider, 1981). However, a recent study on all the courses imparted in the Canton of Vaud, concludes that there is a predominant model with the following features (Bovey, 1992):

- from the point of view of organization, these programmes are offered during the last few years of employment; they last between 12 and 16 hours, either in a two- to three-day seminar form, or in night classes over a longer period. The individual is normally invited to participate by letter. Participation is recommended but optional; spouses are also invited.
- the objective is to provide information on retirement issues and to prepare people for the changes involved.
- the headings discussed cover five main areas:
  i. a general review of retirement and the changes involved;
  ii. financial and tax issues;
  iii. legal issues;
  iv. health education and prevention;
  v. leisure.
  Here and there other points may be added to the list, for instance, questions concerning meaning and values, participation, and pensioners' commitments and responsibilities.
- the courses consist mainly in lectures followed by questions. According to the study's author, these sessions did not really provide training but rather offered information and advise.

RÉUSSIR SA RETRAITE

SOMMAIRE

Connaître tous vos droits

La retraite : hier, aujourd'hui, demain
8 Diversité et complexité telles sont les caractéristiques de notre système de retraite
10 Des principes communs d'une profession à l'autre
13 Les différents régimes de retraite ou vous situez vous ?

Vos revenus
14 Les dates de versements de votre retraite
17 Un revenu revalorisé quand et comment
20 Les conditions pour pouvoir cumuler travail et pension retraite

23 La pension de réversion les conditions sont différentes d'un régime de retraite à l'autre
27 La retraite mutualiste du combattant
29 La cotisation d'assurance maladie

Les aides
31 La majoration pour conjoint a charge en fonction des différents régimes de retraite
33 La majoration pour tierce personne les conditions à remplir
34 Le minimum vieillesse les allocations de base et le Fonds national de solidarité
37 Pour vivre normalement chez soi une aide ménagère
40 Les soins infirmiers à domicile pris en charge par la Sécurité sociale
42 Les aides au logement pour payer son loyer ou améliorer son confort

Votre santé
44 La protection sociale de vos ayants droits, le tarif de la Sécurité sociale
45 La prise en charge à 100 %

Vos avantages
47 Indépendants de vos ressources la carte Vermeil les tarifs réduits
49 En fonction de vos ressources la carte Emeraude les exonérations

Bien vivre au quotidien

Repenser son emploi du temps et ses relations
56 La retraite un cap à franchir
58 Comment éviter les pièges de la solitude
60 Le temps est à vous sachez le gérer
62 Couple une nouvelle intimité à reconstruire
65 Les enfants devenus adultes
67 Des liens privilégiés avec ses petits enfants
69 Nos parents vieillissent comment les aider

La vie pratique
71 Repenser ses dépenses en fonction des nouveaux revenus

73 Rechercher des revenus supplémentaires
77 Placer ses économies
79 Comment gérer son patrimoine
84 Comment calculer et payer ses impôts
91 Transmettre son patrimoine
95 Bien loger sa retraite
98 Un logement adapté a sa situation

Savoir garder la forme

Entretenir son corps
104 Réveiller sa forme
105 Les bienfaits du sport
111 Les points faibles a ne pas négliger

Faire marcher sa tête
113 Un esprit clair
114 La forme n a pas d age

Découvrir de nouvelles activités

Les loisirs
120 La retraite pour le plaisir interview de Danièle Sédières
123 Voyager
128 Index
130 Carnet d adresses

Figure 1. Summary of the Booklet "Successful Retirement" (op cit)

3. Analysis of a Pensioner's Manual

The French retailers' pension fund "Organic" has distributed to its members a booklet edited and produced by the Notre Temps team (a French monthly for retired people). It is called, *"Successful Retirement" (Réussir sa retraite)*, published in a 19 x 24 cm format, with a glossy colour cover; it has 178 pages. The booklet is published every year with updated appendices. The edition being analyzed here is dated 1992.

There is no need to apply sophisticated analytical methods to extract its substance: the gist is clearly enunciated in the *Summary* (pp. 4-5) reproduced in Figure 1, and is also explicitly stated in the blurb (Figure 2). [*]

The booklet is divided into four thematic sections, each introduced by a title on the right-hand side page, a four- to five-line heading and a section summary; on the left page opposite, a photo illustration.

We shall now list the chapter headings and briefly describe the corresponding illustrations.

Knowing Your Rights. The photograph shows a 50-year-old man in a suit, sitting at his desk, pen in hand; clearly, he is a counsellor employed by the Pension Fund to provide information and help people benefit from all the advantages offered by social security and current legislation provisions. He is also there to advise on managing savings and point to further possibilities of support if needed.

Good Daily Living. Three generations gathered in a meadow in bloom: a retired couple, vigorous and smiling, a couple of young parents in ruptures over their two children, a dachshund and two bicycles.

---

[*] "Quotes from the booklet are in italics if they are titles or subtitles, and in between quotation marks if they are extracted from the main body of the text

Figure 2. Cover Page of the Booklet "Successful Retirement" (op. cit)

Keeping Fit. A sexagenarian woman with a thick head of white hair, doing gymnastics on her own.

Discovering New Activities. A couple leaning on a ship's rail. She, arm pointing, shows her companion some object over the horizon. He appears to wish to photograph what she indicates. Are they pensioners or holidaymakers? He might be sixty, she certainly is not, unless aesthetic surgery and photographic touch-up techniques have had a hand in it!

We shall now propose a brief analysis of the booklet focusing on the following categories: the actors involved, the resources associated with them, the recommended activities and the values informing the discourse.

The main actors, the heroes, are retired people. They are the centre of the narrative. Hardly surprising, perhaps, as they are the booklet's intended readers. They are all endowed with attributes concerning, above all, various types of rights (the notion of duty seems singularly absent from retirement vocabulary): right to retirement understood as stopping work, right to a pension, right to a guaranteed minimum income. Moreover, the retired person can also benefit from various allowances, advantages and services. If needed, he has access to different forms of aid: home help, home medical care, housing benefits.

Thus far the material questions. But there is another resource to take into account too: health matters (*Caring for your body; Keeping your mind on its toes*), to which the third section of the booklet is dedicated.

With material and health matters sorted out, the booklet tackles the reorganization of life in retirement. The second section of the work envisages various situations: living alone (*Alone and happy*), couples, relations with "adult sons and daughters", the role of grandparents and, finally, a novel but increasingly encountered situation in the first years of retirement, relations with "aging parents".

The only "others" mentioned are family members and exclusively in direct descent: parents, children and grandchildren. Of collaterals (siblings and cousins) not a word. The family photograph illustrating the section is redundant, as are the statements and omissions of the summary and main text.

The last section is supposed to outline a project for this new life. It is a matter of *discovering new activities*, that is, "to enjoy all those long-sacrificed leisure pursuits". Propositions are grouped into two main headings: *Retirement for pleasure*, and *Travelling*. The introductory illustration suggests, as mentioned above, a cruise.

The retired life is thus organized around two poles: withdrawal into the bosom of the family as the social integration focus for the retired person (with, moreover, a double limitation: a restrictive definition of the family, and a type of integration excluding cohabitation), and a field of activities mostly reduced to leisure pursuits within which the only significant partners are the spouse and the members of the narrowly-defined family.

The main resources are income, the rights enjoyed by retired people, health and free time, a time which should make possible the elaboration of "a life project" (p. 3). But in which social networks is such project inscribed? Which companions and other agents does it include? As we saw, the only interpersonal network mentioned in the booklet is the family, and only the central core. Beyond the family, the pensioner is pictured as an atomised being or, at best, reduced to the dyad of the couple. On the first photograph, we may suppose that the counsellor is talking with a retired person or couple. The confidential nature of the interview legitimates the face-to-face conversation suggested by the picture. But the gymnast, set in a large open-air area, is all alone, the tourists apparently only interested in each other. Apart from the rare moments spent with the family, the retired person or couple is condemned to solitude.

A comprehensive list of useful addresses completes the booklet. It might be argued that, contrary to our contention, these last nine pages, headed Regaining a place in society, propose multiple forms of participation, offering a long list of associations. No doubt. But then the question is why such a call to responsible commitment appears only as an annex, and why it is totally missing from the main body of the text. It is the main text that spells out the doctrine, the ideology of retirement. The list of addresses is a complement, to be consulted if one wishes -at leisure, we should say. In the official discourse, the retired person is never represented as participating in a group, as an active member of an association, but only as a consumer of institutions inasmuch as these may be useful to him.

The photo of the counsellor embodies the presence of institutional actors such as the State, pension funds, the social security system. The retired person is thus linked with "society" as represented by the Welfare-State. This relationship does not characterise a citizen who votes, discusses political decisions, protests, acts or commits himself. Rather, it defines a contributor/services beneficiary type of relationship (cf. Schnapper, 1991). The pensioner's link to society takes the form of rights; duties are nowhere mentioned. The retired person is either alone, in a couple or with his family. He is never in society, although society surrounds him solicitously offering him numerous guarantees which, nonetheless, cannot totally protect him against the ravages of old age and death, which are only discreetly alluded to in the context of medical care services and drafting of wills.

The values expressed in the booklet outline a veritable ideology of retirement; they all gravitate in the universe of modern individualism. At its centre, the individual (the pensioner) and his/her life. What other certitude is universally shared today apart from the truism: "I am alive"? But at the same time, lacking any competition in the modern Olympus, "my life", this new divinity, is in a hypertrophied state from which derive both the obsession of health, as condition for a good life, and the denial of death. In the individual's life course, retirement signals the exit from public space and withdrawal into the private sphere, the "I", the intimacy of the couple, the joys of the family. A sphere conceived as the reign of freedom, choice, pleasure. The exit from public space signals the end of social constraints and duties, the availability to the expression of desire, the possibility at last of doing what, because of obligations, one had hitherto only dreamt about. From constraining, society becomes permissive. Whilst old age is not yet nigh, although looming in the horizon, retirement appears as the last chance to grasp that Grail of modern times: self-fulfilment, happiness.

The booklet *Successful Retirement* is a caricatural example of the projects and hopes our era has vested in retirement. In recent years, a critique of this conception of life has arisen. Critical voices denounce the social exclusion hiding behind the revival of the Golden Age myth, point to intergenerational conflicts and call for the active participation of retired people. We shall outline this approach in the Conclusion. We may well be witnessing the start of a trend. However, as things stand today in the retirement courses field, the attempt to redefine the general guidelines for training programmes is primarily the work of facilitators and educators, and it clashes with the expectations of the vast majority of the people involved. The survey carried out in the University of the Third Age of Neuchâtel also asked participants to assess the preparatory courses they followed (Jeanneret, 1992). Top of the list were items of technical information concerning the workings of pension funds, the upholding of rights, the question of wills and inheritance and, of course, items on health matters. On the other hand, discussions on the meaning of life, voluntary work, community commitment and the place of pensioners in society came bottom of the list.

There is a homology between supply and demand, a correspondence between the already-internalised expectations of future pensioners and the dominant model informing education programmes.

4. Celebration

The preliminal (or "separation") phase of a rite of passage is found in a very structured form in the case of retirement. Preparatory courses constitute future pensioners as a specific social category, distinct from other colleagues at work, often starting a few years before retirement itself. Initiates have their own activities through which are imparted both the knowledge considered necessary for dealing with their new condition and its general ideology.

What about the liminal phase as such? On the one hand, former workers and future pensioners both experience and are very aware of the symbolism of passage. Here is an interview extract:

> That day I crossed the (company's) door and, stepping onto the street, I looked back and thought:
> 'That's it, I have crossed this door for 27 years and now I won't anymore.' It was a strange feeling.

Walls, a door, an inside and an outside, crossing a threshold of no return. All the elements are there.

On the other hand, the latency period, the suspension which forebodes a change of status, that image of death between two lives, is not found in this modern rite. Moreover, although the symbolism of passage is indeed strongly present, it is hardly structured or systematized. Our study shows two typical situations. In the first, the rite is left up to the retired person himself. Up to him to announce or not his leaving, to organize some drinks at the company cafeteria or elsewhere, to draw up a guest list. He cumulates various roles: organizer, butler, officiant, applicant. In the second situation, the celebration is organized by the company, often as part of the annual party. The traditional workers' jubilee appears to provide the model. Retirement is a good opportunity to extol the virtues of the good and faithful collaborator, to show him as an example to young people, to celebrate the company as a great family and to rouse solidarity. A gift confirms the company's gratitude, bears witness to the individual's past life and kindles remembrance. Sometimes the symbolic object is a typical product manufactured by the enterprise (for instance, in watchmaking companies or foundries). The offering of chains, clocks, watches or platters is also common, as these objects are well adapted to commemorative inscriptions.

In the Civil Service this type of homage is contemplated in the administrative regulations. For example, in the Republic and Canton of Geneva, somebody employed for over ten years "has a right to a souvenir-gift" (*"Hommage du Conseil d'Etat"*) on retirement or invalidity superannuation". The person concerned then has a choice between two or three objects!

According to Durkheim, through rites society reaffirms its ideals by expressing them (Durkheim, 1960; Conclusion). Homage paid to retiring colleagues is thus an opportunity to confirm the enterprise culture. However, in these times of economic restructuring and widespread redundancies, the cultural model praised in such celebrations has become dysfunctional and anachronistic. The virtues of the new enterprise culture are no longer faithfulness, hard work, seniority and continuity, but change, innovation, adaptability, mobility and profit at any price. It is no longer appropriate to give a high profile to retirements which come to swell the ranks of departures and which are no longer part of the traditional generational cycle. Discretion is called for. Training is included in redundancy programme packages, but it is better if people leave by the back door, without making a great song and dance about it. Under the circumstances, an official celebration may seem in bad taste and may even provoke some counter-event. Thus, current trends are to leave it up to the person concerned to mark the occasion as he sees fit.

These two typical forms of departure ritualization have an impact on the symbolic domain. Manuals present retirement as a passage from the public to the private space, from an institutionally regulated life to a life guided by free will. When it is the enterprise which presides over the departure ritual, it leads the novice to the place of passage and waves him goodbye from the threshold as the individual goes forth alone towards his new life. In the second typical form, departure is regulated in its more formal aspects, but the celebration is no longer a public event. Although the individual is still within the enterprise's walls, in a sense the company has already withdrawn and does not accompany the neophyte along the departure path but lets him decide whether others will be associated to the event or not. In this case, the whole of the passage, from start to finish, has already been defined as a private affair.

5. The Postliminal Phase: An Belonging-Free Identity?

Up to now, we have only considered the leavetaking part of the passage, the last minutes of working life and of playing a professional role. But a passage, symbolised here by the company door, is by definition both a departure and an entry. According to the theory of rites of passage, the third phase is the outcome of the process, through which the previous ones gain their full meaning. The whole point of separation and the suspension of time associated with the passage is to allow integration into a new social group and, through such incorporation, a rebirth into life and into the group by assignation of a new identity and role. In the case in question, what is the third ritual phase?

Ambivalence haunts the responses collected; they express a sense of strangeness, recount joys and fears, talk of loss and relief, freedom and emptiness, of turning over a new page and of waiting. They tell of returning home that last day sometimes hurriedly, sometimes slowly and stopping over at the pub: "We have the time, now". They portray the spouse waiting at home, a lovingly prepared meal in the background, or getting ready for going out to a restaurant "to celebrate".

The celebration is primarily the couple's affair. Sometimes, a family meal is organized, including children and grandchildren, rarely one or two intimate friends. The celebration may take the form of a more or less extended trip, sometimes presented as an ersatz honeymoon. The respondents' autobiographical narratives reflect the image of the couple leaning on the ship's rail suggested by the manual analyzed above.

The activities reported correspond closely to the ideology of retirement. Retirement consigns the individual to his home and the retired person becomes, sometimes with the spouse's help, the officiant of a new cult. A cult of intimacy, of the couple above all. A cult of "self", of an "I" which can finally affirm itself and do as it pleases; indeed, which not only can but must do so, because henceforth choosing becomes just as much a duty as a possibility.

There is a striking absence of a welcoming group or social reintegration ceremony. To some extent, it is in the nature of things: as retirement commits the individual to his free will, Rosow (1967) had already described the retired person's role as an empty role, it is up to the initiate to organize his new life. Although a lot of information is provided on existing possibilities and on the large number of institutions, groups and associations ready to welcome him (this is one of the tasks of pre-retirement programmes), supposedly nothing should be forced upon him. Nothing is imposed, apart from preparatory courses (which are, in fact, recommended rather than compulsory) which reinforce an ideal of retirement. Nothing, apart from a ritual initiation which accompanies the neophyte to the company door, signalling the breaking of the link binding the individual to society and pointing to a road of freedom through the landscapes of the couple, private life, the family, leisure and consumption.

Applying to retirement the theoretical model of rites of passage produces a surprising result. In the early phases, the preliminal stage, the ritual is precise and strongly structured. But the nearer it gets to the moment of passage, the more it wears thin, becoming -once the transition achieved- an assortment of hazy practices and customs. Retirement is organized as a rite of passage, but the ritual is inconclusive, lopsided. It is a departure not an entry rite, separation without reintegration. But, we would argue, this lopsided structure is not unfinished, it is both coherent and concluded. After all, on retirement, does not society release the individual from his major duty? Does it not withdraw, like an ebbing tide, leaving the individual upon a hindrance-free beach of which he is lord and master? Therefore, this ritual organization, strongly structured in its liminal phase and destructured in its final stage, aptly expresses the modern ideology of retirement.

6. A Rite Between Two Mythical Universes

By myth we mean a narrative or set of images referring to the emergence of a "state of affairs", a foundation, which they endow with meaning and legitimacy (Encyclopaedia Britannica, 1988; Lalive d'Epinay, 1991). We would like to argue that to the split structure of the retirement *rite* there corresponds an articulation of two mythical universes, each applying to a different historical period. Through its celebration of the end of a life cycle (the central phase), the rite expresses the ethos of work and duty typical of industrial society. Whilst through its preparation for later life, it sings the praises of the morality of self-fulfilment, the leading theme of post-industrial society's culture and mythology.

During the first half of the century, social struggles for retirement presented old age, like illness or disability, as a condition justifying stopping work, on the one hand, and the mustering of collective solidarity around retired people, on the other. These struggles expressed a revolt against the tragic and all-too common fate of men and women suffering from the ravages of old age but forced to continue working by the threat of hardship The decrepitude associated with poverty-stricken old age appeared all the more abhorrent that these men and women had toiled throughout their lives, thereby faithfully if anonymously contributing to the great task of industrial society Military metaphors abounded in contemporary texts, like the soldier, the worker was the homeland's child, its servant To die in the mine or in a building site, was like dying in action Some countries exalted labour by establishing Orders of Merit to decorate and celebrate their heroes But, if it was heroic to die "on the battlefield", what about the widows and children? And where was the glory when the hero's strength shrivelled and he was unable to carry out his job, becoming an infirm body, a burden?

Industrial society had made labour into a new adventure, and technique into the ultimate weapon for the domination and transformation of the world At one point in history, it decided to offer its anonymous heroes, the workers, what kings had earlier granted to their faithful servants

Retirement was seen as the reward of deserving workers up to this day, all systems condition pension rights and revenue to the number of years of service (sorry, of work!) The Promethean myth adopts a modern guise when the time comes, if the gods have spared them, the anonymous heroes will withdraw with honours from the scene, and thousands of young champions of humanity will march forward to take up the titanic task

"A right won by a life of toil" this part of the myth retains today all its force There is no retirement preparation course or farewell speech which do not include an eulogy of past working life, a hymn to accumulated merit and hence an apology of legitimate right And no self-respecting debate on raising or lowering retirement age ever fails to intone this refrain

But another part of the original myth would indeed disappear under the impact of the changes brought about by the "thirty glorious years" From the 1930s to the 1960s, retirement was seen as a "right won by toil" and the transition heralded the arrival of the "winter of life" It was a right to rest granted to the elderly Hence the recurrent image of the "old people's bench", the bench symbolising the support granted to people whose legs could not carry them anymore It was the right to shift from the role of actor to that of spectator

But during the following decades there was a growing dissociation between the age of retirement (on a downward trend) and increasingly deferred old age Life expectancy kept going up If we adopted today the rule followed by the early Legislator—i e to make retirement age and life expectancy coincide—men would have to wait until their 75th, and women their 80th, birthday to retire from work A space thus emerged was deployed in between retiring and old age This new stage in the life course was very quickly endowed with the values issuing from the cultural revolution taking place at the time Because, suddenly, after the war, that which industrial society had always promised but nobody was expecting anymore, on that scale at any rate, finally came to pass affluence and a brighter future! Economic growth was reflected in rising purchasing power Home budgets stopped being stretched by mere survival demands and began to assign greater portions to new needs free time activities, leisure, travel The environment changed and so did attitudes A cultural mutation occurred in customs, lifestyles and, more profoundly, values and beliefs This transformation was heralded by four messengers materialising almost simultaneously in the 1960s the car, television, the pill and holidays In its own way, each connoted freedom and proposed a *trip* The fact that these emblematic objects also comprised other aspects and borne their share of darkness, to say the least, was not enough to shake off their imaginary load, the hope which the "spirit of the times" had imbued them with (Morin, 1962 and 1973, cf Baudrillard, 1970, Dumazedier, 1988, Mendras, 1988, Lalive d'Epinay, 1990)

After a century-and-a-half of relentless labour, the individual could breathe at last, discover free time and himself, all his aspirations, drives and dreams which, suddenly, rejoined the realm of the possible. He could now explore territories whose landscapes had until yesterday been but illusion and chimeras. No sooner had industrial society produced the affluence in whose name so much sacrifice by labour had been exacted and legitimised, than the work ethos began to wither away and its values turned upside down (Yankelovich, 1981). Work stopped being seen as a duty, a "social mission". Personal happiness and self-fulfilment, rather than work, became the true vocation of man. Consequently, the relation between individual and society was reversed. Formerly, man fulfilled himself by discharging his social duties, the individual was at the service of the social whole. Now, the vocation of society, especially of the State representing it, was to be at the service of the self-fulfilment of its individual members.

This new ethos developed, to begin with, among the middle classes and the younger generations. At first, the rural areas and older people resisted. The latter were not ready to relinquish the values which had guided their lives; in any case, pensions only slowly followed suit the rise in affluence. But, as years went by, new cohorts reached retirement age having had time to internalise the new spirit of the times. They do not feel old and retirement appears as a "fulltime free time era", the passage signalling the release from constraints, the access to a surprising Thelemes' abbey, to an age of life whose motto is "Do as you wish". Released from the need to work but not yet affected by the tribulations of old age, this phase of life bears most auspicious tidings for the quest for the modern Grail: self-fulfilment.

Obsolete, the original idea of a *right to rest* fades away and the *right to life* replaces it. "Time for living", or stronger still, "time to live", proudly proclaim the two poppies adorning the logo of senior citizens' associations. Indeed, time presses because, even if they do not consider themselves elderly, old age looms on the horizon with its dark clouds: decrepitude, dependence, homes. There is a striking contrast between the eagerness to enjoy life, on the one hand, and the spectre of dependent old age and the proximity of death, on the other. A typical trait of contemporary culture is clearly evinced in such contrast: the idolatry of life and the fear of death.

CONCLUSIONS AND DISCUSSION

1. With the coming of post-industrial society, whose social structure assigns a growing place to organizing relations between "generations",*, The use of ethnological paradigms and categories proves both relevant and fruitful.

Applying Van Gennep's model we have been able to show that the generalisation of retirement has been accompanied by the elaboration of a rite in which the initiate leaves the public space to settle in the private sphere through a separation process. As the process unfolds, the initiate gains a new identity, but an identity without belonging. The rite issues in an opening up, it ends when the individual is restored to himself rather than, as in the

---

* "The increasingly frequent use of the term *generation* (conflicts between generations, intergenerational contract,) is, strictly speaking, incorrect in this context and can lead to confusion We are in fact discussing groups that are distinct due to the socially-defined position their members occupy in the life courses It is this position along the life course, with its social status and definition, rather than individuals' generational membership, that underlies the current debate on the sharing of resources between "active" and retired people The distinction is not without consequences, membership to a generation characterises an individual from birth, from childhood, and it accompanies him throughout life, from one transition to another, In this sense, an "active" person is a potential pensioner, just as a pensioner is a former "active" person, but every one remains a member of the same generation throughout his/her life "

traditional case, to a new belonging group. The rite dramatizes a symbolic script based on elements taken from two distinct mythologies. Some derive from industrial society: the myth of labour as salvation and the figure of the deserving worker. Other elements were added by the cultural revolution of the 1960s: the theme of the Golden Age and the figure of the free and fulfilled "self".

2. Our analysis has pinpointed the three features which, according to Centlivres, characterise the current evolution of rites of passage (cf. *supra*, section 2). Firstly, *emblems* become rarer. If it is the company that bids the employee farewell at a ritual event, then some symbolic insignia are indeed exhibited. But if the farewell is left up to the initiate, no emblematic devices appear, except perhaps the final shedding of working for civilian clothes. Only the parting gift remains. The second significant trait, according to Centlivres, is the withdrawal from the local community into the family. Quite, but we must take the analysis a step further. In the case studied, it is the company that withdraws. Indeed, in contemporary urban society, companies—as working communities—rather than local communities, provide the stronger links between individual and society. During the progress of the retirement rite, the initiate is dispatched from the company to his family without any mediation: the local community plays no role and even the extended family, everpresent in other life cycle ceremonies, is excluded. The pattern here is a most radical leap from the social to the intimate. The third change consists in the effacement of the cleric, the officiant. In this case too, retirement confirms the trend and takes it to extremes. By making the initiate the officiant from mid-rite onwards, retirement conveys the deeper meaning of the message: you are free, your are your own master; up to you to organize the liturgy of your life.

These three trends identified by Centlivres are found in retirement rites in a striking way, revealing the forces at work in such transformations. The disappearance of the local community as the group to which the individual de facto belongs, evinces the establishment of a no man's land between society (embodied in great public institutions: State, business concerns, school) and the private sphere. The liturgical role assumed by the future pensioner expresses the contemporary individualism which makes the "self" the centre of the world.

3. With reference to the canonical theoretical model, retirement appears, in the absence of a reincorporation phase, as an incomplete, lopsided rite. Should we conclude that such incompletion means non-accomplishment? On this issue we beg to differ from the Catalan anthropologist Fericgla (1992, p.119) who defines the retirement rite as "destructuring and distructured" (*"desestructurante y desestructurado"*). In as much as the rite marks the passage from an individual's life course stage characterised by participation in the social project to a new stage in which he is left to his own devices, the rite is destructuring for the individual but not destructured in itself. The rite's incompletion should not be interpreted as non-accomplishment. On the contrary, it thereby evidences a requirement of the ideology it dramatizes. The rite, in this sense, is lopsided; it is structurally incomplete but fully accomplished. Retirement restores the individual to himself, to his free will. Consequently any reintegration ritual would seem prejudicial to that freedom and deny the project vested in this life stage by contemporary culture: society withdraws from the individual and leaves him to his choices, purportedly only reappearing once options have been defined.

4. But has society truly withdrawn? The sociologist must distinguish between the level of ideology, messages and representations, and the level of practices. In fact, society is far from absent. But its presence is channelled through other mechanisms. The *group* has undoubtedly withdrawn from the individual, or rather, the company—understood as a set of groups allowing individual interaction and circulation within the social body. The evolution of modern society reduces the number and role of the groups to which the individual *de facto* belongs and which function to enmesh him in a long chain of exchanges and interactions. The Church, for example, only fulfils this role for a minority; local community institutions yield to voluntary—hence, non general—organizations; the school and companies increas-

ingly gain the monopoly of the de facto articulation between individuals and the social whole. For the retired person, upon leaving the company, the main link is broken.

In other respects, however, society continues to enfold pensioners but in a different way. Society's presence is felt not through human networks, belonging groups, responsibilities and shared activities, but rather through impersonal channels. Such as the State-managed sums which the individual receives monthly; such as the rights he can claim from assuredly competent but definitely anonymous civil servants; such as the permanent and obsessive call to consumption which targets retired people, offering castles in the air and other never-never lands. Retirement is pronounced free, but an anonymous system watches over the way in which freedom is exercised, guiding its choices and decisions. The peculiarity of such system, be it the State through social security or the economy through advertising and consumption, lies in treating the retired person as an atomised individual, without roots or belonging, reduced to an assortment of needs.

5. Retirement, a new Golden Age? We could also speak of the age of golden exclusion. Through the withdrawal into the domestic and family sphere, through the flight into leisure and travel, the retired person exiles himself from society and, seduced by their siren calls, confines himself to those modern limbos, at the frontier between the living and the world of the dead.

Once again, ethnology can shed light upon our own society. In community-based societies, rites of passage into old age are rare, because aging is a slow process. Definite signs and events are required to build up a rite. By choosing a chronological age, modern society assigned a date to old age and made the rite of passage possible. In community-based societies, the infrequent rites of passage of this type concern women, "whose old age is signalled by a clear biological event, menopause" (Belmont, op.cit., p.17). Lupu (1981) has described one of these rites. But in fact, they are less a rite of transition into a chronological age than an "early death rite", an act of mourning. Old age is a marginal or transitional stage in itself, marked off not by true rituals but by social practices which uproot the elderly from the adult age class as well as, to a greater or lesser extent according to specific societies, from the local community as a whole (Belmont, ibid).

In an early death rite, the transition is the final stage. At first, retirement adequately met the anthropological need for an established margin in which the elderly could abide. But in view of the dissociation between pensionable age and senescence, retirement places the prelude of death very early on, whilst covering up this marginalisation with the mirage of the Golden Age.

6. Could we imagine another conception and ritualization of retirement? Many voices call today for a new "intergenerational" pact (Binswanger, 1993; Economic and Social Council, 1993). Could we not imagine local communities and authorities, voluntary organizations and senior citizens' groups arranging an annual neighbourhood retirement party? Retired people from the current and previous years would be invited, as well as representatives from various groups and age classes, to celebrate the potential of the new phase of life. A regularly updated booklet would offer information on the variety of socially useful, though unpaid, activities helping to flesh out a society which would otherwise be mere skin and bone or, at any rate, a dehumanised machine. Without forgetting leisure, and respecting the requirements of the new situation and age group, the contribution that the bounty of retired people's accumulated knowledge could make to society would be vindicated, and the price of wasting it pointed out. Hence, this party would herald a new type of participation no longer based on constraint but on choice, not on the peremptory need for income but on the satisfaction of contributing and demonstrating one's competencies. A utopia? Certainly, if by this is meant a possibility that has not yet found a place in reality. But then, so was retirement at the beginning of the century!

REFERENCES

Atchley, R C , 1982, Retirement as a social institution, *Annual Review of Sociology*, 8 263-287

Baudrillard J , 1970, *"La societe de consommation"*, Gallimard, Paris, in French

Bell D , 1976, *"Vers la societe post-industrielle?"* PUF, Paris (1st ed 1973), in French

Belmont N , 1986, La notion de rite de passage, in *"Naître, vivre, mourir, Actualite de Van Gennep"*, P Centlivres, J Hainard, eds , Musee d'ethnographie, Neuchâtel, pp 9-19, in French

Bernardi B , 1985, *"Age Class Systems Social Institutions and Polities Based on Age"* Cambridge University Press, Cambridge

Binswanger P , 1993, *"Le contrat intergeneration est il encore valable?"* Conference presentee a l'assemblee annuelle de la Fondation suisse Pro Senectute, in French

Bois J -P , 1989, *"Les vieux"*, Fayard, Paris, in French

Bovey G , 1992, *"Les pratiques de preparation a la retraite sous la loupe"*, Memoire de diplôme, Fapse, Universite de Geneve, in French

Centlivres P , Hainard J , eds , 1986, *"Les rites de passage aujourd'hui"*, L'Âge d'Homme, Lausanne, in French

Centlivres P , 1981, Les rites de passage nouveaux espaces, nouveaux emblemes, in *"Naître, vivre, mourir Actualite de Van Gennep"*, J Hainard, R Kaehr, eds , Musee d'Ethnographie, Neuchâtel, pp 161-174, in French

Clark T N , Lipset S M , 1991, Are Social Classes Dying?, *Intern Soc* 6(4) 397-410

Conseil Economique et Social, 1993, *"Les activites d'utilite sociale des retraites et des personnes agees"*, Rapporteur Prof H Thery, Paris, in French

Dumazedier J , 1988, *"Revolution culturelle du temps libre, 1968-1988"*, Meridiens Klincksieck, Paris, in French

Durkheim E , 1960, *"Les formes elementaires de la vie religieuse"*, PUF, Paris, [1ere ed 1912], in French

Encyclopaedia Britannica, art "Myth", 1988, 15th ed , 24 710-727

Fericgla J M , 1992, *"Envejecer Una antropologia de la ancianidad"*, Anthropos, Barcelona, in Spanish

Fourastie J , 1979, *"Les Trentes Glorieuses"*, Fayard, Paris, in French

Glaser B , and Strauss A , 1967, *"The Discovery of Grounded Theory"*, Aldine, Chicago

Gaullier X , 1988, *"La deuxieme carriere'* , Seuil, Paris, in French

Hainard J , Kaehr R , eds, 1981, *'Naître, vivre, mourir Actualite de Van Gennep'* , Musee d'Ethnographie, Neuchâtel, in French

Jeanneret R , 1992, *"La retraite et les cours de preparation a la retraite'* , Universite du 3eme Age, Neucâtel, in French

Kohler P A , Zacher H F , eds , 1982, *"Un siecle de securite sociale* L'evolution en Allemagne, France, Grande-Bretagne, Autriche et Suisse", C R H E S , Nantes, in French

Kohli M , 1985, The World We Forgot A Historical Review of the Life Course in Later Life, in *"Later life The Social Psychology of Aging"*, V W Marshall, ed , Sage, Beverly Hills (CA), pp 271-303

Lalive d'Epinay Chr , 1989, Individualisme et solidarite aujourd'hui, *Cah Int de Soc* LXXXVI 15-31, in French

Lalive d'Epinay Chr , 1990, *"Les Suisses et le Travail Des certitudes du passe aux interrogations de l'avenir"*, Realites sociales, Lausanne, [German translation Die Schweizer und ihre Arbeit, VdF, Zurich, 1991], in French

Lalive d'Epinay Chr , 1991, Les fondements mythiques de l'ethos du travail, *Arch de Sc soc des Rel* , 75 153-168, in French

Lalive d'Epinay Chr , 1992, Beyond the Antinomy Work versus Leisure The Process of Cultural Mutation, *Intern Soc* , 7 397-412

Lalive d'Epinay Chr , 1993, Le contrat intergenerationnel dans la societe de demain, in *"Gerontolologie et Economie"*, Schweizerische Gesellschaft fur Gerontologie / Societe Suisse de Gerontologie, Berne, pp 27-36, in French

Lalive d'Epinay Chr , 1994, La construction sociale des parcours de vie et de la vieillesse, in *'Une histoire de la vieillesse"*, G Heller, ed , Editions d'en Bas, Lausanne, pp 127-150, in French

Lupu Fr , 1981, Les passages a la mort chez les Tin dama du Sepig (Nouvelle Guinee), in *"Naître, vivre, mourir Actualite de Van Gennep"*, J Hainard, R Kaehr, eds , Musee d'Ethnographie, Neuchâtel, pp 149-160, in French

Mendras H , 1988, *"La seconde Revolution française, 1965-1984"*, Gallimard, Paris, in French

Morin E , 1962, 1973, *"L'esprit du temps"*, Grasset, Paris, in French

Roheim G , 1942, Transition Rites, *The Psychoanal Q* , 11 336-376

Rosow I , 1967, *"Social Integration of the Aged"*, Free Press, New York

Schnapper D , 1991, *"La France de l'integration"*, Gallimard, Paris, in French
Schneider H -D , 1981, Selbstverstandnis, Ziel, Inhalte und Formen der Vorbereitung im Alter, in *"Vorberei-
    tung auf das Alter im Lebenslauf"*, Pro Senectute ed , Paderborn, Schoning, pp 39-61, in German
Tornay S , 1988, Vers une theorie des systemes de classes d'âge, *Cah d'Et Afr*, 110 XXVIII(2) 281-291, in
    French
Touraine A , 1969, *"La societe post-industrielle"*, Denoel, Paris, in French
Turner V , 1969, *"The Ritual Process Structure and Antistructure"*, Routledge & Kegan, London
Van Gennep A , 1908, "Essai d'une theorie des langues speciales", *Rev d'Et Ethno et Soc* , 1, Paris, in French
Van Gennep A , 1969, *"Les rites de passages"*, Mouton, Paris-La Haye, [lere ed 1909], in French
Winterthur-Vie, 1991, *"Le nouveau troisieme âge"*, Winterthur, in French
Yankelovich D , 1981, *"New Rules Searching for Self-Fulfilment in a World Turned Upside Down"*, Random
    House, New York

6

THE ROLE OF A PREPARATION FOR RETIREMENT IN THE IMPROVEMENT OF THE QUALITY OF LIFE FOR ELDERLY PEOPLE

René Jeanneret

Université de Neuchâtel
CH-2000 Neuchâtel
Switzerland

INTRODUCTION

In fact, the question is a simple one: are courses in preparation for retirement useful or not? Are they of real benefit to those who have attended them?

Opinions differ. For some people: "A preparation for retirement is a very positive initiative of a nature to play a decisive role when one's active life comes to an end, and, as such, deserves to be widely encouraged and developed" (Rageth, 1982).

For others: "The survey reveals a very critical attitude towards a preparation for retirement, almost stressing its uselessness. On the contrary, paradoxically, those who have not attended such a course expect a great deal from it and frequently express a need to be able to discuss this key period of life..." (Micheloni, 1989).

These contradictory opinions led us to question directly those concerned, and first of all the students—all retired persons—of the Senior Citizens' University of Neuchâtel, a small university town in Switzerland. Having been privileged to direct this institution over a period of 15 years, I had often been struck by the vitality shown by those persons, by their open-mindedness, and their desire to keep on learning and to remain active, in short, by the quality of their lives. For such persons, retirement had clearly been a success, and I wondered whether or not this success was due to the fact that they had attended a course designed to prepare them for retirement.

My talk is therefore mainly built around the results of a survey organized in the early 1990's. Given the limited time at my disposal, however, it will not be possible to deal systematically with all of these results.

I should like simply to present a few of the global results of this research work, and I must point out that my aim is merely to demonstrate certain general trends, without paying too much attention to the absolute value of the figures. Moreover, the notion of a quality of life cannot be defined and quantified precisely. It depends essentially on individual cases considered separately, and one can meet an elderly, disabled person who will declare that he

Preparation for Aging, Edited by E. Heikkinen *et al.*
Plenum Press, New York, 1995

is perfectly content with his lot, whereas others, less tried by fate, may complain of conditions of life that they appear to put up with rather than to appreciate.

We merely wish to say that the aim is to provide an aging person with as good a quality of life as possible from the point of view of health, housing, security and financial resources, with a caring environment for heart and mind, with a life free from the worries that come from advancing years and its side-effects of all kinds, in short, a life worth living. Such a list, of course, appears utopian. It is nevertheless an ideal to aim at, and its attainment may be furthered, at least in part, by courses designed to prepare people for retirement.

THE SURVEY

I drew up a questionnaire on the basis of information provided by several institutions or organizations that had set up such courses. The different themes dealt with, which were very comparable from one organization to another, served as a basis for 50 questions to be answered by ticking one of the following columns: very useful; useful; not very useful; useless. In addition to the members of the Neuchâtel Senior Citizens' University and its annexes, the survey was extended to four organizations inside and outside the canton.

In all, 547 persons filled in the questionnaire out of just under a thousand, of whom 33 had attended no course whatsoever and 214 had benefitted from a preparation (154 men and 60 women). In our analysis, we took into account the following variables, apart form the sex: the social classes based on occupations, namely: lower working class, salaried working class, salaried middle class, self-employed middle class, upper class and house-wives. The period of time elapsed since retirement was taken into account, as well as the place or organizing body of the course, classified under the term "origin" (Jeanneret, 1993).

RESULTS OF THE INQUIRY

As regards the general question: "Did you find this preparation useful?" 186 persons answer in the affirmative, i.e. 87.3% (very useful or useful) while 3 (1.4%) consider it to have been utterly useless. These figures are encouraging, and the case appears to be proved. But it is wise to beware of taking this assertion at its face value, since it concerns a general average, and the opinions expressed by the persons questioned with regard to the usefulness of the course attended and the positive after-effects they have noted (or believe they have noted) in the more or less short term following their retirement are not homogeneous, being contingent on the above-mentioned variables. Then too, not all the topics dealt with are judged to be equally beneficial and interesting.

Roughly speaking, these topics can be classified under three major categories that are fairly well typified, but which, I hasten to add, are not to be thought of as constituting watertight compartments.

The first category is confined to technical or utilitarian information concerning the status of the retired person, within the social fabric, his rights and his duties.

The second category consists of advice concerning rather the individual in particular and aiming at a readjustment of his attitude to life in the future.

Finally, the third category is given over the themes designed to make an aging person reflect on the significance of work, on the meaning of life and on the problem of solitude.

Each of the sections forming our three categories is sub-divided into precise themes on which our questions had a bearing.

ANALYSIS BY CATEGORIES

1. Technical or Utilitarian Information

In fact, we note that this category is considered as being of the most immediate use, with 87.76% of favourable opinions (very useful or useful). It comprises four sub-groups which can be set out in order of importance:

- In the first place, the sub-group *Retirement* comprising: the retired citizen, rights and duties; models of early retirement and the status of retired persons (91.5%).
- In the second place, the sub-group *Succession*, i.e. making one's will, death duties, and marriage settlements.
- In the third place, the sub-group *Institutions* comprising: services and institutions at the disposal of the general public: Pro Senectute; Pro Infirmis; home care and home helps (90.6%) (there was naturally no question of passing judgement on the usefulness of these institutions as such, but solely on the usefulness of the information obtained concerning them).
- In the fourth place, the sub-group *Finance*, composed of annuities and superannu-ation schemes; l'Assurance Vieillesse et Survivants (AVS); insurances, taxation, savings, balancing a budget (77%).

There is nothing surprising about this kind of information being preferred, for retirement sets a number of concrete problems, and it is not only indispensable but it also provides a feeling of security to know what practical consequences will follow the end of one's active life: what will the pension amount to? Where must one apply in order to benefit from the *Assurance Vieillesse et Survivants* (State Pension Scheme)? How long is there to wait? How much will taxation amount to? In the event of difficulties arising which institution must one apply to? How ought one to deal with one's inheritance problems?

In all these spheres, ignorance engenders anguist, and the answers given during courses certainly help to smooth the path towards old age.

But one must beware of drawing overhasty conclusions, for, as is only to be expected, these themes do not all arouse the same interest. In addition, we note that the lower working class is more appreciative than others of the usefulness of this type of information (93.25%), while the upper class, composed of people who are obviously better acquainted with these problems, comes out bottom among the social classes, with 81.35% of positive opinions. If one takes as one's criterion the number of years elapsed since retirement, it is clear that, the older one grows, the less interest one takes in the information described in this category, with the exception of the problems of inheritance, which remain of vital concern: 92.8% usefulness rating among persons who have been retired for 9 to 12 years.

But one cannot really talk here of an improvement in the conditions of life for an aging person, an aspect dealt with more specifically in the advice provided in our second category.

2. Advice Concerning a Readjustment of the Individual's Future Attitude Towards Life

If retirement is a punctual event, marking a sharp change in a person's social and working life, the fact remains that every human being inevitably undergoes the progressive effects of passing years, so that, inescapably, courses designed as a preparation for retirement must, to a great extend, take into account the need to prepare people for growing old. The aim must therefore be to provide advice and to issue warnings, in order to bring to people who are advancing in years an awareness of several factors that may make it easier for them to adapt to a new kind of life and render their conditions of existence as pleasant as possible. In particular, it is vital to maintain and stimulate their intellectual capacities, to assist them to preserve their health and their security, to prevent them from withdrawing into a life of isolation and loneliness, in short, to help them to achieve a greater wellbeing.

With a score of 83.65%, the category *Advice* is placed globally after the technical or utilitarian information in the interest shown by the senior citizens consulted.

We find here the following sub-groups, classified in order of importance:

- In the first place, the sub-group *Intellectual Capacities* i.e. intellectual capacities and advancing years; memory-training (90.7%).
- In the second place, the sub-group *Security*: security and prevention; domestic security; road security (88.3%).
- In the third place, the sub-group *Leisure*: leisure activities; how to use the free time; hobbies; sports; the Senior Citizens University (83.2%).
- In the fourth place, the sub-group *Retired Persons' Testimonies* (82.9%).
- In the fifth place, the sub-group *Health*: medical questions: how to preserve one's health; how to avoid illness; how to compose one's diet; dietetics (80.6%).
- In the sixth place, the sub-group *Voluntary work* (76.2%).

It is in these various fields that a direct relationship can be established between the advice given during the courses and their consequences. However, the figures given for each sub-group must be interpreted to a certain extent, for, in examining our survey point by point, we note that the themes concerning health are regarded as being more useful than all the others: in the first position among the 50 points of the questionnaire: how to preserve one's health (96.3% of positive answers); in the second position: how to avoid illness (94.8%). To enjoy good health is obviously one of the key-factors for a good quality of life, and it is one of the major preoccupations of aging persons, who are more and more prone to illness. It therefore comes as no surprise to note that this interest not only remains constant, but becomes even more marked in the category of the senior citizens who have been retired for 9 to 12 years.

The problems of security come next; 100% of the women regard these topics as useful, compared with 90.4% of the men. This difference is easily explained by the fact that women are more frequently exposed to conning or mugging.

Without denying the interest of these attitudes, we feel it is more important to know whether, on precise points, the participants have modified certain aspects of their behaviour as a result of the courses in preparation for retirement.

It is obvious that the end of a working life brings in its wake a modification of the conditions of existence. But 27.6% of the people questioned recognize the influence of the courses. According to them, these changes affect the organization of the daily routine, with the pleasure of being able to plan their time and their personal activities, and of making good use of their new-found freedom.

On the other hand, we were led to wonder whether certain eating habits had been modified. The results are relatively disappointing, since only 28.8% seem to pay heed to the recommendations, i.e. to have a healthy, balanced diet appropriate to their age, factors which help to preserve good health and improve the conditions of life. Women are more alive to this necessity than men (respectively 33.3% and 26.3%).

As regards the question: "Do you spend more time on physical exercise and sport?" The result is more encouraging, since 63.1% of the persons consulted declare that they practise more sport than before retirement. Walking heads the list, followed by gymnastics, swimming and skiing, and, to a lesser degree, cycling, yoga and tennis.

A clear renewal of interest is also shown in cultural activities. For most of the persons concerned (50.8%) this consists in attending lectures and courses at the Senior Citizens' University or the *Université populaire* (Open University), not to mention club-schools. Interest is shown in the study of languages and history, in reading and in radio, television and the cinema.

Finally, the question was asked whether, as a result of the course, the time spent on artistic activities had also increased. On this point, 33.7% of positive answers were given. We note, on the one hand, a fresh interest in concerts, the theatre, visits to museums or exhibitions, while, on the other hand several persons themselves practise artistic activities (painting, sculpture, sketching, photography, instrumental or choral music) or various crafts (embroidery, weaving, knitting, pottery etc.).

3. Reflection

Our third category comprises topics designed to arouse reflection and discussion (81.2%).

- In the first place, the sub-group *Generalities*, comprising the themes: thinking ahead; my future retirement; constructing one's retirement; the retirement adventure (this being generally the same subject, presented as an introduction to the courses) (86.1%).
- In the second place, the sub-group *Philosophy*: the meaning of life; a philosophy for later years; suffering and death; modifications in mental outlook and loneliness (81.9%).
- In the third place, the sub-group *Work*: the significance of work; reflection on work; work and the space of life; bridging the gap between work and retirement (81.7%).
- In the fourth place, the sub-group *Life*: living alone; living as a couple; living with several generations; housing and environment (76.7%).

In this category, if we place ourselves in the perspective of the lapse of time since retirement, we note two contrary movements.

As regards the sub-groups *Generalities* and *Work,* the interest tends to flag, since, with the passing of the years, people feel less and less concerned by themes in which they are no longer personally involved. On the other hand, reflections on life, which have little appeal for newly-retired persons, are much more sought after by their elders (87.6%) as is the sub-group *Philosophy*, which is also greatly favoured by the more elderly.

This fact in itself is scarcely to be wondered at, and confirms the appeal of spiritual values and reflection among the older members of the population, who lean to relativize things and attribute less importance to "concrete problems".

THE ROLE OF SENIOR CITIZENS' UNIVERSITIES IN THE PREPARATION FOR RETIREMENT

These few remarks lead us to define the role that might be played by Senior Citizens' Universities in the sphere of advancing years and, why not, in that of the courses in preparation for retirement and their follow-up.

A first remark must be made: the themes forming our first category are not of a specifically academic character. They are designed to give information of a practical or technical nature, the level being that of vulgarization and imparting of information.

On the contrary, the second category, reserved for advice, can be included within the sphere of training processes, part of which, to our mind, appears to fall within the domain of University. Thus, the problems of health in general, avoidance of illness, dietetics, intellectual capacities, disablement due to advancing years, could be dealt with by our Senior Citizens' Universities in the form of lectures or talks given by specialists.

On the other hand, certain cultural or artistic activities already occupy a place of honour in the Senior Citizens' Universities, whose role in opening their students' minds to such activities has proved to be of long-term benefit, and we would like to put in a plea for the attribution of a greater and more responsible role to the older students in these mind-training processes.

In addition, we ought increasingly to seek the collaboration of members of the Senior Citizens' Universities in the preparation for retirement. We have seen that the testimonies of retired persons are regarded as being useful by 84.6% of those consulted in our survey, so true it is that the experience of persons who have succeeded in bridging the gap cannot but be beneficial, especially, as has already been noted, for the younger retired persons who have not yet solved all the problems set by their working life having come to an end.

Finally, the third category, thanks to its aspects of reflection on life and its philo-sophical approach to the problem of existence, might become the privileged domain of the Senior Citizens' Universities, by encouraging exchanges in the form of workshops, group research with and by the older students, resulting in a new balance of life, a new "savoir-être".

These few considerations confirm the idea that the Senior Citizens' Universities, strictu sensu, can and should play a part in the prolongation (or the follow-up) of the courses in preparation for retirement by concentrating on domains in which their high-level contri-bution would complete and enrich the practical information given initially by ad hoc organizations.

The life of a retired person is indeed far from forming an immutable unit. It demands constant readjustments, dependent as it is on phenomena linked to advancing years. Not so long ago, the age of 65 was considered to mark a human being's life span. This is no longer the case today, when the life expectancy is far more favourable. This implies that the role of the Senior Citizens' University is not restricted to teaching, but should consist in stimulating and encouraging the activity of its members, in improving the quality of their lives, and, to sum up, in playing its part in the struggle against aging.

COURSES IN PREPARATION FOR RETIREMENT FOR ELDERLY UNEMPLOYED

We now come to the last part of this talk. The part played by courses in preparation for retirement in the improvement of the quality of life among the middle-aged unemployed.

At first sight, being made redundant can be assimilated to early retirement. But its effects are far more pervasive. On reaching the legal retirement age, all individuals can look back on their past and assess his working life, with its attendant happiness and sorrows, its successes and its failures. Admittedly, retirement, successful or not, constitutes a great upheaval in the life of every human being, but this change is foreseeable, and can be programmed, prepared several months in advance. Redundancy, on the contrary, strikes a much harsher blow, generally without any middle-term or long-term warning, and its victims, whether they be 30, 40 or 50 years of age, still have a future, or part of a future before them. They therefore see themselves as being not only deprived of a livelihood, but also of the social status due to their occupational situation. The man or woman who finds himself or herself suddenly deprived of the right to work cannot help thinking that his or her dismissal will be interpreted as being due to inability, incompetence, laziness or some hidden flaw. Even if all this is untrue, such persons feel they have to redeem themselves, so to speak, in the eyes of society, of their neighbours, their friends, to regain their self-respect, and when all attempts to find new employment prove fruitless, the tendency is to loose confidence in oneself- in one's abilities, to withdraw into one's shell, to isolate oneself and even to acquire a feeling of guilt. This impression of failure is all the more difficult to bear when redundancy comes close to the legal retirement age, for it is a sad truth that to find a regular, lucrative occupation when over the age of 50 is something of an utopia in the present-day economic situation. We must therefore assist these women and these men to overcome their discouragement and to persevere in spite of the refusals they encounter in their search for an occupation. We must help them to avoid despair, to take heart again and adapt to new conditions.

Faced with such problems, the Swiss Foundation Pro Senectute, specialized in the organization of courses in preparation for retirement, carried out a most interesting experiment in the cantons of Vaud and Neuchâtel (Calande, 1993/1994). Last year they set up courses in preparation for retirement designed for unemployed men and women aged 58 to 62 or 65 (for younger persons there exist other institutions that cater for people up to the age of 57). Officially recognized, these courses which are non-fee-paying, are held for 3 hours a week over a period of 15 weeks. They are attended by 10 or 15 participants who contract to attend them regularly from start to finish, and who themselves propose the themes to be presented and discussed in the form of workshops. The subject, generally speaking, is identical to that dispensed in the "traditional" courses in preparation for retirement, i.e. the three categories previously mentioned. But another factor has to be taken into account: the highly emotional atmosphere in which these meetings are held. Those who attend them have often become easily upset, destabilized by this enforced "early retirement". It is indispensable to help them to deal with this break in their lives, with all its after-effects on the personal, relational and social planes.

The organizers of the courses had the following aims in mind: the unemployed must be helped

- to live through the crises without hiding from themselves the inevitable hardships;
- to regain confidence in themselves and their personal values;
- to deal with administrative situations that can sometimes by very personal;
- to care for the family and social relationships that have suffered from the unemployment situation;
- to redefine a project for a balanced life.

The participants are therefore encouraged to look after their health by walking, swimming and observing a correct diet. They are also urged to remain open to the world by means of the radio, television and cinema. Within the group, social contacts are made and deepened, while links of confidence - which are indispensable - are forged with the group leaders, whose role is vital.

They have the delicate responsibility for giving these persons psychological support and working out, with their cooperation, the instruments that will provide them with a solid basis in their search for a new occupation, for it has been noted that, after, several setbacks, the applicants for a post end up by adopting a defeatist attitude. The aim is, therefore, to persuade them not to resign themselves, but to keep on fighting, and to stress certain abilities and skills that may be different from their basic job qualifications. Some of the students hold cards they do not know how to play or which they can only discover with help. At the present time, flexibility and adaptability prove more indispensable than ever, and it is the task of the group leaders to point this out. Simulations of interviews are organized for potential applicants, who are also taught how to set out an application or a curriculum vitae, and how to make up a dossier that will give a positive image of the applicant.

In the event of the unemployed persons being too close to the age of legal retirement to have any hope of finding another stable post, the important thing is to provide them with enough incentive to build new life projects and to assist them in achieving them, just as any "normal" retired person is led to do, in order to live a life of quality.

How can one assess the results of these courses? The participants in the first workshop give the following evaluation of these sessions:

- a restoration of one's self-respect;
- an improvement in conjugal, family and social relations;
- a better emotional balance;
- an ability to come to terms with events;
- a resumption or not of various occupations.

As regards this last point, the "concrete" results are significant, since, out of a score of unemployed persons having attended the two workshop courses in 1993, nine have since found full-time or part-time posts, two have good prospects of finding employment, six have launched personal projects and two are due to receive hardship benefit.

One the social plane, the former students from these groups continue to meet and help one another out. They have formed a network of friendly relations, they ring one another up and assist one another in the event of illness.

CONCLUSION

My conclusion will be brief. It is clear that courses in preparation for retirement can contribute to an improvement in the lives of retired persons, as well as of those approaching retirement. But we must be prudent in assessing this impact and beware of expecting miracles, as other figures prove. The fact is that the needs and expectations are extremely variable, according to the criteria used in our survey. To gain in efficiency, these courses should be given to homogeneous audiences (as was the course for unemployed persons) and they ought to provide the participants with opportunities for expressing their wishes, in as unacademic atmosphere as possible, enabling participants to join in a free discussion. In spite of all the information and advice he may receive, it remains up to the individual to shoulder his responsibilities and settle his own particular problems, as many have done long before their time of retirement.

REFERENCES

Calande, C , 1993/1994, *"Ateliers-chômeurs Premier bilan, Rapport atelier-chômeurs 2e semestre 1993"*, *["Workshop on Unemployment First Results, Report from the Workshop on Unemployment, Second Semester, 1993"]*, Pro Senectute, Neuchâtel, in French

Jeanneret, R , 1993, *"La retraite et les cours de preparation à la retraite" [Retirement and Courses to Prepare One's Retirement"]*, Université de Neuchâtel, Université du 3e âge, Neuchâtel, in French.

Micheloni, M , 1989, *"Etre âge dans le canton de Neuchâtel", ["The Elderly in the Canton of Neuchâtel"]*, Université de Neuchâtel, Cahiers de l'ISSP no 11, Neuchâtel, p 30, in French

Rageth, J -P , 1982, *"L'apprentissage de la liberte, chances et dangers de la preparation a la retraite Vieillir aujourd'hui et demain ", ["Learning to Be Free The Advantages and the Dangers of Preparing One's Retirement Growing Old Today and Tomorrow"]*, Realites sociales, Lausanne, p 369, in French

7

IN SEARCH OF THE MEANING OF EDUCATION

The Case of Educational Generations in Finland

Ari Antikainen, Jarmo Houtsonen, Hannu Huotelin, and Juha Kauppila

Department of Sociology
University of Joensuu
Box 111
FIN-80101 Joensuu, Finland

WHAT IS THE MEANING OF EDUCATION?

In the research project "In Search of the Meaning of Education", we are studying the meaning of education and learning in the lives of Finns (Antikainen, 1991). In addition to formal education, we are interested in adult education and other informal and nonformal ways of acquiring knowledge and skills. Our conception of meaning refers to our method and theory both.

We are investigating intersubjective social reality by means of a qualitative logic, not of statistical representativeness. We are using a biographical method, namely a life-history approach with a life-story or narrative interview and semistructured thematic interview (Thomas and Znaniecki, 1918-1920; Denzin, 1989; Huotelin, 1992). According to our theory, the meaning of education can be analyzed on three levels, as the following three restricting questions:

1. How do people use education in constructing their life-courses?
2. What do educational and learning experiences mean in the production and formation of individual and group identity?
3. What sort of significant learning experiences Finns have in the different stages of their lives? Do those experiences originate in school, work, adult study or leisure-time pursuits? What is the substance, form and social context of significant learning experiences?

In this article we approach these questions from the point of view of generational analysis looking for differences concerning cohorts and generations. What are the most essential differences between older and younger generations in contemporary Finnish society?

Preparation for Aging, Edited by E Heikkinen *et al*
Plenum Press, New York, 1995

DATA AND METHOD

We collected our data by means of biographical and thematic interviews. In the initial interviews the interviewees related their life stories orally. As needed, each interviewee was also posed more specific questions about education, self-definition, and areas of knowledge important in his or her life. An interview typically lasted three to four hours. We then picked out a list of significant learning experiences from each life story and presented it to the interviewee to be accepted or changed.

In the second interview we considered each significant learning experience and its social and biographical context in greater detail. Assuming that education can also destroy identity, we asked, finally, for the interviewee's most negative education-related experience. The second interview usually lasted about as long as the first did.

In accordance with our purpose, we interviewed many kinds of people: women and men, representatives of different social classes and ethnic groups, and persons of various ages. Most of the 44 interviewees were Finnish speakers (n=28), but the group also included Swedish-speakers, Samis (Lapps), Romanies (Gypsies) and individual members of immigrant and refugee groups. Beforehand, we classified those to be interviewed into four age groups or cohorts whose representation we thus wished to guarantee. In accordance with our grounded-theory approach, we ended the collection of the data when we reached the saturation criterion.

THE EDUCATIONAL GENERATIONS

In recent decades Finnish society, and education as a part of that society, has changed swiftly. By the concept of generation we were able to examine the processes by which historical and social time shape the individual's life-course (Mannheim, 1952). A generation consists of group of people born during the same period of time and united by similar life experiences and a chronologically coherent cultural background. In the present study we have been looking for Finnish educational generations. At this stage of the analysis, ethnic minorities are not involved (Huotelin and Kauppila, 1994). Figure 1 outlines the course and outcome of our analysis.

We classified the data according to the following educational cohorts, on the basis of the structure of opportunities - that is, on changes in social mobility, in the educational system and in participation in education (cf. Pöntinen, 1982; Roos, 1987; Järvelä, 1992):

1. Cohort with little education (persons born before 1936)
2. Cohort of educational growth and inequality (persons born in the years 1936-1945)
3. Cohort of educational growth and welfare (persons born in the years 1946-1965)
4. Cohort of young people (persons born after 1965).

By applying the life-course perspective to the data thus classified into cohorts, we attempted to clarify the experiential base of the education, as well as education-related actions in the various stages of life. From this perspective we were able to take note of the individual's actions under specified sociohistorical conditions. (Huotelin and Kauppila, 1994).

On the basis of this very intensive reading of the life stories we found the following core-categories of education and life-course, corresponding to the educational cohorts:

1. Education as an ideal, life as a struggle
2. Education as a means to an end, work as the substance of life

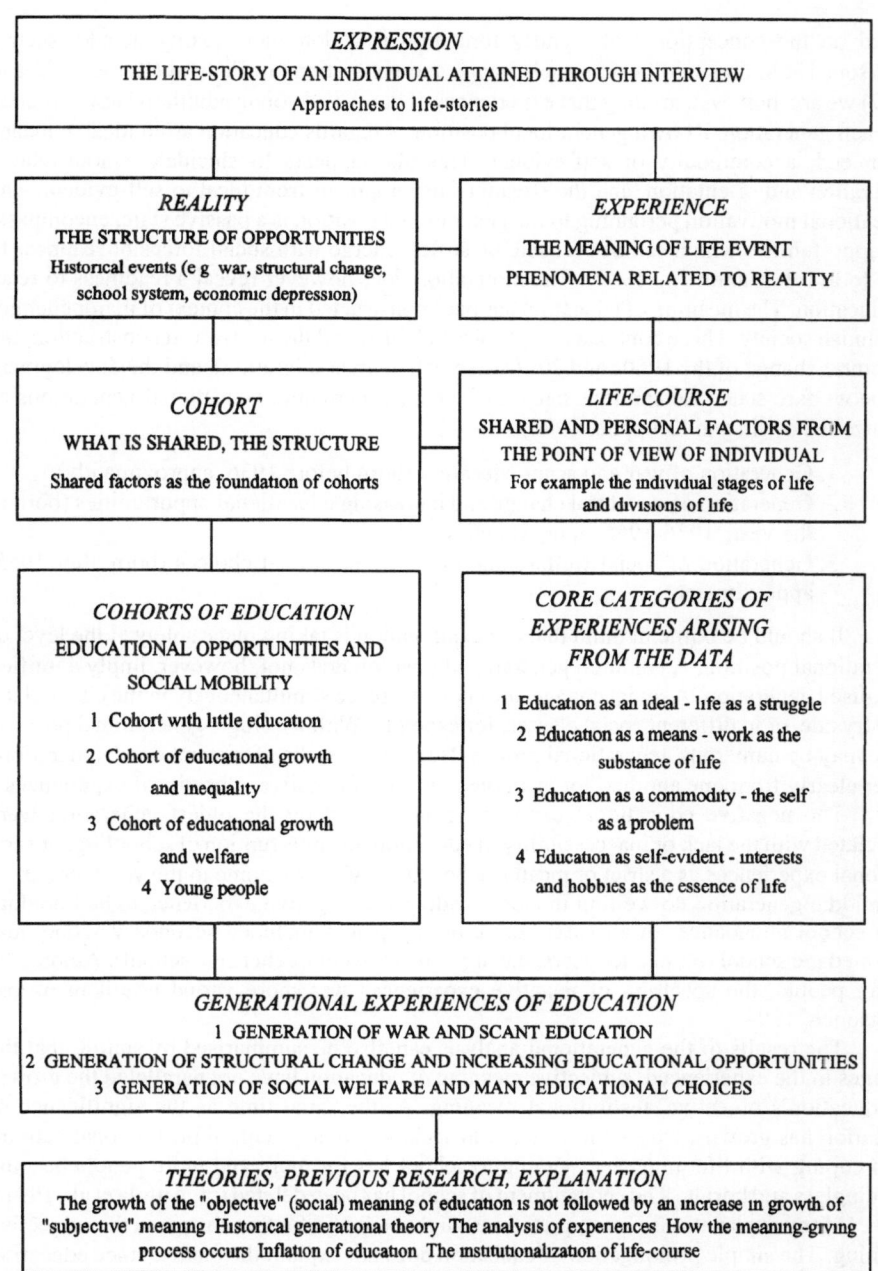

Figure 1. Towards generational experiences of education

3 Education as a commodity, the self as a problem
4 Education as self-evident, personal pursuits as the substance of life

It's impossible to isolate the influence of age and the influence of history in this kind of generational analysis This means by some critics a conceptual confusion (Kertzer, 1983)

Based on the conception that identity formation is a dominant quality in adolescence (Erikson, 1968) and studies on social change and generations in Finland (Roos, 1985 and 1987) we are, however, arguing that experiences of youth and young adulthood have created, for each generation, its own generational position as regards education as an ideal, a means to an end, a commodity or self-evident. This also appears to elucidate school-related motivation and orientation, and the stream of development from ideal to self-evident. The educational motivation pertaining to the generational position is a passive state, encompassing opportunities which can be realised, be stifled, merge with social forces or reappear in new form (cf. Mannheim, 1952). The motivation does however reveal a readiness to react to education. The meaning of education has been constructed in the context of major changes in Finnish society. These contexts include World War II and the post-war reconstruction, the structural change of the 1960s and 70s (the so-called great migration), and the development of the welfare state. We can thus name the actual, experiential educational generations as follows (Huotelin and Kauppila, 1994):

1. Generation of war and scant education (born before 1936, approximately)
2. Generation of structural change and increasing educational opportunities (born in the years 1936-1955, approximately)
3. Generation of social welfare and diverse educational choices (born after 1955, approximately)

It should be borne in mind that our examination is taking place solely at the level of generational position. A common generational position does not, however, imply a unified actualised generation. A social change does not take place simultaneously in the city and the countryside, or in different social classes, for example. Within a single generational position there may be numerous generational groups. Nevertheless, the aforementioned generations differ clearly from one another, for example in terms of negative educational experiences.

The negative educational experiences mentioned by the oldest generation were associated with the lack or inaccessibility of education, the interruption of schooling, or such personal experiences as a strict or mean teacher. Only when we come to the younger half of the middle generation do we find the most widespread negative experience to be boredom with school attendance. In all cases, these boring places included secondary school and intermediate school (or, in later years, the upper level of comprehensive school). Among the young people, the spectrum of negative experiences was more varied (Antikainen and Houtsonen, 1994).

The results of the generational analysis can also be summarised by stating that the changes in the experienced, subjective meaning of education have not paralleled the growth of education's objective, institutional meaning. At the same time as the significance of education has grown vastly—for example in seeking employment, in professional activity or in coping with life in general—going to school has for more and more people become meaningless and boring. The enchantment of school has been deleted (cf. Aittola et al., 1991).

Many explanations or interpretations might be found for this change in the subjective meaning. The simple prolongation of education does not explain the matter, since education was an ideal even for the members of the oldest generation who had received most extended education. Second, a partial explanation for the change may be found in the change in education's status in distribution of knowledge - the "loss of monopoly"—and in the inflation of degrees—the "educational inflation"—which developed as the increase in the number of educated persons outstripped the expansion of the job market. The subjective meaning of education began to crumble first among members of the middle generation, even although their professional careers were by no means being obstructed. A third explanation is associated with the institutionalisation of life-course and with the role of education in that process (Kohli, 1985; Meyer, 1987; Buchmann, 1989; Dannefer, 1991; Lindroos, 1993).

Modernity has meant the standardisation of products, culture, social organisation and human life itself. Increasingly, one's life-course proceeds in accordance with formal rules. At the same time, life has become more individualised—we act by making our own individual choices. The dynamic of these two seemingly opposite developments has institutionalised life-course: the "self"—the individual's life-course—has become an institution. We all make our own choices, but in a very similar way or within specified boundaries. Education and the student or pupil's role are assigned to a certain age group. Schooling which proceeds "normally" produces no significant experiences.

Age-related norms and roles have in recent years weakened in modern society, but in our educational life-stories this development is not visible. Unless this weakening of age-related norms and roles outside of school has contributed to the feeling that school is insignificant.

EDUCATION AND IDENTITY

In telling their life stories, people express and explain themselves. Thus according to our working definition, a person's identity is composed of the meaningful parts of his or her life story (Antikainen, 1991, 6). Identity is the individual's socially constructed definition, formulated by using available cultural meanings. The person's identity can exist only through the system of linguistic and cultural codes which people use in defining their identities as objects (Weigert et al. 1986, 30-36).

In this analysis we may define identity as the typified self in each stage of the life-course, in the context of social relationships. Through the typification process the identity is termed a meaningful social object. Typification is a conceptual process which helps people to organise their information concerning the world. The organising takes place on the strength of the typical features—rather than the unique characteristics—of people, things and events (Starr, 1983, 162).

We have distinguished among four dimensions or manifestations of identity: social identity ("objective"), personal identity ("biographical"), self-identity ("feeling of identity") and cultural identity ("meaningfulness"). Of these four, the first three correspond to the definition used by Goffman (1963). In this paper we pick up some remarks from our longer report (Houtsonen, 1994; Antikainen et al., 1994a and 1994b).

Education and Social Identity

In all the generations, education produced social status either directly or via a vocation or profession. The younger the generation, the less a degree or vocational certification tended to create status directly. It is a more indirect process than earlier working through vocational career. Vocational or academic education was perceived as an irreplaceable producer of the basic or theoretical skills needed in working life. All the same, representatives of the oldest, little-educated generation acquired their knowledge and skills through self-training, primarily in agriculture. For the young people, summer jobs and personal pastimes were important sources of knowledge and skills. Especially those members of this generation who were receiving vocational education stressed that "real skills" are learned in practical working life, while education primarily offers "theory" (Antikainen and Houtsonen, 1994).

Education and Personal Identity

For most of the interviewees, the timing of their education followed a cultural script typical to their generation, gender and social class. Vocational courses, vocational school, college or university-level education followed immediately upon completion of either compulsory schooling or academic secondary school. Some other sorts of scripts did however come to light. For example, some of the interviewees sought education later in life, to get new jobs. Internal or external constraints led these interviewees in that direction. The external constraint may for example have been illness or the death of a close relative, whereas the internal constraint or challenge may have been a desire for social advancement or meaningful work.

The meaning of vocational education following basic schooling varied from generation to generation. For the oldest generation vocational education meant a way to cope with life's difficulties; for the middle generation, a vocation, a career, and social advancement; for the youngest generation, often, the realisation of a personal dream. The same trend seemed to prevail when we examined the meaning of education in terms of educational level. A vocational course after basic schooling was linked to the need to cope with life's difficulties; an advanced university education to the fulfilment of a personal hope or dream. By the "personal dreams" of the youngest generation we refer to matters related to personal values or lifestyles, such as creativity, ecological consciousness or sexual marginality.

As noted earlier, this generational change in the subjective meaning of education was thus connected to a change in the overall themes of the life stories. By the same token, in our examination of educational generations we noted how primary educational experiences have changed. Representatives of the oldest generation respected education in general, although they did not describe their own time spent in compulsory school attendance as at all pleasant. Representatives of the youngest generation found secondary school and the upper forms of comprehensive school (or the former intermediate school) boring, tiring and oppressive. Regardless of the generation, the most favourable experiences were linked to personal relationships and friendships developed in school. In like fashion the absence of such relationships was viewed very negatively (Antikainen and Houtsonen,1994).

Education and Self-Identity

Lack of education seemed to bother men of the oldest, limited-education generation, but it did not to any extent bother women of that generation. The young people seem in particular to see themselves as either "practical" or "literate" in their orientation. The upper level of comprehensive school—and, sometimes, secondary school—were pictured as boring, confining and old-fashioned, while the interviewee's self was depicted as "nice", "free" or "creative". Very seldom did an interview include a self-definition as a pupil, for example as a "good", "average" or "poor" pupil, but an explanation for so few cases could be sensitiveness of these definitions in Finnish society. Certain interviewees said that education had instilled in them such characteristics as "creativity", "leadership", "empathy" and "self-confidence" (Antikainen and Houtsonen, 1994).

Education and Cultural Identity

Our observations on cultural identity are very tentative: the analysis is continuing. Anyway, it seems that the victory of education has meant a standardization of culture and life-course. In our data this trend is visible most in regard to ethnic minorities. Finnish school—or any modern western school—does not encounter Romany culture; in fact, for Romanies (Gypsies) school attendance can mean rejection of their own culture. For Samis (Lapps), vocational or higher education often represents a comparable threshold; on the other hand, it appears that the mutual accommodation of education and Lappish culture has succeeded. For representatives of the mainstream culture, education often seems to be associated with the continuation of the culture, lifestyle and values represented by the family, but the data also reveals some "defections" over the borders of cultural and social structures. Changes of vocation or profession during adulthood, as mentioned earlier, furnish examples.

SIGNIFICANT LEARNING EXPERIENCES

In the context of a life story, we defined as significant learning experiences those learning experiences which appeared to guide the interviewee's life-course or had changed or strengthened his or her identity (Antikainen, 1991). It makes sense to assume that a possible change in identity concerns secondary areas or the person's relationship to his or her identity, rather than the core of identity.

In terms of the experience's duration, two types of significant learning experiences came to light in our research: clearly definable events and vaguer, cumulative experiences such as the development of self-awareness as an outgrowth of certain events related by the interviewee. In our pilot study we coded, in a very theoretical and thus largely deductive way, five cases which represented all four educational cohorts (Antikainen et al., 1992 and 1993). Some observations from the pilot study follow.

In many respects the oldest generation's experiences were in general more numerous and intensive than those of the younger generations. We can interpret this difference firstly by the retrospectiveness of the educational setting and secondly by maturation, but a genuine generation gap also exists, and plays a role here. Reference should also be made to the increasing role of the media and to the aforementioned institutionalisation of life-course.

As early as in the 1930s, Dewey (1938) stated that the quality and continuity of experiences constituted key factors in learning and human development. Our data likewise provide a picture of how a significant learning experience gives a person strength to cope with future problems. Without knowing of the powerful, antecedent significant learning experiences involved, it would for example be difficult to imagine what might motivate Anna, a 66-year-old woman living in a rural village, to learn a foreign language for the first time at her advanced age, in spite of her relatives' negative view of the matter (Antikainen, 1993a); or to conceive of Ville, a 50-year-old employed as a building contractor, continually taking university studies or evening-school classes (Antikainen, 1993b, p. 132).

It was easy to note the supportive human relations—"significant others"—in every significant learning experience. These relations were described as egalitarian or encouraging. So, it is possible to analyze learning as concrete human and social relations even in a technological society.

In our pilot study we also examined the situations which led to significant learning experiences in terms of where and in how formal an environment the learning had taken place (Jarvis, 1987). We first noted the variety of such situations. Nonetheless, not one interviewee had a single significant learning experience solely in a pupil's role in a school attended on a compulsory basis. On the other hand, school learning may be involved, along with other environments, in longer-term, cumulative experiences of the sort mentioned earlier.

Our analysis thus far indicates that, for representatives of the oldest generation, the typical context of the significant learning experience is a situation we refer to as a constraining situation, which demands new ways of action and thought. Taken literally, the term may be an exaggeration; "demanding challenge" might be more descriptive, since the interviewees reacted to such situations as challenges which included possibilities for choice. For the young people, the first typical context is the pastime and hobbies; the next is self-definition and the search for identity. We have not as yet adequately analyzed the extent to which learning experiences vary according to gender, social class or cultural membership.

Finally, let's look at our first informant Anna's life story as an illustration to significant learning experiences. Anna is aged 66 and married with two children. She was born in those part of Karelia which later became the Karelian Soviet Republic and from where she was evacuated when she was 16. For years she lived in the west of Finland, but after her marriage in the early 1960's she moved to North Karelia (on the Finnish side of the border). She has been working as a housekeeper, caterer and family mother. Anna's formal education consisted of 4 primary school grades plus 1 upper grade; the second upper grade was interrupted by the war. During the war Anna also took a 7-month course in housekeeping. So her formal schooling was exceptionally modest. But she has made the best of what education she has had, and she has continued her non-formal studies throughout her adult life. In fact Anna's story epitomizes some Finnish or Nordic ideals of traditional liberal adult education. The following experiences of personal skills and the learning experiences behind them were identified in Anna's life story:

1. Household work and needlework. These experiences go back to her childhood home, where in particular her aunt taught her household work and needlework. She started needlework at the age of 7 and has continued throughout her life (see 3.).
2. Plant-growing. During the war the 18-year evacuee was assigned by the official local crop manager to grow tomato, turnip and cabbage plants. This was not a well-known skill to the villagers, but it was very important in that food supply situation. Anna's efforts were very successful, and she was duly proud of her success. She has continued growing plants throughout her life.
3. Independent house-keeping. Anna was 31 when she was hired as a housekeeper by two farmer brothers. Her duties included all household work from baking and vegetable gardening to cattle tending. She was given a free hand in her work, and the farmers treated her as their equal. Anna learned to trust her own knowledge and became aware of her quite versatile skills. In the interview she told us about cases of cattle diseases where her action saved the lives of the animals. With the experience that she gained from housekeeping she could later take on catering jobs, and she continued to tend cattle until the last few years.
4. Sewing and other needlework. All her adult life Anna has been attending study circles at adult education centres, and she still belongs to some sewing circles. She is so highly skilled in so many kinds of needlework that she herself claims she has at least "completed secondary education in sewing".
5. Elementary English. Anna had her children at an advanced age. When they began learning English at the comprehensive school, Anna—58—was inspired to join the English circle at the local adult education centre. Her English studies really got off to start when she was hospitalized for a short time. In Anna's home village the news about her English studies was first received with astonishment and even laughter. But she went on, obviously supported by the self-confidence she had gained from her previous positive learning experiences. Later several villagers have told her that they regret they did not follow her example—they could even have shared rides to the adult education centre. After five years of language studies she visited London, which was her first journey abroad. She found she could understand and even speak some English. In the interview she described London as almost "like a favourite child to her".

DISCUSSION

The meaning of education in the lives of Finns was studied by a biographical method at three levels: life course, identity and significant learning experiences. By generational analysis based on cohort and experiences through life course the institutional, "objective" meaning of education has enormously grown, but the perceived, "subjective" meaning of education has not followed the development. For the oldest generation education is an ideal, for the middle generation a means to an end and for the youngest generation a commodity or self-evident. Remarks on educational identity and significant learning experiences both strengthened the results of the generational analysis and at the same time made the portrait more complex by referring to social class, cultural membership and gender. The tentative analysis of significant learning experiences demonstrated creativity and participation being hidden in life stories. Lived significant learning experiences are like solved problems—resources in constructing the future life course.

REFERENCES

Aittola, T, Jokinen, K, and Laine, K, 1991, Koulun haihtunut lumous [Disenchantment of school], Nuoriso-tutkimus [The Finnish Journal of Youth Research] 9(3) 9-17, in Finnish

Antikainen, A, 1991. *"Searching for the Meaning of Education A Research Proposal"*, University of Joensuu, Bulletins of the Faculty of Education, 38, University of Joensuu, Joensuu

Antikainen, A, 1993a, Life course, education, identity Anna's learning experiences, *LEIF,* no 2 28-29

Antikainen, A, 1993b, *"Kasvatus, koulutus ja yhteiskunta" ["Education, School and Society']*, WSOY, Juva, in Finnish

Antikainen, A, and Houtsonen, J, 1994, In Search of the Meaning of Vocational and General Education Preliminary Remarks on Identity and Education in Life-histories of Finns in *"Vocational Education and Culture—European Prospects from History and Life-History"*, A Heikkinen, ed, Tampereen yliopiston Hameenlinnan opettajankoulutuslaitos, ammattikasvatussarja 9 University of Tampere, Tampere

Antikainen, A, Houtsonen, J, Huotelin, H, and Kauppila, J, 1992, Auf der Suche nach der Bedeutung von Bildung Eine Soziologische Fallstudie uber Lebenslauf und Lernerfahrungen von Finnen Einfuhrende Hinweise von Klaus Hurrelmann, *BIOS, Zeitschrift fur Biographischen Forschung und Oral History*, 5(2)

Antikainen, A, Houtsonen, J, Huotelin, H, Kauppila, J, and Turunen, A, 1993, In Search of the Meaning of Education A Pilot Case Study of Education and Learning in the Finnish Life Course in *"Developing Education for Lifelong Learning"*, E Ropo, and R Jakkola, eds, University of Tampere, Department of Education, Series A, 52, Tampere

Antikainen, A, Houtsonen, J, Huotelin, H, and Kauppila, J, 1994a, Elamankulku, sukupolvet ja koulutus nyky-Suomessa, [Life-Course, generations and education in contemporary Finnish society] *Aikuiskasvatus, [The Finnish Journal of Adult Education]*, 14(2) 102-112, in Finnish

Antikainen, A, Houtsonen, J, Huotelin, H, and Kauppila, J, 1994b, In Search of the Meaning of Education A case study of education and learning in the life of Finnish young generations, paper presented at the XIII World Congress of Sociology, July 19, 1994, Germany, Bielefeld

Buchmann, M, 1989, *"The Script of Life in Modern Society Entry into Adulthood in a Changing World'*, University of Chicago Press, Chicago

Dannefer, D, and Duncan, S C, 1991, Homogeneity vs Heterogeneity Social and Developmental Perspectives on the Life Course in *"Jatkuva koulutus ja elinikainen oppiminen", ["Continuing Education and Lifelong Learning"]*, R Raivola, and E Ropo, eds, Tampereen yliopisto, Kasvatustieteen laitos, Tutkimuksia A 46, Tampere

Denzin, N K, 1989, *"Interpretive Biography"*, Newbury Park, Sage

Dewey, J 1963, 1938, *"Experience and Education"*, Collier, New York

Erikson, E H, 1968. *"Identity Youth and Crisis"*, Faber and Faber, London

Goffman, E, 1963, *'Stigma Notes on the Management of Spoiled Identity"*, Prentice-Hall, Englewood Cliffs, NJ

Houtsonen, J, 1994, *"How to Approach Education, Culture and Identity?"* Upublished Licentitate Thesis, University of Joensuu, Department of Sociology, Joensuu

Huotelin, H, 1992, *"Elamakertatutkimuksen metodologiset ratkaisut Esimerkkitapauksena "Koulutuksen merkitysta etsimassa"-projektin menetelmalliset valinnat", ["Methodological Choices of the Biographical Research The Methodological Choices of the Project 'Searching for the Meaning of*

Education"], Joensuun yliopisto, Kasvatustieteiden tiedekunnan tutkimuksia, A 46, Joensuu, in Finnish

Huotelin, H , and Kauppila, J , 1994, From the Ideal to the Self-Evident Education in the Life-course of Finns in *Vocational Education and Culture—European Prospects from History and Life-history"*, A Heikkinen, ed , Tampereen yliopiston Hameenlinnan opettajankoulutuslaitos, ammattikasvatussarja 9 University of Tampere, Tampere

Jarvis, P , 1987, *Adult Learning in the Social Context"*, Croom Helm, London

Jarvela, M , 1992, ' *Palkkatyo ja koulutustarve", ["Waged Labour and Need for Education"]*, Tutkijaliitto, Helsinki, in Finnish

Kertzer, D I , 1983, Generation as a sociological problem, *Ann R Soc ,* 9 125-149

Kohli, M , 1985, Die Institutionalisierung des Lebenslauf, *Kolner Z So ,* 37 1-29

Mannheim, K , 1952, The Problem of Generations, in *"Essays in Sociology of Knowledge"*, P Kecskemeti, ed , Routledge & Kegan Paul, London

Meyer, J W , 1987, Self and Life Course Institutionalization and Its Effects, in *"Institutional Structure"*, G M Thomas, et al , eds , Sage, Newbury Park, CA

Mezirow, J , 1981, A critical theory of adult learning and education, *Adult Ed ,* 32(1) 3-24

Pontinen, S , 1982, Sotasukupolvi, sotanuoret, sotalapset ja suuret ikaluoka, [War generation, war youth, war children and baby booms] *Sosiologia* 3 153-162, in Finnish

Roos, J P , 1985, Life stories of social change Four generations in Finland, *International Journal of Oral History* 6(Nov)

Roos, J P , 1987, *Suomalainen elama' , [' The Finnish Life"]* SKS, Helsinki, in Finnish

Starr, J M , 1983, Toward a social phenomenology of aging Studying the self process in biographical work, *Int J Aging,* 16(4) 255-270

Thomas, W I , and Znaniecki, F , 1918-1920, *"The Polish Peasant in Europe and America'* , Volumes I-II, University of Chicago Press, Chicago, Volumes III-V, Badger Press, Boston

Weigert, A J , Teitge, J S , and Teitge, D W , 1986, *"Society and Identity Toward a Sociological Psychology'* , Cambridge University Press, Cambridge

8

LIFELONG LEARNING EXPERIENCES FROM NORWAY

Reidun Ingebretsen and Tor Endestad

The Norwegian Institute of Gerontology
Oslo
Norway

INTRODUCTION

The last decades' international studies on aging have focused on the possibilities of elderly to learn and participate in education programs. In the United Nations' declaration of human rights, the educational rights of the elderly are particularly mentioned (Recommmendation 45, 1983).

The growth in the share of elderly, as well as the increasing longevity and better health of elderly, are reasons to highlight this topic. Elderly people are seen as potential contributors on social arenas, and it is therefore important that they are given the possibilities for further education and development. More than dwelling on the topic of whether old dogs are able to learn new tricks, the question is how old tricks and knowledge are to be sustained, developed and renewed according to the interests of the elderly themselves and the society.

A document from the Ministry of Health and Social Affairs (1993) emphasizes the fact that many Norwegians will live as much as one third of their lives in retirement. This fact is a challenge to the policies of aging. Studies and participation in courses are some of the possibilities of development as well as a way of strengthening the position of the elderly in a changing society.

The capacities of the elderly to learn may be an important aspect of adaptation in later years (Baltes and Baltes, 1990). Knowledge helps to gain control and influence and is of crucial importance for active participation in community-roles. Successful aging is a result of the transaction between the elderly and society. "Successful aging is individuals' learning to plan and society's planning to learn". (Featherman et al., 1990, p. 84)

Educational gerontology refers to the study and practice of instructional endeavours for and about the aged and aging (Peterson, 1990). Different models for categorizing and understanding the motivation of the elderly are developed. (Houle, 1961; McClosky, 1975; Peterson, 1986). Cross (1981) takes into consideration variables related to personality, attitudes and expectations of the elderly in addition to factors related to the actual situation of the elderly learner and information about availability and organization of courses.

In spite of increasing interest and initiative, course-attendance among elderly is, in most studies, found to be rather low compared to younger generations. American, English,

German and New Zealand studies show an attendance between 1 and 6 percent in samples of elderly above the age of 65 (Fisher, 1986; Glendenning, 1989; Karl, 1991; Tobias 1991).

THE STUDY

The aim of the study is to map the participation in educational activities, motivating factors and barriers. Main questions are:

- Who participates in courses?
- What are the interests of the elderly?
- Why are elderly interested in courses?
- What are the barriers to participation?

This study is a part of a Nordic project with Jan-Erik Ruth, Finland, as the research coordinator. The Norwegian contribution to the study is based on information from 710 randomly selected 67-85 years old persons living in their homes. This paper is based on the Norwegian data only (Ingebretsen and Endestad, 1993).

Participation in Courses

Ten percent of the respondents had attended a course of a certain duration during the last three years, eight percent during the last year. In line with results from similar studies, women (11% course-attendance) were found to be somewhat more active than men (9%). One fifth of the respondents (21%) had planned to participate. Most elderly (77%) wanted to learn more about at least one topic.

High socioeconomic and educational level increase the probability of attending courses. The participants more often look upon their school years as a positive experience and most of the them (83%) attended courses before retirement age. This seems to reflect a family-pattern. The parents and the children of course-participants have considerably higher education level compared to the families of non-participants.

Participation rates were equally high among the age group 80-85 as among those below 80. The participants evaluate their financial status, their ability to learn and to memorize more positively than the non-participants, but there are small differences related to subjective evaluation of health.

The use of a senior centre seems to increase the probability of attending courses (23% vs. 8% attendance), so also does a short distance between the dwelling and a community centre. Hence, the combination of interest and accessibility is crucial.

The respondents were asked to plot their level of activity on 17 areas during the last week and their interest in learning more about 28 different topics.

Figure 1 shows that participants in general are more active than non-participants (mean 9.4 vs. 7.8 activities).

Figure 2 shows that participants more often than non-participants have a lot of topics they want to learn more about (mean 6.8 vs. 3.8 topics of interest).

Both with respect to level of activity and scope of interests, the median was used to split the high and low scores and the two measures were combined. The two variables are strongly correlated ($\chi^2 = 34.64$, p<.0001).

The informants with a high level of activities and a broad scope of interests have the highest participation rate. Interest to learn more about many topics is, however, highly motivating for course-attendance even for persons with a relatively low level of activity.

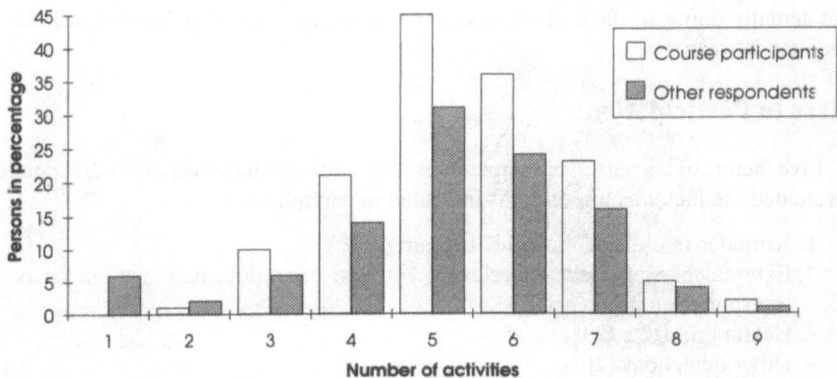

Figure 1. The number of activities last week among course participants and other responderts

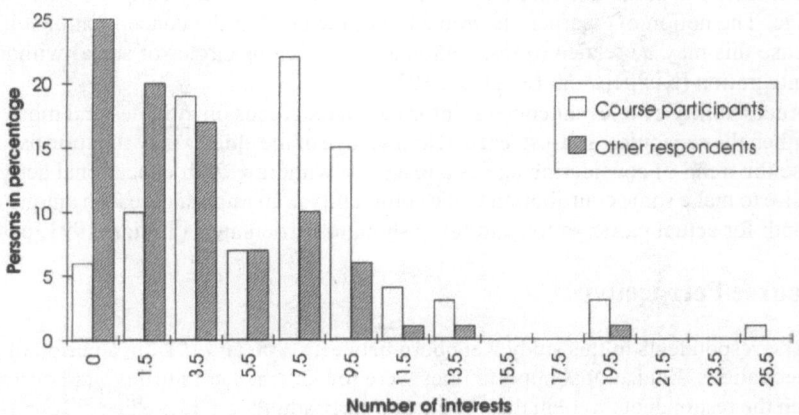

Figure 2. Scope of interest among course participants and other respondents

Table 1. Course attendance related to level of activity and
scope of interests (percentage of participants)

Scope of interest	Level of activity	
	High	Low
High	17	16
Low	9	2

Motives

The most important motive for participation is the wish to learn. The wish to learn, the desire for personal development and for social contact are, however, highly intercorrelated (r=.84, r=.75, r=.65). The last two factors were more important for women than for men.

The respondents often agree that there are many good reasons to attend courses without actually doing so themselves. «Good reasons» are necessary, but not sufficient to explain attendance.

Barriers to Participation

Five factors of barriers to participation were identified (the percentage of respondents who evaluated the factor as important is indicated in parenthesis):

1. Attitudes to old age, "too old" to learn (45%)
2. External barriers: lack of relevant courses, poor information, high costs and transportation problems (25%)
3. Health problems (23%)
4. Other objections (20%)
5. Negative self-concept (16%)

Factor 2 and 5 are more important for women than for men.

It is worth noting that regardless of chronological age, feeling «too old» is the most significant barrier to participation. Percy (1989) relates the factor of feeling too old to a poor self-image. The notion of "learner" is probably not included in the conception of self. In a wider sense this may be related to resignation and the vicious circles of social withdrawal and disintegration (Kuypers and Bengtson, 1973).

Accessibility of relevant courses, information and focus on attitudes and motivation to strengthen the capacities, self-esteem and knowledge of the elderly may function as means to reverse the trend of considering age as a reason to withdraw from educational activities. "The desire to make some contribution to the community is an important reason among older people both for actual participation and for wishing to participate." (Tobias, 1991, p. 420).

Life Course Perspective

The respondents in this study were born between 1906 and 1924 in a period of great social inequalities. Educational opportunities were few in "the hard thirties" and during the war, when the respondents were in their youth and early adulthood. In studies of life reviews of the elderly, a wish for more education is often highlighted (Thorsen, 1992; Ingebretsen and Solem, 1994; Antikainen, see in this volume;). In our study, we wanted to find out how the respondents evaluated their interests and opportunities for education throughout life. We asked the respondents to mark their interest and possibilities for education and studies at four stages in their life; at 20, 40, 60 years of age and today.

The methodological problems related to this kind of retrospective study are to be taken into consideration. It is difficult to "compare" the information given at different stages of life, because both the interests and the actual opportunities may have a different content and be evaluated differently during the life span. The questions are seen through the glasses of an old person looking back on his or her life, and the information will be coloured accordingly. A longitudinal study might therefore have shown another picture. Moreover the non-response rate on this question is high, but some results and patterns are still worth mentioning.

By and large the informants report higher interests in education when they were twenty than today. Following the normal life course expectations this seems logical. The mismatch between great interests and poor opportunities for education were most evident at the age of twenty. The majority of the respondents (86%) reveal great interest for education at that time, however nearly half of them (47%) evaluate their opportunities to fulfil their aspirations as poor. Later in life the mismatch is decreasing. At 60 years of age and after the age of 67, the group reports modest interest as well as poor opportunities increasingly. Today

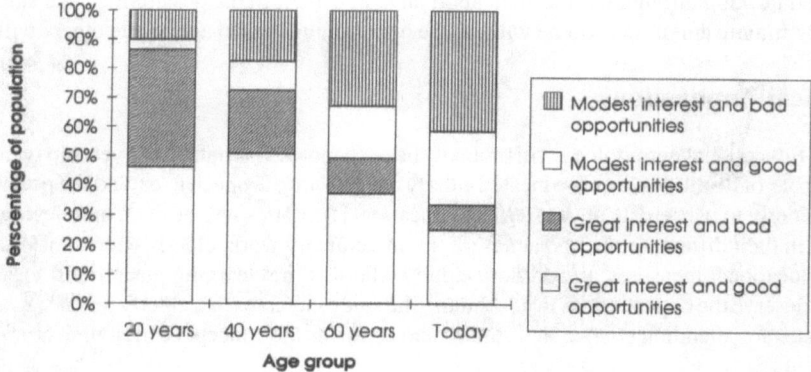

Figure 3. Combinations of interest and opportunities to education on different stages of life

this is the largest group, followed by the persons with the combination of strong interest and good opportunities (25%), and those with little interest but good opportunities (25%). A few respondents (8%) are still keeping up their great interest regardless of perceiving the opportunities for educational activities as poor.

Groups with different combinations of interest and opportunities for education in their careers at the age of twenty, are followed up till to day. We find both continuity and discontinuity regarding interests and opportunities for educational activities across the life course. Persons reporting little interest and poor opportunities at age twenty most often do so at all stages of life. The responderts with great interest and good opportunities show continuity to some extent, but quite a few of them change motivation and/or perception of opportunities. The others have more varied responses, and show partly a continuous and partly a discontinuous pattern.

If we follow the careers of the respondents who told about great interest but poor opportunities when they were twenty, we can still see some continuity until the present day. After the age of forty, however, they are more likely to be found in one of the other groups. In old age, some of the respondents that regarded themselves as interested, but unlucky learners in their youth, eventually have formed positive opportunities, others have lost their interests regardless of whether the opportunities are good or poor. As already mentioned, a small group continues to be interested in education in spite of poor opportunities.

In this study we had no opportunity to go into further details about the kind of internal and external barriers the respondents have experienced throughout life, but one of the respondents made a note telling:

"I was twenty years of age when the war started The life was turned upside down After 5 years of active service the ones of us who survived, came home to the challenges of rebuilding the country This situation had a crucial influence on plans and educational opportunities and may influence the answers to the questions of education for this generation"

CONCLUSION

Interests and perception of opportunities will work together in a complex manner throughout life. We find most continuity in the group of little interest and poor opportunities and the one of great interest and good opportunities. Lack of opportunities can discourage motivation and dreams of education. Some, however, keep their strong interests regardless of lack of gratification. Wolf (1985) reports about similar results. When given the possibilities to learn and develop throughout life, students at all ages can grasp the possibilities that suit their interests and life conditions. Some will use the opportunities in old age, while others will not.

Practical Implications

Educational gerontology will probably be even more important in the years to come As the number of healthy and well-educated elderly are growing, the need to extend the possibility of the elderly to participate in training and educational activities will be even more evident

In their different roles and activities, f.i. in voluntary work, elderly people may benefit from educational measures. In addition, elderly who just find learning meaningful for themselves, deserve the opportunities to education. The study indicates that elderly people are actual candidates for attending courses and studies and confirms the concept of lifelong learning.

REFERENCES

Baltes, P M , and Baltes, M M , 1990, Psychological Perspectives on Successful Aging Models of Selective Optimization with Compensatio, in *"Successful Aging"*, P B Baltes and M M Baltes, eds , University Press, Cambridge, Cambridge

Cross, K P , 1981, *'Adults as Learners"*, Jossey-Bass Publishers, London

Featherman, D L , Smith J , and Peterson, J G , 1990, Sucessful Aging in a Postretired Society, in *"Successful Aging"*, P B Baltes and M M Baltes, eds , Cambridge University Press, Cambridge

Fisher, J C , 1986, Participation in educational activities by active older adults, *Adult Ed Q* , 36, 202-210

Glendenning, F , 1989, Educational gerontology in Britain as an emerging field of study and practice, *Educ Geron* , 15, 121-131

Houle, C O , 1961, *"The Inquiring Mind"*, University of Wisconsin Press, Madison

Ingebretsen, R , and Solem, P E , 1994, Tilbakeblikk på livet, [Life Review] in *"Alderdoms vekst Festskrift til Eva Beverfelt i anledning 70-årsdagen"* *["Age and Development Volume of Honour to Eva Beverfelt on her 70th anniversary"]*, A Helset, ed , Norwegian Institute of Gerontology, Oslo, in Norwegian

Ingebretsen, R , and Endestad, T , 1993, *"På skolebenken i eldre år En kartlegging av interesse for kurs og studievirksomhet* , *["Back to School in Old Age A Survey of the Interest in Courses and Other Forms of Education']*, NGI-rapport 6-1993 Norwegian Institute of Gerontology, Oslo, in Norwegian

Karl, F , 1991, Outreach counseling and educational activities in a district *Educ Geron* , 17, 487-493

Kuypers, J A , and Bengtson, B L , 1973, Competence and social breakdown A social psychological view of aging, *Human Dev* 16, 37-49

McClosky, H Y , 1975, Education for Aging The Scope of the Field and Perspectives for the Future, in *"Learning for Aging"*, S M Grabowski, and W D Mason, eds , Adult Education Association of the USA, Washington D C

Ministry of Health and Social Affairs, 1993, *"Trygg og aktiv i eldre år"*, *[' Safe and Active in Old Age']*, Sosialdepartementets tiltak for eldre i 1993/94, [Social Department], Oslo, in Norwegian

Percy, K , 1989, Participation of older people in learning activities in the United Kingdom, *Educ Geron* ,15, 133-150

Peterson, D A , 1990, A History of the Education of Older Learners, in *"Introduction to Educational Gerontology"*, R H Sherron , and D B Lumsden, eds , Hemisphere Publishing Corporation, New York, pp 1-22

Peterson, D A , 1986, Aging in Higher Education, Older Students, Older Faculty and Gerontology Instruction, in *"Education and Aging"*, D A Peterson, J Thornton and J Birren, eds , Prentice Hall, Englewood Cliffs, N J

Thorsen, K , 1992, *Livsløp og Aldring [Lifespan and Aging]*, NGI-rapport 6-92 Norwegian Institute of Gerontology, Oslo, in Norwegian

Tobias, R M , 1991, Participation by Older People in Educational Activities in New Zealand Survey Findings, *Educ Geron* 17 409-421

United Nations, 1983, *"Universal Declaration of Human Rights"*, United Nations Publications

Wolf, M A , 1985, The experience of older learners in adult education, *Lifelong Learning*, Washington, DC, 8 8-11

9

IF I HAD MY LIFE TO LIVE OVER AGAIN...

T. I. Tikkanen and J. Kuusinen

University of Jyväskylä
Department of Education
P.O. Box 35
FIN-40351 Jyväskylä
Finland

ABSTRACT

The study was a replication of DeGenova's (1992) study, aiming at identifying what elderly people would do differently if they had their life to live over. The Ss were 174 retired women (56%) and men (44%) born in 1927 or 1929, residing in urban areas. The subjects completed a questionnaire, Life Revision Index (LRI), developed by DeGenova (1992), assessing life revision in the areas of friends, family, work, education, leisure, religion, and health. The results show that education and leisure enjoyment are the areas with the greatest amount of desired change. More than 70% of the subjects would spent more time in education, if they had their lives to live over again. More than sixty percent (61.8%) would spend more time in doing things they enjoy in life. Three quarters of the subjects (75%) would spend less time with worrying about job. The high value of education and leisure enjoyment, together with low relative value of work, will be discussed in regard to the levels of formal education of this age-cohort, as well as to the possibility of changing values of work and leisure in society in general.

"WHAT BETTER WAY TO LEARN WHAT LIFE IS ABOUT THAN TO ASK SOMEONE WHO HAS LIVED IT?"

This question was asked by Mary Kay DeGenova (1992) in her article. The question in itself gives high significance to the educating value of life experiences—a lot more so than what we have used to in our information loaded society. Though, the world has changed tremendously during the last century, in the fundamental issues about life and what it is all about, the older people could still be a great deal more of a highly valuable resource to younger people than what they are today. They do not only carry with them a rich collection of personal life experiences, but also a part of the collective history of a mankind. Among the most complicated issues in life is the fact that, when we finally have learned something about it, it is time for us to go.

Preparation for Aging, Edited by E. Heikkinen *et al.*
Plenum Press, New York, 1995

In addition to the educating value of life-review of those less experienced others, life-review has been stated to play a significant role for the personal lives of older people. The issue of life review has been connected with the multifaceted meaning of regret in life (Landman, 1993) as well as with life satisfaction (DeGenova, 1993). Also, in Erikson's theory life-review plays a significant role in the last developmental phase, integrity vs. despair. In this article the focus is in what might be learned about the life review of older people, and how to enhance the quality of their lives according to that knowledge.

WHAT WOULD YOU DO DIFFERENTLY IF YOU HAD YOUR LIFE TO LIVE OVER AGAIN?

The study was based on life revision, which was referring to "a desire or a wish to change certain past thoughts, feelings, actions, or accomplishments relative to some object, person activity, or situation, if it was possible to relive one's past life" (DeGenova, 1992, 136). We asked people, what would they do differently, if they could live their lives over again.

METHOD

The Life Revision Index

The data used in this study is a part from the research project 'Retirement and retirement preparation' (University of Jyväskylä, Department of Education). The Life Revision Index (LRI) was included in the last phase of data collection, accomplished as a mailed questionnaire. The LRI consists of 35 questions dealing with what people would do differently, if they had their lives to live over. The 35 questions include five questions of all of the following seven life-domains: family, friends, work, religion, health, education, and leisure. For example, "If you had your life to live over, how much time would you spend with good friends?". The participants were asked to indicate on a five-point scale, whether they would spend time on each activity much more (5), more (4), as much (3), less (2), or much less (1) than what they had actually done. For the purpose of this paper the scale has been recoded to indicate more (3), as much (2) or less (1) of desired change. The internal consistency reliability (alpha coefficient) for the life revision was 0.85.

Subjects

The participants (56% women), who were born on 1927 or 1929, were 64 or 66 years old, having been retired for 1½ years on average. Out of the 205 respondents 174 (85%) completed the Life Revision Index. Half of the participants had completed only basic level of education, 38% had vocational training, and 12% had higher education. The majority (70%) of the respondents were married, 13% widowed, 6% divorced, and 11% single. More than half (55%) rated their current health-status as good or very good.

MORE EDUCATION AND ENJOYMENT, LESS WORRYING ABOUT JOB

Whether more or less change was desired was examined for each of the items. The results are shown as percentage distributions in table 1.

Table 1. Results of the Life Revision Index[1] (N = 174)

If you had your life to live over again, how much time would you spend	Would spend time (%)		
	more	as much	less
With good friends	47 3	51 5	1 2
Keeping up with good friends (letter writing, phone calls, visits)	41 4	56 2	2 4
In social activities	28 2	68 2	3 5
Developing friendships	36 8	59 6	3 5
Getting to know more people	32 7	62 5	4 8
In family activities	56 6	40 2	1 2
Keeping up with family members (letter writing, phone calls, visits)	44 7	54 1	1 2
At home with your family	40 0	57 0	3 0
Developing close relations with your children	53 8	45 5	0 6
Developing close relations with your siblings	33 9	64 8	1 2
Keeping up with the demands of work	34 5	60 8	4 7
In work activities	3 6	69 0	27 4
Worrying about job	3 5	21 1	75 4
Financially preparing for the future	36 5	58 1	5 4
Developing your career	36 5	56 9	6 6
Keeping current on topics that interest you	60 6	38 2	1 2
In learning activities	76 6	21 1	2 3
Studying	70 8	26 9	2 3
Developing your mind or intellect	70 8	26 9	2 3
Pursuing your education	71 8	26 5	1 8
Travelling	37 1	58 8	4 1
Making sure you fit in leisure time or time for fun	18 7	75 4	5 8
Doing things you enjoy	61 8	33 5	4 7
Relaxing	42 7	54 4	2 9
Developing hobbies	40 9	55 0	4 1
In devotion to a religion	18 3	71 0	10 7
In charitable activities	22 8	66 1	11 1
Developing your spirituality	32 0	59 8	8 3
In prayer	25 1	65 9	9 0
Studying a religion	24 3	66 9	8 9
Developing good eating habits	55 6	43 9	0 6
Exercising	50 3	48 5	1 2
Taking good physical care of your body	59 6	38 6	1 8
On personal appearance	40 4	57 3	2 3
Visiting the doctor	8 8	84 7	6 5

[1]DeGenova (1992)

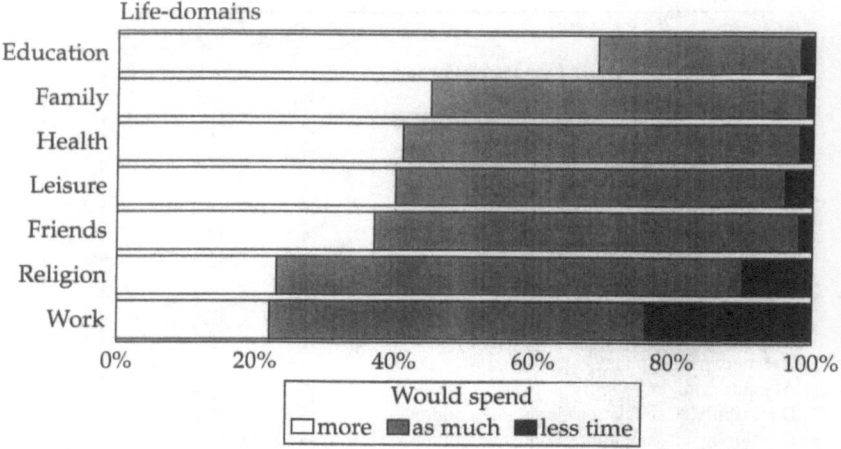

Figure 1. How would people spend their time differently, if they had their lives to live over.

Table one shows that in general people would spend more time with the issues especially in learning and education (generally over 60% of the respondents), but also in doing things they enjoy (61.8%), and with health related issues (more than 50%), if they had their lives to live over. The only area where most (75%) of the respondents would have spent less time was worrying about one's job.

Since the results were quite consistent among the five activities concerning the same life domain, we summed up the results within each of them. The results of this sum-indicator, which shows the average percentages of desired change for more, for less, or for no change on the part of each domain, are presented in figure 1.

Figure 2 shows the percentages for the change-options only, i.e. those who would spend more or less time.

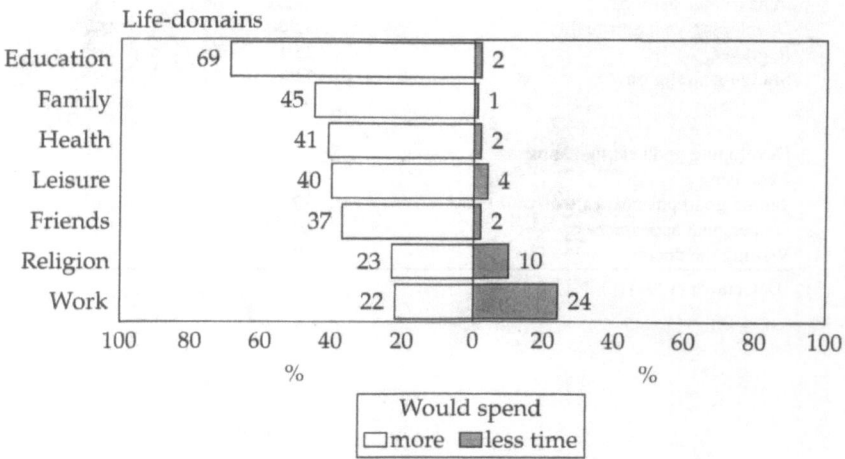

Figure 2. Desired changes in time spent in different life domains.

Figures one and two show how education indeed would be one life domain significantly different from all the others in regard to spending more time with it. Almost half of
the subjects (45%) would spend more time with their families. Religion seems to be the
life-domain where the least changes (the biggest proportion of 'would spent as much time')
would be made. These figures also show, how work is the life domain where the most of
people would spend less, and the least more of their time.

The differences between men and women within the seven life-domains are shown
in figure 3. Figure three presents only the percentages of the option 'would spend more time'.

Women and men differ most in regard to religion, health, and education. Whereas
almost a third (30%) of the women would spent more time in religion related issues, only
17% of the men would do so. There were also more women (47%) than men (35%), who
would spend more time concentrating on health. Although education was the life-domain
within which the majority of the respondents would spend more time, women tended to think
even more so (74%) than men (64%). The only life-domain where men, compared to women,
would spend more time than what they actually have, is family. Figure three also shows how
among men there are less of those who would spend more time with religion related issues
(17%) than with work (22%).

Since the research was a replication of the research of DeGenova (1992), who did her
research in the United States, the findings were finally compared to her original findings. For the
comparison, presented in figure 4, the results of DeGenova were summed up and reclassified from
the five point scale on the part of each life domain, to match the three point scale of presentation
used in this paper. Figure 4 shows only the 'would spend more time' -option.

The comparison showed that the major findings in both countries were similarly
related to education and work, with the same direction in regard to the desired change. The
most salient differences between the two samples where related to the proportion of desired
change within education and to religion. Education was regarded as far more important
among the Finns (69%) than among the Americans (47.5%). Actually, among the Americans
there were slightly more of those who would invest more of their time on leisure (48.3%)
than on education (47.5%). In this regard they differ from the Finns, of whom 40% would
spend more time on leisure, compared to 69% in education. On the other hand, much more

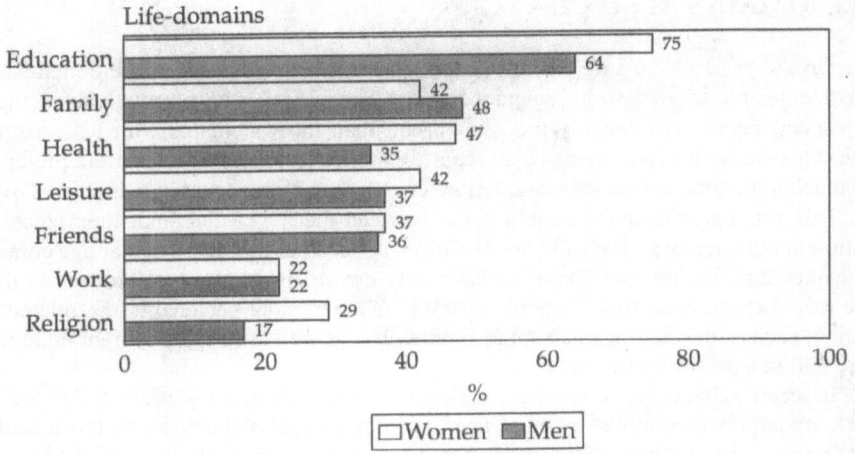

Figure 3. Gender differences in desire to spend more time within different life-domains.

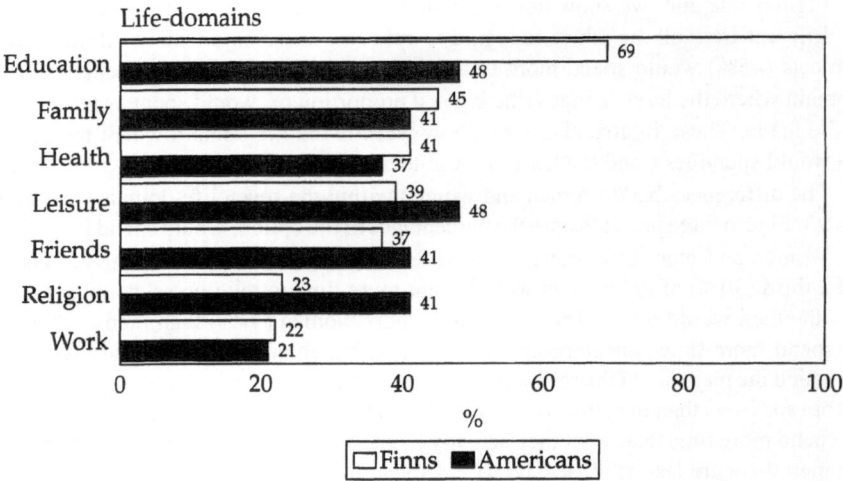

Figure 4. Differences between the Finns and the Americans in desire to spend more time within different life-domains.

of the Americans (41.2%) than of the Finns (24%) would spend more time on religion related issues.

The differences in the results between these two studies may partly be due to the age difference of these two samples; the subjects in the DeGenova's study were almost ten years older on the average, 54-91 years of age (mean 72.1 years), than in our Finnish sample. Partly the differences are due to socio-cultural differences. The Finnish welfare society, together with state-church system may well have produced more secularized life-style, resulting in diminished role of religion in the everyday lives of people.

INFORMATION AND KNOWLEDGE NEEDED FOR BETTER COPING, BUT ARE THE VALUES OF WORK AND OTHER LIFE-DOMAINS ALSO CHANGING?

On the basis of the Life Review Index we can assume that those life domains, in which older people say they would spend more time than what they have actually done, they, in retrospect, regard as relatively more important than those domains with less desired change. This study showed that based on the life experiences, older people would prefer to spend much more time in education and leisure enjoyment, if they could live their lives over again. This finding reflects the contradiction between the high value and importance of education in our society and the low level of formal education of this particular age cohort. What makes this finding significant, is that it was not solely instrumental education that people would spend more time with, but also all the other, more personal issues related to education, such as developing one's mind and intellect as well as keeping current on topics that are within a person's interests.

It seems self-evident that when one's job as well as one's resources for living in general, are largely determined by one's level of formal vocational education, those adults who do not have it, say they would spend more time on it, if they could live their lives over again. Also, as information needs for coping with everyday social life have become more demanding in our more complex societies than still a few decades ago, older people may very concretely feel the lack of knowledge and information

However, if education would be regarded important only for instrumental purposes, the other more personal issues related to education that according to this study, are as important, would not necessarily be regarded as such This suggests that in what comes to information needs of older people in our society, we should invest more on educational options available to them

Learning in later years can enhance the quality of life of older people Especially for people, who for any reason have been excluded from educational opportunities in earlier life, participation in older ages has the potential to fulfil a lifelong wish In general, the findings give support to the idea of lifelong learning and education, suggesting we should have educational opportunities available through the adult years

However, it needs to be taken into account that though not every adult, younger or older, is willing to participate in institutional forms of education, most of them seem to be willing to develop their mind and intellect within topics of personal interest Thus, in addition to the educational opportunities to be available, we can conclude from the findings that through the adult years it is important to have more balanced match between work and leisure, enabling adults to devote their time and energy to personal intellectual interests, if they wish to do so

The generations currently in retirement ages, have lived their lives mainly for two general purposes, family and work In older ages they, however, seem to recognize also other as important values in life, which spending more time e g in developing one's mind might be referring to The idea of changing values gets also support by the finding showing it is work above all the other major life domains, where people would spend less time with if they had their lives to live over As well as the finding may support the idea of life-span developmental changes in personal values, it may also reflect the socio-historical changes which are accompanied by value changes in society Concluding his study Alkula (1990) writes that "the concept of work as a central moral duty (often connected to the Weberian Protestantic ethic) has been eroded Thus, centrality of work is largely determined in comparison to the amount of positive experiences leisure can offer Work has lost this competition (p 157)" Learning and education also in later life may well be a means for positive leisure experiences, which is shown for example by the rapidly growing popularity of the third age universities

What younger people can learn from these results can be stated like DeGenova (1992) did "Education matters!" Regrets of low levels of education are common not only for less educated older people, but also for younger less educated people, and even for relatively highly educated people (Landman, 1993) The educational regrets of the aged as well as their perceived willingness to participate in educational activities poses a moral question to the society The aged of today, in the industrialized societies belong to those generations that, after the second world war, contributed by their lifelong work periods and taxes to the building of an educational system and learning opportunities for the younger generations that we now refer to as a "learning society" This point of view stresses the moral duty of the younger generations in society to take care of the educational needs of the aged to the same degree that educational services are developed for the other age groups of the population

REFERENCES

Alkula, T , 1990, *Work orientations in Finland A conceptual critique and an empirical study of work related expectations"*, Societas Scientiarum Fennica, Commentationes Scientiarum Socialium 1990 42, Finnish Society of Sciences and Letters, Helsinki

DeGenova, M K , 1993 If you had your life to live over again What would you do differently? *Int J Aging Human Dev* 34(2) 135-143

DeGenova, M. K., 1992, Reflections of the past: New variables affecting life satisfaction in later life, *Educ. Geron.* 19: 191-201.
Landman, J., 1993, *"Regret. The Persistence of the Possible"*, Oxford University Press, New York.

10

THE NEW PUBLIC HEALTH APPROACH TO IMPROVING PHYSICAL ACTIVITY AND AUTONOMY IN OLDER POPULATIONS

John B McKinlay

New England Research Institutes
9 Galen Street
Watertown, Massachusetts 02172

INTRODUCTION

A whole population (or purportedly "new") public health approach is emerging which is characterized by (a) the promotion of health through upstream planned sociopolitical change, (b) appreciation of the ontologic and epistemologic limitations of risk-factor epidemiology to inform this new approach, (c) application of more appropriate methodologies (including both quantitative and interpretive methods), and (d) the incorporation of multi-level outcomes (including broader measures of qualities of life) Confronting this new public health, established epidemiology and community health methods are limited in three key ways a dominant paradigm which produces atheoretical studies and the equation of infinite risks with causation, a restrictive focus on downstream individual behavior and interventions, and methodologies that produce a rigid adherence to positivist approaches which devalue complementary interpretive approaches This paper considers the potential of the "new" public health approach to promote increased physical activity (and thereby autonomy and improved quality of life) among sedentary older adults

Thomas Kuhn in *The Structure of Scientific Revolutions* (1962) described the history of science as a chronicle of the rise and fall of "paradigms"—which are overarching viewpoints or prevailing conceptions which, for some time, dominate a discipline or field of inquiry A paradigm serves as a guide for all activity in a particular field It determines what topics of inquiry are appropriate, what methods are most desirable, the proper way things ought to be done, and finally how support and recognition are awarded Scientific revolutions and change result from a breakdown of the prevailing paradigm—internal inconsistencies emerge, anomalous findings persist, and alternative viewpoints promise greater explanatory utility Kuhn traces the paradigmatic shift from Ptolemaic to Copernican astronomy and he shows how theoretical anomalies emerged that were not explicable by the Ptolemaic system The resulting Copernican view point was, in turn, characterized by its own guiding paradigm This new paradigm then directed science, focused attention, dictated methods and asserted and reinforced its own priorities From this viewpoint, science is essentially conservative and resistant to change, while new views and methods are ordered and institutionalized by the system supporting the prevailing paradigm (Watson, 1973; Gholson and Barker, 1985). Using a Kuhnian approach, Sloan (1987) believes a new paradigm may be emerging in health promotion, fostered by recognized limitations of the prevailing paradigm, including misfocused outcomes and ethical quandaries. There is increasing disappointment with the failure of efforts which derive from the dominant paradigm, concern about inequalities in its application and frustration over the current resistance to alternative promising approaches and methods.

This paper concerns the improvement of autonomy and the quality of life of older people and focuses specifically on increasing physical activity as one means of promoting such independence. Physical activity can be viewed as a necessary but not sufficient condition for increasing autonomy or independence in elders: it improves functional status, it increases beneficial social interaction with others and improves subjective perceptions of well-being and qualities of life (Grimby et al., 1992; Paffenbarger et al., 1993; Stewart et al., 1993). It is reasonable to assume that increasing physical activity levels in elders is one important contributor to improving autonomy or independence, yet there are obviously many other psychological, environmental, socioeconomic and health factors which also contribute to levels of physical activity in elders.

Recognizing that increased physical activity has many beneficial outcomes, particularly among older populations, most interventions remain focused on "downstream" tertiary treatments or individual one-on-one interventions (The Lancet, 1994). It is argued that these well-intentioned (but questionably effective) efforts have their origins in the prevailing biomedical paradigm and the risk factor epidemiology and public health research methods that currently serve as its handmaiden. Like good servants, these epidemiologic approaches and methods are always readily available, do whatever is asked of them, but seldom question the underlying reasons. Research methods with origins in behavioral science are being narrowly and inappropriately applied to measure the effectiveness of tertiary and secondary prevention activities to promote physical activity in selected high risk groups. We must move beyond these "downstream" efforts towards a more appropriate whole population public health approach to physical activity—what may be termed a social policy approach to healthy lifestyles (that include physical activity) rather than the current lifestyle approach to social policy. Appropriate research methods must be developed and applied to match these emerging, appropriate levels of whole population intervention. Viewed from the history of public health this refocus on a whole population approach to physical activity suggests a journey back to the future, rather than the development of a brand new public health. After reviewing some limitations of the prevailing biomedical paradigm, the promise of an upstream approach to physical activity and its role in the development of independence and autonomy in older populations is illustrated.

PHILOSOPHICAL LIMITATIONS OF THE PREVAILING PARADIGM

The current debate over the most appropriate approaches to promoting physical activity in older people has traceable origins in divergent social philosophies and different conceptions of disease and health. Attention, however, is focused not on these underlying origins, but rather on their more immediate manifestation in health interventions and the methodologies which measure their impact. Inviting colleagues to move discussion to a philosophic level, Nijhuis and Van der Maesen (1994) suggest:

most theoretical debates about the pros and cons of public health approaches are confined to the methodological scientific level Philosophical foundations such as underlying ontological notions are rarely part of public health discussions, but these are always implicit and lie behind the arguments and reasoning of different viewpoints or traditions (1994, p 1)

Wrestling with terms like "public" on the one hand and "health" on the other, they make crucial distinctions that facilitate understanding of the consequences of these different social philosophies and conceptions of health for intervention activities

With respect to divergent *social philosophies*, Nijhuis and Van der Maesen (1994) identify two major types as follow

- *Individualistic* (or "individualistically oriented social philosophy") Here the emphasis is on the individual and following for example, Pareto (1963) and Weber (1947), "the total (the Gestalt) is considered to be the outcome of the actions and motives of distinct individuals" (1994, p 2)
- *Collectivistic* (or "collectivistically oriented social philosophy") Here the emphasis is on "the social constellations of which individuals are part " From this perspective and following the views of Marx (1964) and Durkheim (1938), "the Gestalt is primarily the social constellations of which individuals are part" (1994, p 2)

Regarding different *conceptions of health*, two general types can be identified

- The *natural science (mechanistic) view*, which is the dominant orientation of allopathic medicine, focuses on disease states and factors which predispose, are associated with, or increase the chances of entering into one of those states This *pathogenic view* treats people as biopsychosocial and neurophysiologic systems, while disease represents a perterbator which produces disequilibrium, dysfunction and disease Apart from its mechanistic approach, this view results in a conception of health as "non-disease", an exclusionary state, or one that is "intrinsically residual in nature" Accordingly, "because health is seen as non-disease it can only be viewed as a condition brought into being through causal mechanisms" (1994, p 2)
- The *holistic view of health*, originally associated with the Goddess Hygeia in classical Greek thought, appears to be undergoing a renaissance in the new public health and upstream health promotion strategies of today This *salutogenic view* considers health "as an expression of the degree to which an individual is capable of achieving an existential equilibrium This equilibrium is not static but constantly in motion" (1994, p 2)

Even though thinking in terms of dualities or binary opposites may, in itself, be a consequential limitation of the prevailing paradigm, we nonetheless (and only for the purpose of convenience) combine these dimensions into a conventional 2 x 2 array in order to derive a conceptual device to permit identification of general categories or classes of phenomena (McKinney, 1954) This enables us to locate the origins of different levels of intervention and epidemiologic research methods in different social philosophies and conceptions of health Accordingly, discussion can actually advance from disparaging evaluations of the advantages and disadvantages of different approaches/methods, or from futile discussion of "the best" approach, to appreciation of the underlying philosophies and views of health which manifest themselves in everyday health activities and their measurement I encourage all my colleagues to accept the invitation of Nijhuis and Van der Maesen (1994) to "reflective philosophical discussion" of these issues

This typological differentiation invites several observations

Conceptions of Health

		Natural Science	Holistic View
Social Philosophies	Individualistic	1 Downstream Curative Focus	3
	Collectivistic	2	4 Upstream Health Promotion Policies

Figure 1. Typology Social philosophies and conceptions of health

a it permits us to understand some international differences in types of public health studies and activities In Europe, for example, where a more collectivist/holistic orientation is evident, there is great interest in upstream public health policies, or the purportedly "new" public health In the US, with its more individualistic/natural science orientation, there is heavy investment in individual knowledge and behavior change and also in the reduction of disease in identifiable categories (high risk individuals) Some examples of this include the National Cholesterol Education Programs whose goals are that every American ought to "know their number" by the year 2000 and the National Heart Attack Alert Program which seeks to teach people the signs and symptoms of a myocardial infarction Individualized and group "rights" are given such prominence in the US that enactment of utilitarian public health measures often meets strong resistance

b It also permits us to understand the dominance and resilience of different methodologies in different national settings In the US and Great Britain, two settings in which I have both lived and worked, Popperian logical positivism prevails (Popper, 1968, 1974) In other settings (e g , groups in Canada, Europe, Australia and England) there is a refreshing interest in qualitative, interpretative, inductive methodologies which are more appropriate to the programs suggested by a collectivistic/holistic orientation These alternative approaches (I prefer the term "complementary") have their origins not in dissatisfaction with the limitations of positivistic methods, or the inherent superiority of one over the other according to some standard of science, but rather in the collectivistic/ holistic philosophies of their proponents Until these divergent origins are widely appreciated we will continue, as David Mechanic (1989) warns, to talk past each other

c Most of the erudite and interesting debates among devotees *within* a particular orientation have little appeal to the proponents of divergent philosophical views The utility of Popperian views and new derivative falsificationist criteria for deciding causes (Weed, 1988), while important contributions within the scientific materialist tradition (Whitehead, 1985), have very little appeal to collectivistically oriented interpretists This is not to disparage the valuable contributions of Greenland (1988) Rothman (1986, 1988) Petitti (1991), or Susser (1988, 1991), instead, it emphasizes their irrelevance *(not error)* to those driven by a fundamentally different social philosophy and conception of health These contributions are as dissimilar as two farmers with divergent views on crop production—one applies chemical sprays to kill weeds and prevent harmful insects, the other applies natural fertilizers and waters the crops Depending on one's philosophy, either approach may be considered appropriate and will produce acceptable yields

d This analogy prompts the question asked by Peter Rossi (1994)—can the quals and the quants ever live together in harmony? For some, like Foucault (1973) Feyerabend (1987) and Habermas (1981), there appears to be little hope—their different methodologies are all derived from distinct philosophical perspectives, and furthermore, each includes its own ultimately irreconcilable presuppositions While the explanations for it are perhaps necessarily divergent, the two groups nonetheless continue to suffer from the same malady—hardening of the categories

APPROACHES THAT FOLLOW FROM THE PREVAILING PARADIGM

No one can question the remarkable contribution of epidemiology to understanding the causes and consequences of illness, disability and death From the early public health activities of 17th and 18th century Europe to the initiatives of today, the range of problems tackled, the exquisite methods developed, and the programs and policies attributable to specific findings have been remarkable Some commentators, while acknowledging this remarkable progress, question the current direction of the field of epidemiology and its emergent theoretical assumptions In marked contrast to its origins, the established epidemiology of today appears hamstrung by the assumptions that follow from its adherence to the individualist/natural science paradigm and also by the results of that paradigm (Miettinen, 1985, Rothman, 1986)

Established epidemiology is characterized by the following features

- *Biophysiologic Reductionism* This involves a process in which phenomena, whether primarily physical or primarily behavioral are explained by tracing their causes back to some bacteriological, genetic or molecular origins Even sociological phenomena such as socioeconomic, racial and gender differences in heart disease are presumed to have biophysiologic explanations Plausible structural explanations in social deprivation as well as biases in treatment are overlooked in preference for identifying physiological risk factors and biomedical interventions,
- *Absorption by Biomedicine* Moving from its origins in public health and its status as an independent discipline, epidemiology is simply becoming an adjunct to clinical medicine Some reduce it to a body of expertise that is only useful for improving clinical decision making among practicing physicians (to check that they are being good Bayesians) I can well understand how my good friend John Last (1988a, 1988b), with his background in public health at Edinburgh University, has come to regard the term "clinical epidemiology" as an oxymoron!
- *It is atheoretical* Established epidemiology can explain very little because, unlike most disciplines, there is little interest in the field in developing theories which can be tested Lamenting the absence of theory development, Alwyn Smith of Manchester likens the products of today's epidemiology to "a vast stock-pile of almost surgically clean data untouched by human thought" (Smith, 1985),
- *Limitations of Dichotomous Thinking* Even though it is now widely accepted that the response curve is continuous and smooth for most risk factors and conditions, dichotomous thinking nonetheless prevails and still determines our actions The well demonstrated fact that most illness conditions and risk behaviors (for example, physical activity) are normally distributed, still appears to escape most

researchers. Using hypertension as one example, Rose (1992) has described the markedly different activities that logically follow from either dichotomous or continuous thinking. He observes "Paradoxically, it is epidemiologic research which has now repeatedly demonstrated that in fact disease is nearly always a quantitative rather than a categorical or qualitative phenomenon, and hence it has no natural definitions" (Rose, 1992, p. 8). Whole population approaches to public health which follow from acceptance of the continuous nature of risk are precluded "because it is a departure from the ordinary process of binary thought to which they are brought up. Medicine in its present state can count up to two but not beyond" (Pickering, 1968 in Rose, 1992, p. 7).

- *A Maze of Risk Factors*. Established epidemiology is analogous to a person trapped in a maze (of risk factors) in which there is no opening or exit in sight. Researchers enter this maze with great enthusiasm but are quickly diverted to the left or to the right by new, exciting and endless risk factor openings. Every new turn produces ever more promising openings but finally results in exhausting and frustrating disputes over which, among the numerous possibilities, is the "correct" direction in which to proceed. Often, after expending large amounts time, effort and resources, researchers often return to the same place they started but without the added knowledge base that is required for action.

- *Observational Associations are Confused with Causality*. Disregarding the inferential superiority of randomized controlled trials, even when such trials are feasible, there is a preference for weaker observational studies. When elevating simple associations to causal status as occurs in most risk factor epidemiology, important qualifications for membership into the causal club are disregarded. Bradford Hill (1965) listed five conditions, all of which must be fulfilled before observed associations can even begin to qualify for consideration as cause and effect variables and hence as candidates for interventions. These included magnitude, consistency, specificity, dose-response and biologic plausibility. Using these criteria, what proportion of observational reports qualify for membership in the causal club? Wider acceptance of Susser's discussion of levels of causality encompassing types of social organization, individuals, organ systems and molecular contributions may lead researchers beyond the obstacles of single risk factor studies to social action (Susser, 1973).

- *Dogmatism by Design*. There is a belief, often incanted by the epidemiologic faithful, that certain designs are purer than others. For example, many claim that cohort studies are inherently superior to case control studies is common dogma. Of course, each of these observational designs have their own particular strengths and weaknesses yet both are still observational sinners! One may be more superior than the other in different circumstances, yet neither has an intrinsic advantage, nor is one more appropriate than the other (Schlesselman, 1982; Kelsey, 1986). Appropriateness is a contingent status. Advocating that one method is inherently superior to other methods betrays a shallow understanding of research methodology as opposed to research techniques.

- *More of the Same is the Answer*. Even while recognizing some of these ontologic and epistemologic limitations, many researchers believe that the solution lies in ever more of the same—bigger observational studies, better measurement techniques and more sophisticated multivariate manipulation rather than in improving the basic structure of these designs. Phillips and Smith (1993) recently proposed an improvement to observational epidemiology. They recommended more measurements of risk factors in order to overcome the limitations of reduced sample sizes. Skrabanek (1993) responded to their idea with the old Irish saying, "you can't make a pig grow by weighing him". The point is that improved measurement techniques and statistical manipulations are no cure for the wasting condition which is now afflicting established epidemiology.

Yet another logical consequence of the dominant paradigm is the current individual risk factor approach to solving population based health problems. For approximately 20 years now (McKinlay, 1974; 1979; 1980; 1992; 1993) I have been describing limitations associated with an individual level risk factor intervention approaches. Namely, such policies (a) *divert limited resources* away from upstream healthy public policies; (b) *blame the victim*; (c) produce *a life style approach to health policy*, instead of a social policy approach to healthy lifestyles; (d) *decontexualize risk behaviors* and overlook the ways in which such behaviors are culturally generated and structurally maintained; (e) seldom assess the relative contribution of *nonmodifiable genetic factors* and modifiable social and behavioral factors. (In this regard, socioeconomic reductionism among social scientists is as myopic as bio-physiologic reductionism among natural scientists); and (f) can actually be harmful to the health of the targeted populations. Marshall Becker (1986) reminded us that not all health efforts are benign, an observation given added weight by recent reports that programs to lower total cholesterol in children may have actually worsened the HDL/LDL ratio. Further-more, when one wishes to publish such negative results, existing publication bias makes it difficult to do so (Rosenthal, 1979; Begg and Berlin, 1988; Esterbrook et al., 1991). Successes tend to get published; failures seldom do.

Even more problematic, many downstream efforts to reduce risks and to improve quality of life at the level of the individual have unfortunately produced disappointing results. One report which reviewed the best designed risk factor intervention trials for the primary prevention of cardiovascular disease (CVD) concluded that the major interventions to date have had no effect whatsoever on total mortality (McCormick and Skrabanek, 1988). My colleagues and I are undertaking the first formal meta-analytic assessment of the effective-ness of community interventions, and about one-half of all published reports lack even the basic data required to calculate an effect size! We have furthermore all been eagerly awaiting the results of the three most sophisticated, well-designed community interventions ever conducted in the US (Stanford, Minnesota, Pawtucket) (Blackburn et al., 1984; Elder, et al., 1986; Farquhar et al., 1990). The results are now in, and they too add to the list of failures.

The failure of these promising community level interventions highlights a tension and inconsistency in the field of public health. Moving beyond downstream approaches in interventions also requires that we move beyond the limited quantitative methods currently employed to measure individual behavior change. In other words, more appropriate upstream interventions require appropriate research methods. Currently, more appropriate community interventions are being assessed with inappropriate methods. Interventionists are setting themselves up for failure. When programs appear to fail, as most do, there is then a defensive McCawberish search for some other positive outcome *at the same level*, rather than a move to more appropriate levels, methods and outcomes (Green and Lewis, 1986; Fries et al., 1989).

Faced with failure on the very grounds upon which these interventions were mounted, some dispute whether they are truly failures since they may have produced improvement in other areas such as morbidity and quality of life. However, it is never clear that morbidity improvements were due to the voluntary behavior changes which the trials were specifically designed to promote. Instead, resultant improvements were almost always an unanticipated side effect of more the aggressive drug treatment of those persons identified as high risk individuals (McKinlay, 1993). One can understand why interventionists and public health researchers select the outcomes they do—these workers have humanistic interests and want to prevent death and sickness. They also know that, without behavior change, most untoward outcomes cannot be avoided. But are the most widely accepted outcomes the most appropriate ones? Are those interventions bound to fail? Perhaps they are successful, but their success is not immediately observable, nor easily measured.

Prochaska's Transtheoretical Model integrates current behavioral status with a person's intention to maintain or change his/her pattern of behavior. Five stages of change are hypothesized: precontemplation, contemplation, preparation, action and maintenance (Prochaska and DiClemente, 1983; Prochaska et al., 1990). In the precontemplation stage an individual is not engaged in the behavior of interest and does not plan to become involved in the behavior. In contemplation, an individual is not engaged in the behavior of interest but is thinking about becoming involved in the behavior. In preparation, an individual has taken some steps towards behavior change. During the action phase, an individual has initiated behavior change. Finally, in the maintenance stage, an individual has been actively engaged in the behavior of interest for some time. Consider this model in relation to the promotion of physical activity among an older people. Even among U.S. adults who are motivated to join exercise programs, 50 per cent drop out during the first three to six months (Carmody et al. 1980; Dishman, 1988). The Transtheoretical Model has been adapted to physical activity by Marcus and her colleagues (Sonstroem, 1988; Barke and Nicholas, 1990; Marcus et al., 1992a; Marcus et al., 1992b; Marcus et al., 1992c; Marcus and Simkin, 1993). As applied by these researchers, precontemplation includes individuals who undertake no physical activity and do not intend to start. People in the contemplation stage do not participate in physical activity but intend to start. Subjects at the preparation stage participate in some physical activity but not regularly (Regular physical activity is defined as exercises that are engaged in three or more times per week for a total of 60 minutes or longer). The action stage includes subjects who currently participate in regular activity but have done so for less than six months. Maintenance entails the participation in regular physical activity for six months or longer (Marcus et al., 1992b). With these definitions, Marcus et al. (1992c) then utilized the Transtheoretical Model to design an intervention to increase the adoption of physical activity among 236 community volunteers in which they successfully moved a sizable proportion of the study population along the transtheoretical continuum, from unobservable contemplation to observable action. Moving a large number of people a small distance (for example, from contemplation to preparation) may generate more overall social benefit than moving a small number of identifiably high risk people a large distance along the Transtheoretical Model's continuum. Furthermore, a paradox of community interventions is that, while they may eventually bring large benefits to the community by increasing everyone's kilocalory expenditure a modest amount, they will seldom afford any real benefit to individuals (Rose, 1992).

I have never joined those colleagues who argue that if ineffective interventions are run long enough they are likely to become effective. It is possible, however, that an unobserved, but not unmeasurable, outcome will eventually work for those who reached the preparation stage during the first trial, if that intervention is conducted again. The initial intervention may make the target group more receptive to the purposes of that intervention when it is repeated. However, we still need to avoid double standards in the assessment of results. Public health researchers often chastise clinicians for becoming so invested in a procedure that they will not abandon it, even when consistent evidence from several well-designed trials shows that it is ineffective (McKinlay, 1981). Those in the field of public health must likewise abide by those same standards.

WHERE TO FROM HERE

Having focused so far on inappropriate policies, let the discussion now turn to appropriate methods. Arguing for a refocusing of efforts does not imply that all resources should be invested in upstream interventions That can never occur, since if it did, resource allocation would be as distorted as it is at present A balanced distribution of effort and resources across the whole range of possible points of intervention is required to accommodate the continuous distribution of physical activity in the general population

For purposes of convenience one can distinguish three levels of public health interventions that could be instituted to improve physical activity among elders *Downstream efforts* comprise treatments, rehabilitation counseling and patient education for those already experiencing some disease and disability This is the level which, while consuming most of the available resources, encompasses a very small segment of the general population—probably under 5% of those already occupying the sick role *Mid-stream prevention efforts* to improve physical activity involved two main areas (a) secondary prevention efforts which attempt to modify the risk levels of those individuals and groups who are very likely to experience some untoward outcome, (b) primary prevention actions to encourage people *not* to commence sedentary behaviors that may unnecessarily increase their changes of experiencing a negative health event Even further upstream are healthy public policy interventions which include governmental, institutional, and organizational actions directed at entire populations which require adequate support through tax structures, legal constraints and reimbursement mechanisms for health promotion and primary prevention

Geoffrey Rose in the *Strategy of Preventive Medicine* (1992) provides eloquent arguments for an upstream or whole population public policy approach to increasing physical activity in the general population He moves beyond the traditional paradigm and shifts thinking from the level of statistical association as in relative risks, odds or rates to the absolute levels of risk in populations Of this new emphasis Marmot (1994) says

> By shifting attention away from *relative risk* (how many times more likely is this exposed person to succumb than someone not exposed?) to *absolute risk* (what is this exposed person's increase in absolute level of risk?), and even further to some measure of *population attributable risk* (how much of the disease in the population can be attributed to this level of exposure?) the notion of what constitutes an important risk can change dramatically (Marmot 1994 p 3)

Such distinctions lead to what may be called the Rose Theorem, one of the most important insights in modern public health "A large number of people exposed to a small risk may generate many more cases than a small number exposed to a high risk" (Rose, 1992 24) This theorem has dramatic implications for public health policies and resource allocation With necessarily limited resources, large investments in questionably effective attempts to sustain the few leave little to promote the health of the majority If utilitarian principles guide resource allocation then small improvements in the level of physical activity among the majority are a better bet than dramatic *attempts* to mobilize the sick and to prevent illness in a high-risk minority Appreciation of the continuum of risk (the dose-response curve) suggests that small and perhaps even imperceptible improvements in everyone's physical activity (including those at low risk) will yield greater overall gains for a society than very perceptible improvements in the level of physical activity among a minority of high risk individuals This harsh reality must be coupled with an equally harsh certainty— society has necessarily finite resources Therefore, what is invested in attempts to improve the sickness levels of the minority diverts resources away from promoting the health of the majority We are confronted with what Rose terms the prevention paradox "a preventive measure that brings large benefits to the community affords little to each participating individual" (1992 3) Alternatively, downstream measures that yield possibly large benefits to sick or at risk individuals affords little to the overall health of our community which is where the real benefit lies

Our continuing misfocus on downstream individual risk factors and the greater promise of upstream approaches is well illustrated by the dominant approach to hypertension, a prevalent condition which may affect up to 25 per cent of the adult population

Hypertension is the most pervasive cardiovascular disease, the most critical stroke risk factor, and the leading cause of heart attack, kidney failure, eye diseases and congestive heart failure. It strikes about 35 per cent of people without their knowledge and disproportionately afflicts minorities and the poor. Therefore, controlling this condition may produce improved outcomes in many different areas. Hypertension is especially challenging for public health workers because it is subject to the rule of halves—only half of hypertensives are known, only half of those known are under treatment, and only half of those being treated are managed effectively. As a result, a whole population public health approach to hypertension must involve all three levels simultaneously—upstream public policy, mid-stream secondary prevention and downstream tertiary treatments.

Figure 2 (from Ashton and Seymour's *New Public Health*) depicts the normal distribution of blood pressure in the population, dividing it into segments based on standard deviations from the overall mean. The area within two standard deviations above and below the mean includes 95 per cent of the entire population. Many physiological phenomena can be modeled using such a Gaussian curve. With respect to hypertension, for example, those with a diastolic BP greater than100 mmHg, constitute about 2.5 per cent of the total population at the greatest risk for heart disease and stroke. Most resources are devoted to the identification, treatment and modification of risk factors in this 2.5 percent of the population, mostly by pharmacologic measures. While these individuals are at great risk, they represent only a very small proportion of all deaths from heart disease and stroke. Ashton and Seymour (1988) argue:

> ...a public health approach to the problem of hypertension involves achieving a shift to the left of the complete population distribution through general measures to reduce risk factors among the entire population, none of whom individually may ever know whether they would have developed problems had they not adopted the proposed changes (Ashton and Seymour, 1988, p. 28-30).

The measures envisioned, many deriving from the work of established epidemiology are not totally dependent on voluntary individual behavior change. Rather, they are largely determined by social policies, macroeconomic structures and the prevailing cultural milieu (that is, upstream healthy public policy). A similar approach to the one advocated here for hypertension offers considerable promise in regards to the promotion of increased physical activity among the adult population. My colleague Alan Jette has recently described the

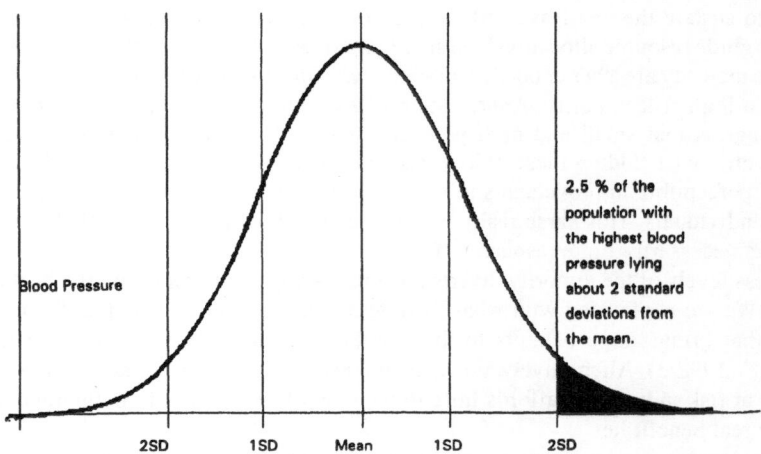

Figure 2. The normal distribution of blood pressure in a human population. The public health approach involves a shift in the entire distribution to the left. (Source: Ashton, and Seymour, 1988)

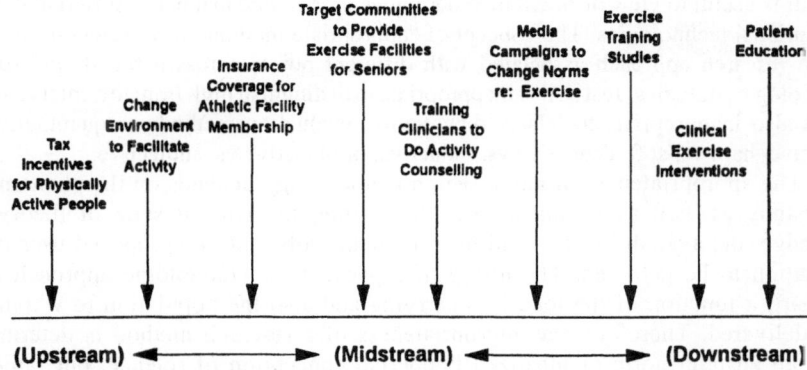

Figure 3. Points of intervention for physical inactivity. (Source: Jette, 1994).

importance of physical activity for older adults (Jette, 1994). He reports that approximately 70 per cent of people in the US over the age of 45 years do no regular exercise. This lack of exercise ranks with tobacco use as one of the leading preventable contributor to US mortality. Regular physical activity reduces the risks of mortality and morbidity. Finally, the level of activity required to beneficially affect health is actually quite modest.

Building on an appreciation of the continuous distribution of physical activity levels in the general population, Figure 3 (Jette, 1994) summarizes some points of intervention that can be used to improve physical activity levels in older adults. To my knowledge this is the first attempt to apply the principles of whole population public health to the challenge presented by physical inactivity among elders.

TOWARDS MORE APPROPRIATE EVALUATION METHODS AND MEASURES

I have argued that the prevailing paradigm, with its inherent assumptions and orientation, results in a disproportionate emphasis on downstream, individually-oriented interventions which have limited effectiveness for whole population public health. I extend my concern to the research methods currently employed to quantitatively measure either the success or the failure of these downstream, one-on-one, secondary prevention and tertiary treatment efforts.

As defined by the Oxford Dictionary, the term "appropriate" denotes something that is "specifically fitting or suitable," or phenomena that are "proper." The term "appropriate technology" supersedes the high-low continuum. Depending on the problem of concern, so-called "low technology" may be appropriate or inappropriate—likewise with so-called "high technology". "Appropriate" health technology does not conform to some idealized national or international standard, nor is it necessarily optimal or even "simple" (Newell, 1977). Instead, it serves as a suitable approach for some purpose at a particular point in time, taking into account the nature and magnitude of the problem as well as the available resources. Obviously, what is appropriate in one setting may be quite inappropriate in another setting. Moreover, even within a particular setting there are often differences over time in what is deemed appropriate. As a result, appropriateness is a Herculean notion: it connotes fluidity. Appropriateness is not a state that is achieved, nor is progress easily measured against some gold standard.

It is useful to view different methodologies in the same manner as different types of interventional technologies. The concept of "appropriate methodology" refers to the most suitable research approach associated with different points across a broad spectrum of methodologic strategies. Just as it is inappropriate to distinguish high from low interventions, so it is also inappropriate to falsely dichotomize evaluation methods as quantitative vs. qualitative, hard vs. soft, deductive vs. inductive, or objective vs. subjective.

The appropriateness of any research methodology depends on the phenomenon under study as well as its magnitude, the setting, the current state of theory and knowledge, the availability of valid measurement tools, and the proposed uses of the information to be gathered. The utility of a particular methodologic approach is, in large part, a function of the load it is carrying and also the population to whom it is being delivered. Therefore, the appropriateness of a research method is determined, not by an abstract norm or idealized Popperian conception of science, but rather by the nature of the problem under consideration, the community resources or skills available, and the prevailing norms and values at the national, regional or local levels. Acceptance of the notion of "appropriate methodologies" requires adaptation and refinement of traditional quantitative research methods such as social surveys and conventional experimental designs in order for these methods to remain applicable to the perspective of the "new public health." Moreover, well-designed and carefully conducted qualitative studies, including ethnographic interviewing, participant observation, case studies, and focus group activities, are now required not only to complement quantitative approaches, but also to fill gaps where quantitative techniques are suboptimal or even inappropriate (Spradley, 1980; Sanday, 1983; Yin, 1984; Basch, 1987; Bernard, 1988; Morgan, 1988; Grunig, 1990; Kitzinger, 1994; Kitzinger, in press). One problem is that quantitative and qualitative methods are viewed by their more rigid adherents as fundamentally incompatible rather than as mutually enriching partners in a common enterprise. Most quantitative researchers view qualitative approaches as inductive, subjective, unreliable and "soft." These advocates of quantitative methods constitute the dominant force in biomedical research (and control the purse strings). Investigators employing qualitative methodologies see quantitative researchers as positivistic, mindless data dredgers who suffer from hardening of the categories.

Generally speaking, quantitative methods as developed by epidemiologists including randomized controlled trials, or case control studies tend to be employed exclusively to measure outcomes of downstream interventions in which individuals are the unit of analysis (Kleinbaum et al., 1982). These individual level experiments could almost certainly benefit from judicious integration of appropriate qualitative methods. As one moves upstream, the utility of quantitative methods becomes problematic, not because they are intrinsically defective or flawed, but because the phenomena to which they are applied, the units of investigation, are of a *qualitatively different type*. Rigorous experimental control and manipulation are not always possible at the level of sociopolitical intervention, especially when change is unexpected or unplanned. Thus, different design approaches, measurements and data collection techniques must be employed. Quite often, egregious methodologic errors result from confusing an upstream unit of random assignment such as a community or school with a downstream unit of analysis such as an individual student. When an intervention program is applied to an aggregate unit such as a community, school or worksite, and the analysis is based on individual level observations, the residual error is deflated by intracluster correlation which leads to overstatement of the statistical significance, and also includes the problem of measuring the wrong outcome.

Diverse methods can complement and enrich each other, leading to a better understanding and appreciation of the phenomena under investigation (Strange and Zyzanski, 1989). The application of qualitative methods can provide further insight into the meaning of quantitative findings at both the individual and system level. While quantitative techniques can elucidate statistical significance, qualitative methods can reveal substantive significance (Reichardt and Cook, 1979; McGraw et al., 1989). Similarly, quantitative methods can be used to improve the generalizability and inferential strength of findings from qualitative approaches. An ethnographic study was recently conducted at NERI as an essential component of a larger AIDS community intervention experiment. This study employed purposive sampling schemes, stratified in various ways to ensure the development of a picture of the whole community and to guard against the danger that the ethnographer would end up with informants who, while conveniently available, did not represent all groups of interest. Incidentally, this ethnography was not an afterthought but actually served as the source of specific components of the subsequent intervention; it was the very foundation for the entire two-community experiment and informed the content of the pre- and post-intervention surveys.

The concept of process evaluation is relatively recent and still has not been widely applied in health promotion research (Flay, 1986; McGraw et al., 1989; McKinlay et al., 1989; Rossi and Freeman, 1993). Increasingly, researchers are recognizing that this type of evaluation may be as important as outcome measurement because outcomes of health or other social programs become uninterpretable without it (Nutbeam et al., 1993). Process evaluation permits the systematic exclusion of competing explanations for an observed experimental result. Most of the disappointing community interventions already discussed had no process evaluation so it is impossible to know why they failed, or whether they succeeded on some other level. When no effect is observed for an intervention, process evaluation can answer the following questions:

- Is there no effect because the program was not properly implemented or the implementation was not fully effective?
- Is there no effect because the program could not be fully implemented for some subjects, or compliance levels were variable?
- Is there no effect because of barriers to program access?

If a beneficial effect is observed from the outcome measures, then process evaluation can answer the following questions:

- Is the effect actually due to the program or is it due to the receptivity of selected subjects or target groups?
- Is the effect actually due to the experimental program or is it due to other competing, uncontrolled interventions, recognizing the reduction in experimental control at the aggregate level?

In order to answer these questions adequately it is necessary to monitor potentially competing programs, to measure exposure to the experimental programs, and to observe implementation of these experimental programs. With the use of case studies or video- or audio-taped interactions, qualitative methodologies are particularly appropriate for monitoring program implementation. Traditional quantitative approaches cannot measure aspects of group interaction that determine successful implementation. This situation provides a clear example of the appropriateness of qualitative over quantitative methods. In this instance, qualitative techniques serve as a necessary complement to traditional quantitative evaluation methods.

In summary, my central argument is quite simple. For a variety of reasons, health promotion efforts to increase physical activity levels in older populations need to move from the level of the elderly individual to the level of the social system (healthy public policy). Although tried and true quantitative methods generally work when the focus is limited to voluntary lifestyle changes at the individual level, they are not always useful or adaptable when the emphasis shifts to the social system level. Some techniques are misapplied, and others are inherently inappropriate. The notion of "appropriate methodology" emphasizes the match between the level of intervention and the most suitable evaluation approach in which the choice of approach is contingent on the problem, the state of knowledge, the availability of resources, the audience, and so forth. There is no right or wrong methodological approach, rather appropriateness given the purpose of the intervention must be the central concern.

Any reorientation of efforts to organizations, communities or national policies, requires the development of measurements and indicators appropriate to that upstream level of focus. In contrast to measurements of individuals, systemic interventions must be assessed through the use of systemic outcomes, that is, how have you improved the community, *independent of individuals* and their risky behaviors. In other words QOLs become QOCs (quality of community) or QORGs (quality of organizational environment). The interest is not in whether an individual quits smoking or lowers his or her cholesterol level, but whether there is improvement in the quality of the organizational environment, whether and how many restaurants add heart healthy items to their menus, and whether or not the air quality, measured by CO_2 concentrations or particulate matter, shows observable improvement. What proportion of schools change the way school meals are prepared? How many exercise facilities become available and what proportion of the elderly population utilize them? Is there a change in the availability of healthful products in stores, and if so, what proportion of space is devoted to them? How many different voluntary organizations devote time to healthful activities? How often do local leaders devote themselves to health promotive activities in fulfilling public responsibilities? What added revenues are generated from the imposition of taxes on harmful products? Is there a reduction in the overall rate of avoidable death? The list of system outcomes is extensive and the appropriateness of any is largely a function of the problem being addressed.

Upstream policy approaches to pressing public health challenges have advantages and disadvantages and these should be recognized. The advantages are as follows: (a) they are theoretically grounded, and based on a collectivistic orientation and a holistic conception of health; (b) they reflect the continuous distribution of disease and risk as distinct from artificially dichotomous categories; (c) they are cost effective and utilitarian since admittedly small health gains for the many bring greater results than even dramatic gains for the few; (d) they are egalitarian since they contribute to health improvement in all socioeconomic groups; (e) they have broad outcomes since a single act may affect many conditions whereas screening for hypercholesterolemia and subsequent treatment with Lovastatin may measurably affect only cardiovascular risk for a few; (f) they make team work in health a reality rather than promoting only physician-based public health; and (g) all research methods are given equal status based on their appropriateness to the problem, the available resources and the cultural group targeted. There are doubtless other advantages.

But there are also arguments against the upstream approach and these too should be recognized. They include the following: (a) paternalism and/or authoritarianism will increase since the government would be coercively telling people what to do and how to act; (b) they are frequently regressive as risk behaviors are disproportionately concentrated among the poor and minorities so legal and fiscal prohibitions disproportionality affect them; (c) people will disobey legal injunctions; (d) individual rights are comprised; (e) lowering everyone's risk may adversely affect the otherwise healthy; and (f) it is idealistic and not feasible because the medical industrial complex has too much to lose.

CONCLUSIONS

Because of their underlying paradigm health studies in general and established epidemiology in particular are limited by three types of inappropriateness. These are, in logical order, an inappropriate, atheoretical approach which is producing the misfocus on individual behavior change which, in turn, is producing inappropriate methodologies. If we are to move beyond individual behaviors or attributes, it is necessary to move beyond the currently limiting paradigm. The public health challenges of increasing physical activity levels and autonomy in older populations present an exciting opportunity to go back to the future through whole population epidemiology and healthy public policy. In pursuing these new directions, however, we must avoid any disjunction between more upstream interventions and the methods used to measure their effect—appropriate unto the intervention level must also be the research method thereof.

REFERENCES

Ashton, J , and Seymour, H., 1988, *"The New Public Health"*, Open University Press, Buckingham

Barke, C R , and Nicholas, D R , 1990, Physical activity in older adults the stages of change, *J Applied Gerontol* , 9(2) 216-223

Basch, C , 1987, Focus group interview an underutilized research technique for improving theory and practice in health education, *Health Educ Q* , 14 411-448

Becker, M H , 1986, The tyranny of health promotion, *Publ Hlth Rev* , 14 15-25

Begg, C B , and Berlin, J A , 1988, Publication Bias A problem in interpreting medical data, *J Roy Statist Soc A* , 151 419-445

Bernard, H R , 1988, *"Research Methods in Cultural Anthropology"*, Sage Publication, Newbury Park, CA

Blackburn, H , Luepker, R , Kline F G , et al , 1984, The Minnesota heart health program a research and demonstration project in cardiovascular disease prevention in *"Behavioral Health A Handbook of Health Enhancement and Disease Prevention"*, J D Matarazzo, S M Weiss, J A Herd, N E Miller, and S M Weiss, eds , John Wiley and Sons, New York

Carmody, T P , Senner, J W , Manilow, M R , and Mattarazzo, J D , 1980, Physical exercise rehabilitation Long-term dropout rate in cardiac patients, *J Behav Med* , 3 163-168

Dishman, R K , 1988, Overview, in *"Exercise Adherence Champaign"*, R Dishman, ed , Human Kinetics Books, IL, pp 1-9

Durkheim, E , 1938, *"Rules of Sociological Method"*, University of Chicago Press, Illinois

Esterbrook, P J , Berlin, J A , Gopalan, R , and Matthews, D R , 1991, Publication bias in clinical research, *Lancet*, 337 867-872

Elder, J , McGraw, S , and Abrams, D , 1986, Organizaitonal and community approaches to community-wide prevention of heart disease the first two years of the Pawtucket heaert health program, *Prev Med* , 15 107-117

Farquhar, J W , Fortmann, S P, Flora, J A , et al , 1990, Effects of communitywide education of cardiovascular risk factors the Stanford five-city project, *JAMA*, 264(3) 359-365

Feyerabend, P , 1987, *"Farewell to Reason"*, Verso, New York, pp 162-91

Flay, B R , 1986, Efficacy and effectiveness trials (and other phases of research) in the developments of health promotion programs, *Prev Med* , 15 451-474

Foucault, M , 1973, *"The Birth of the Clinic An Archeology of Medical Perception"*, Tavistock, London

Fries, J F , Green, L W , and Levine, S , 1989, Health promotion and the compression of morbidity, *Lancet*, 1 (8636) 481-483

Gholson, B , and Barker, P, 1985, Applications in the history of physics and psychology, *Am Psychol* , 40 755-769

Green, L W , Lewis, F M , 1986, *"Measurement and Evaluation in Health Education and Health Promotion"*, Palo Alto, California Mayfield Publishing Company

Greenland, S , 1988, Probability versus Popper An elaboration of the insufficiency of current Popperian approaches for epidemiologic analysis, in *"Causal Inference"*, K J Rothman, ed , Epidemiology Resources, Chestnut Hill, MA, pp 95-104

102   J. B. Mckinlay

Grimby, F, Grimby, A, Frandin, K, et al, 1992, Physically fit and active elderly people have a higher quality of life, *Scand J Med Sci Sports*, 2 225-230

Grunig, L, 1990, Using focus group research in public relations, *Publ Relat*, 1 36-49

Habermas, J, 1981, *Theorie des kommunikatieven Handelns*, Band 1 und 2, Frankfurt am Main, Surhkamp Verlag

Hill, A B, 1965, Environment and disease association or causation? *Proc R Soc Med*, 58 295-300

Jette, A, 1994, *"Designing and evaluating psychosocial interventions for promoting self-care behaviors among older adults"*, National Invitation Conference on Research Issues Related to Self-Care and Aging, NIA

Kelsey, J L, Thompson, W D, and Evans, A S, 1986, *"Methods in Observational Epidemiology Monographs in Epidemiology and Biostatistics"*, Vol 10, Oxford University Press, New York

Kitzinger, J, 1994, The methodology of Focus Groups the importance of interaction between research participants, *Soc Health Illn*, 16(1) 103-121

Kitzinger, J, in press, Focus groups method or madness? in *"Challenge and Innovation Methodological Advances in AIDS Research"*, M Boulton, ed, Falmer Press, London

Kleinbaum, D G, Kupper, L L, and Morgenstern, H, 1982, *"Epidemiologic Research Principles and Quantitative Methods"*, Van Nostrand Reinhold Company, New York

Kuhn, T S, 1962, *"The Structure of Scientific Revolutions"*, University of Chicago Press, Chicago

Lancet Editorial, 1994, Population health looking upstream, *Lancet*, 343(8895) 429

Last, J, 1988a, What's clinical epidemiology? *J Pub Health*, (Summer), 9(2) 159-163

Last, J, 1988b, *"A Dictionary of Epidemiology"*, Oxford University Press, New York

Marcus, B H, Rakowski, W, Rossi, J S, 1992a, Assesing motivational readiness and decision-making for exercise, *Health Psychol* 11(4) 257-261

Marcus, B H, Rossi, J S, Selby, V C, et al, 1992b, The stages and process of exercise adoption and maintainence in a worksite sample, *Health Psychol* 11(6) 386-395

Marcus, B H, Selby, V C, Niaura, R S, et al, 1992c, Self-efficacy and the stages of exercise behavior change, *Res Q Exerc Sport*, 63(1) 60-66

Marcus, B H, Simkin, L R, 1993, The stages of exercise behavior, *J Sports Med Phys Fitness*, 33(1) 83-88

Marmot, M, 1994, Cardiovascular Disease, *J Epidemiol Community Health*, 48 2-4

Marx, K, 1964, *'Selected Writings in Sociology and Social Philosophy'*, McGraw-Hill, New York

McCormick, J, and Skrabanek, P, 1988, Coronary heart disease is not preventable by population interventions, *Lancet*, 2(8615) 839-842

McGraw, S A, McKinlay, S M, McClemments, L, Lasater, T M, Assaf, A, and Carleton, R A, 1989, Methods in program evaluation the process evaluation system of the Pawtucket Heart Health Program, *Eval Res* 13 459-483

McKinlay, J B, 1974, A case for refocussing upstream - the political economy of illness In Applying Behavioral Science to Cardiovascular Risk, Proceedings of American Heart Association Conference, Seattle, Washington

McKinlay, J B, 1979, Epidemiological and political determinants of social policies regarding the public health, *Soc Sci Med*, 13A 541-558

McKinlay, J B, 1980, Evaluating medical technology in the context of a fiscal crisis the case of New Zealand, *Milbank Mem Fund Q*, 58(2) 394-443

McKinlay, J B, 1981, From "Promising Report" to "Standard Procedure" seven stages in the career of a medical innovation, *Milbank Mem Fund Q*, 59(3) 374-411

McKinlay, J, 1992, Health Promotion Through Healthy Public Policy The Contribution of Complementary Research Methods, *Can J Public Health*, 83 11-19

McKinlay, J B, 1993, The promotion of health through planned sociopolitical change challenges for research and policy, *Soc Sci Med*, 36(2) 109-117

McKinlay, S M, Stone, E J, and Zucker, D M, 1989, Research design and analysis issues, *Health Educ Q*, 16(2) 307-313

McKinney, J C, 1954, Constructive typology and social research, in *"An Introduction to Social Research"*, Stackpole Co, Harrisburg

Mechanic, D, 1989, Medical Sociology Some Tensions among Theory, Method, and Substance, *Health Soc Behav*, 30(2) 147-160

Miettinen, O S, 1985, *"Theoretical Epidemiology Principles of Occurrence Research"*, Wiley, New York

Morgan, D, 1988, *'Focus Groups as Qualitative Research"*, Sage, London

Newell, K W, 1977, Research for an appropriate health technology, Annual Address ANZSERCH Conference

Nijhuis, H G J, and van der Maesen, L J G, 1994, The philosophical foundations of public health an invitation to debate, *J Epidemiol Community Health*, 48 1-3

Nutbeam, D , Smith, C , Murphy, S , and Catford, J , 1993, Maintaining evaluation designs in long-term community-based health promotion programmes heartbeat Wales case study, *J Epidemiol Community Health*, 47 127-133

Paffenbarger, R S , Hyde, R T , Wing, A L , et al , 1993, The association of changes in physical-activity level and other lifestyle characteristics with mortality among men, *New Engl J Med*, 328 538-546

Pareto, V , 1963, *"The Mind and Society"*, Dover, New York

Petitti, D B , 1991, Associations are not effects, *Am J Epidemiol*, 133 101-102

Phillips, A N , and Smith, G D , 1993, The design of prospective epidemiological studies more subjects or better measurements? *J Clin Epidemiol*, 46-1203-11

Pickering, G W , 1968, *"High Blood Pressure"*, 2nd edition, Churchill, London

Popper, K , 1968, *"The Logic of Scientific Discovery"*, Harper and Rose, New York, pp 276-81

Popper, K , 1974, *"Conjectures and Refutations"*, Routledge and Keegan Paul, London, p 339

Prochaska, J O , and DiClemente, C C , 1983, Stages and processes of self-change in smoking towards and integrative model of change, *J Consult Clin Psychol*, 51 390-395

Prochaska, J O , Rossi, J S , and Velicer, W F , 1990, A comparison of four minimal interventions for smoking cessation An outcome evaluation Paper presented at the 7th World Conference on Tobacco and Health Perth, Australia

Reichardt, C S , and Cook, T D , 1979, Beyond qualitative versus quantitative methods, in *"Qualitative and Quantitative Methods in Evaluation Research"*, T D Cook, C S Reichardt, eds , Sage Publications, Beverly Hills, CA

Rose, G , 1992, *"The Strategy of Preventive Medicine"*, Oxford University Press, Oxford

Rosenthal, R , 1979, The "file drawer problem" and tolerance for null results, *Psychol Bull*, 86 638-641

Rossi, P H , 1994, The war between the Quals and the Quants Is a lasting peace possible? *New Directions Program Eval*, 61 23-36

Rossi, P , and Freeman, H , 1993, *"Evaluation A Systematic Approach"*, Fourth edition, Sage Publications, Beverly Hills, CA

Rothman, K J , 1986, *"Modern Epidemiology"*, Little Brown and Company, Boston, p 11

Rothman, K J , ed , 1988, *"Causal Inference"*, Epidemiology Resources, Chestnut Hill, MA pp 15-32

Sanday, P , 1983, The ethnographic paradigm(s), in *"Qualitative Methodology"*, J van Maanen, ed , Sage Publications, Beverly Hills, CA

Schlesselman, J J , 1982, *"Case Control Studies Design, Conduct, Analysis"*, Oxford University Press, New York

Skrabanek, P , 1993, The epidemiology of errors, *Lancet*, 342 1502

Sloan, R P , 1987, Workplace Health Promotion A Commentary on the Evolution of a Paradigm, *Health Educ Q*, 14(2) 181-194

Smith, A , 1985, The epidemiological basis of community medicine, in *"Recent Advances in Community Medicine 3"*, A Smith, ed , Churchill Livingstone, Edinburgh, pp 1-10

Sonstroem, R J , 1988, Psychological models, in *'Exercise Adherence Champaign"*, R Dishman, ed , Human Kinetics Books, IL, pp 125-154

Spradley, J P , 1980, *"Participant Observation"*, Holt, Rinehart and Winston, New York

Strange, K C , and Zyzanski, S J , 1989, Integrating qualitative and quantitative research methods, *Fam Med*, 21(6) 449-451

Stewart, A L , King, A C , and Haskell, W L , 1993, Endurance exercise and health-related quality of life in 50-65 year-old adults, *Gerontologist*, 33 782-789

Susser, M , 1973, *"Causal Thinking in the Health Science"*, Oxford University Press, New York

Susser, M , 1988, Falsifcation, verification and causal inference in epidemiology Reconsideration in the light of the philosophy of Sir Karl Popper, in *"Causal Inference"*, K J Rothman, ed , Epidemiology Resources, Chestnut Hill, MA, pp 33-58

Susser, M , 1991, What is a cause and how do we know one? A grammar for pragmatic epidemiology, *Am J Epidemiol*, 7 635-648

Watson, R I , 1973, Psychology A prescriptive science, in *"Historical Conceptions of Psychology"*, M Henle, J Jaynes, and J Sullivan, eds, Springer, New York

Weber, M , 1947, *"The Theory of Social and Economic Organization"*, Oxford University Press, New York

Weed, D L , 1988, Causal criteria and Popperian refutation, in *"Causal Inference"*, K J Rothman, ed , Epidemiology Resources, Chestnut Hill, MA, pp 15-32

Whitehead, A N , 1985, *"Science and the Modern World"*, Free Association Books, London, p 22

Yin, R K , 1984, *"Case Study Research Design and Methods"*, Sage Publications, Beverly Hills, CA

11

HEALTHY AGING

Utopia or a Realistic Target?

Eino Heikkinen

University of Jyväskylä
The Finnish Centre for Interdisciplinary Gerontology and
Department of Health Sciences
P.O. Box 35
FIN-40351 Jyväskylä
Finland

INTRODUCTION

Both health and aging are multidimensional and complex concepts. When they are used in combination the situation becomes even more difficult to study and interpret. The basic biological mechanisms of aging are not yet known and the numerous theories or metatheoretical notions focusing them have appeared to be difficult to test by experimental research. Consequently there are as yet no specific ways of modifying the process of aging. It seems to be increasingly evident that there are multiple causes of aging and that almost all theories of aging may have some relevance (Baker and Martin, 1994).

The concept of the intrinsic factors of aging refer to the basic biological mechanisms of the aging process. Because they remain unknown they only can be defined as residual factors which remain when the effects of extrinsic factors are excluded (Grimley Evans, 1988). The study of external factors, on the other hand, is rendered difficult by great variation in defining the dependent variable, the process of aging. This definitional variety is related to differences between research frameworks, and methodologies, and the epistemological and ontological assumptions underlying them. The only common denominator in these definitions is the concept of time, which is usually seen as a linear, quantitatively measurable category. Aging is usually defined as changes that occur in an individual over time (see Birren and Birren, 1990; Medvedev, 1990), the nature and the rate of such changes being partly dependent on the membership of a birth cohort, the characteristics of the historical period, and the ecological context.

Despite the prominent role of health in gerontological research and practice, there is no clear consensus regarding its conceptualization and measurement (e.g. Liang and Whitelaw, 1990). The definition of health by the World Health Organization, i.e. health as a state of complete physical, mental, and social well-being, and not merely the absence of disease and infirmity, is difficult to operationalise, and it may be argued that health is not a

Preparation for Aging, Edited by E. Heikkinen *et al.*
Plenum Press, New York, 1995

state but rather a dynamic, lifelong process. Furthermore, there are no unanimous interpretations of the content and meaning of the various aspects of well-being.

A major challenge for gerontological research has been to describe the differences between the aging process that is conditioned by chronic diseases and so-called normal aging (Grimley Evans, 1988; Svanborg, 1988; Wallace, 1992). This has lead to a number of epidemiological studies, both cross-sectional and longitudinal, attempting to identify the changes that occur with age in individuals who have no clinically diagnosed conditions as compared to the changes that are found in people with various chronic diseases. In many cases the differences seem to be of a quantitative rather than qualitative nature (e.g. atherosclerosis, hypertrophy of the prostate gland, loss of bone mass, changes in the ocular lens, degeneration of joint cartilages, involution of the cerebral cortex).

In consequence of the above-mentioned difficulties in attempts at defining aging and health, it becomes understandable that the concept of healthy aging can be regarded more as an empty abstraction than as a tool for scientific inquiry. The sympathetic attitudes connected with the concept may, however, make it suitable for health policy purposes, where healthy aging is seen as the target of policies and programmes which are built up on the basis of probabilities and associations rather than on verified causal chains.

In any scientific approach focusing healthy aging it is necessary to divide the global issue into its constituent parts and select the definitions and the perspectives used for both creating theoretical models and adopting methodologies. I believe that gerontological research would benefit from the observation of Karl Popper (1965) who argued against the empiricist view that in scientific research theory results from observation, proposing instead that pure observation does not exist; according to Popper observation is always selective. It needs a choice of object, a definite task, an interest, a point of view, a problem.

THE THEORETICAL MODEL

Figure 1 illustrates the key concepts and suggested causal chains proposed as the theoretical model for this paper. The starting point in the development of the theoretical model was that it must be structured on a rational logic of interference, on an empirically logical sequence of events which renders understandable the phenomenon in question. It was assumed that there are risk factors common to both aging and disease. The risk factors may cause some pathological condition, or they may have a direct impact on the process of aging. Furthermore, it is assumed that the physiological damage caused by different risk factors may be partly identical in relation to both the process of aging and pathological conditions. Aging and diseases may interact such that the disease may accelerate the process of aging or that the process of aging predisposes the individual to the disease (Heikkinen, 1995). Furthermore, aging, life-expectancy, diseases, disability and self-rated health interact with each other. Levels of self-rated health and disability also depend, on the one hand, on the characteristics of the living environment and the preconditions for activities and, on the other hand, on the quality and quantity of external support (or stress) together with goals, objectives and compensatory mechanisms the individual has been able to develop and adopt. It is now widely recognized that self-rated health and disability are important concepts in relation to the quality of life of elderly people (e.g. Schipper et al., 1990).

An important element of the model is the question about "causes of causes" (Rose, 1992). It is assumed that behind the identified risk factors differences between individuals and groups of people exist in living conditions, life styles and inherited biological properties. Medical technologies may also play a role in modifying the course of diseases.

Self-rated health and functional capacity are used in this paper as the indicators of health. The aim is to describe how they change with age and to identify the factors causing

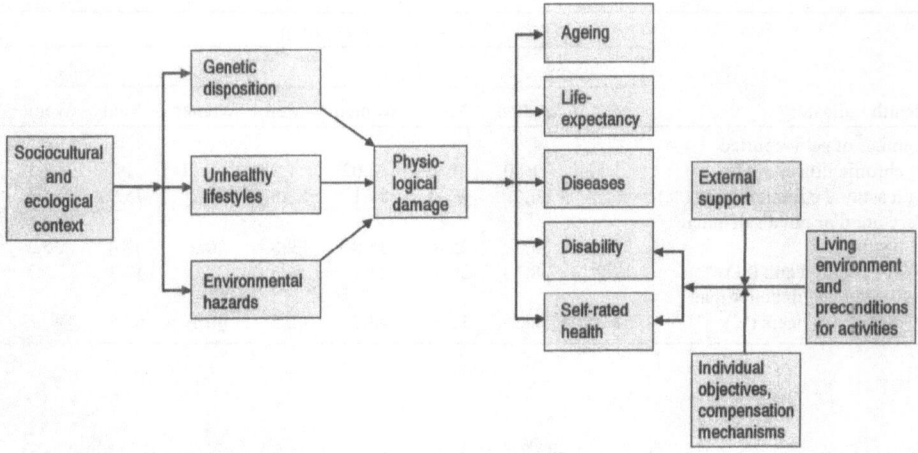

Figure 1. A theoretical model describing the suggested pathways from sociocultural and ecological context through physiological damage to variation in aging and health. Additional factors affecting disability and self-rated health are also presented.

these changes. Furthermore, the possibilities of maintaining good self-rated health and satisfactory functional capacity with advancing age are elaborated, focusing in particular on population-based studies.

SELF-RATED HEALTH AND ITS MODULATORS

The concept of subjective health has been gaining increasing importance in gerontological research. It has been shown that self-rated health or perceived health is an important independent predictor of mortality (Kaplan and Camacho, 1983; Kaplan et al., 1988; Idler et al., 1990; Idler and Kasl, 1991; Pijls et al., 1993), whereas the incidence of chronic diseases among older adults is not. Subjective self-ratings of health are designed to capture the respondent's interpretations of their own objective medical and functional health (Liang and Whitelaw, 1990). The most common measure of self-rated health is a simple global item, for example, "How would you evaluate your health in general, is it very good, good, average, poor, very poor?" This item may be modified to include a reference point such as health at an earlier time or the health of age peers.

The independent effect of self-rated health on mortality suggests that people have "tacit knowledge" about their health, knowledge based on a body which has lived and with the experience of its reactions in various situations. An elderly person may thus feel disturbances in his/her bodily reaction at the preclinical stage of chronic illnesses that are not yet responsive to diagnostic measures or that would merit comprehensive clinical and laboratory examinations (Idler and Angel, 1990).

The determinants and dimensions of the self-evaluation of health have become an increasingly popular topic in gerontological research. Other health indicators which have been reported to correlate with self-assessments of health include e.g. number of symptoms, functional health status, and use of health care, but also activity, social status, and several indicators of psychological well-being and morale associate with self-rated health (see Jylhä, 1994). Other factors which have been suggested to predict self-rated health among elderly

Table 1. Health status of 65 to 84-year-old residents of Jyväskylä in 1988 (N=1224)

Health indicator	65-69 Men	65-69 Women	70-74 Men	70-74 Women	75-79 Men	75-79 Women	80-84 Men	80-84 Women
Number of self-reported chronic illnesses ($\overline{X}$)	1.53	1.80	1.99	2.02	1.98	2.11	1.76	2.17
At least one chronic illness (%)	83.1	80.0	87.8	81.1	83.5	85.2	75.9	85.6
Very good or good self-rated health (%)	39.6	35.0	25.8	35.8	29.8	30.2	47.4	35.3
Worry about health (%)	17.9	28.3	26.3	23.1	14.3	27.0	12.3	35.3
Self-rated health better than that of age peers (%)	39.3	38.6	32.3	46.2	42.5	46.1	65.5	49.6

people include social support and social networks (Mellström et al., 1982; Welin et al., 1985). The factors associated with self-rated health may vary with age as suggested by the findings among men of different age groups (Jylhä et al., 1986). Among younger men symptoms and index of physical fitness were the most important predictors of self-rated health whereas among older men chronic diseases showed the highest associations with their ratings of health. Studies among middle-aged people have suggested that an important dimension reflected by the self-evaluation of health is the individual's perception of his/her own physical performance and sufficiency in general (Fylkenes and Førde, 1991). A positive health-related lifestyle also appears to affect self-rated health positively, while a preoccupation with health has a negative impact (Fylkenes and Førde, 1992).

In the interview study carried out among representative samples of 65 to 84 year-old residents of Jyväskylä (a small town in central Finland, population about 70.000) in connection with the Evergreen project (Heikkinen et al., 1990), it was observed that only about 10-20 percent in the oldest age groups had no diagnosed diseases, whereas almost a half of the persons interviewed reported their health to be very good or good. A large proportion of the subjects in all the five-year age groups estimated their health to be better than that of their age peers. Men in the age group 80-84 years reported surprisingly high levels of health, presumably representing the survival of the fittest. A significant proportion of those who reported chronic illnesses did not worry about their health status (Table 1).

The growth in the delivery and use of health services in the industrialized societies together with the progress made in the diagnostic technology of medicine have presumably increased the number of diagnosable diseases in elderly people which, among other things, affect people's own health ratings.

The factors which might be associated with self-rated health among elderly people in the city of Jyväskylä was examined using multiple regression analysis among the 75-year-old people who participated in an interview (93% of the target population) and in laboratory examinations (77% of the target population). It was observed that among the women a small number of symptoms and a good ability to manage daily activities were the most important predictors of good self-rated health (Figure 2). In addition, the lack of depressiveness, good cognitive performance, and good walking speed were associated with good ratings of health among the women. Among the men the most important predictors of good health were the small number of chronic illnesses and good PADL-performance, but good hearing ability and moderate drinking also predicted good self-rated health.

The variation in self-rated health explained by the models was about 40%. The survey data do not, however, enable us to identify how self-rated health is socio-culturally con-

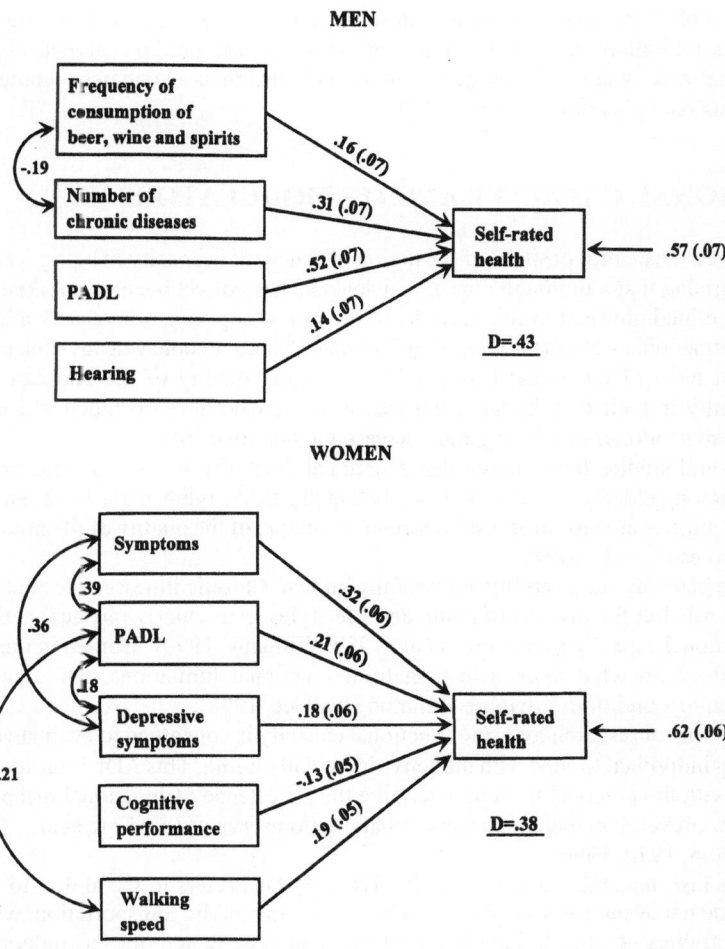

Figure 2. Predictors of self-rated health among 75-year-old men (n = 94-112) and women (n = 179-224) living in the city of Jyvaskyla, central Finland. Multiple regression analysis (LISREL) were used to examine the predictors D = total coefficient of determination

structed or understand the meanings of various factors identified by the model as important predictors of self-rated health in individuals' minds. The models described above suggest that there are differences between elderly men and women in the way they construct their subjective evaluations of health.

Self-rated health is a multidimensional concept. It varies with age and between the sexes. It is a socio-cultural construct and can, therefore, be expected to show cross-cultural variation between ethnic groups and between people living in different living conditions. It presumably also takes on a contextual shape despite the relatively good correspondence that exists between the interviewee's reports and physician's assessment performed in ambulatory settings. Most of the findings reported above were, however, obtained from cross-sectional studies which do not tell about the stability of the self-assessments or about the predictors of possible changes in them over time. The results from the Ontario Longitudinal Study of Aging (Hirdes and Forbes, 1993) suggest that the most important predictors of remaining in good health are an "advantaged" socioeconomic status, not smoking, and moderate alcohol use. The method used in surveys to assess people's own ratings of health has serious limitations in that it requires answers that are unidimensional, abstract and decontextualized, whereas in people's minds their health is a concrete, contextual and multidimensional phenomena (Jylhä, 1994).

FUNCTIONAL CAPACITY AND ITS MODULATORS

Given the central position of the notion of functional capacity in aging research, it is rather surprising that a thorough conceptual analysis has not yet been undertaken. Brook et al.(1979) defined physical health in terms of functional capacity, whereas Twaddle (1974) identified functional status as the social definition of health. People who are able to maintain a sufficient level of functional capacity have a higher quality of life and can live more independently in their own homes compared to those who develop functional limitations which threaten autonomous living and decrease the quality of life.

Several studies have shown that functional disability is an important predictor of mortality among elderly people (e.g. Campbell et al., 1985; Jylhä et al., 1992; Parker et al., 1992). Disabilities are also the most important predictors of the quality of life among elderly people (Parkeson et al., 1992).

The pathways to disability are not fully known. Chronic illnesses presumably play a significant role but the process of aging and life styles (particularly physical activity) also affect functional capacity (Verbrugge et al., 1989; Williams, 1990). Women in the older age groups suffer somewhat more than men from functional limitations, while among men life-threatening conditions are more common (Wallace, 1992).

One line of research focusing functional capacity is concerned to evaluate the ability of the aging individual to cope with the activities of daily living. This ADL tradition is closely bound up with the practical need to determine the prevalence of functional disabilities and the need for preventive, therapeutic, and rehabilitation services (e.g Katz et al., 1963; Katz, 1983; Lawton, 1970, 1990).

The instrumental activities of daily living (IADL) refers to the ability to cope with tasks outside home such as shopping, banking, and using public transportation, whereas the physical activities of daily living (PADL) denotes such basic activities as indoor mobility, dressing and undressing etc. One problem with the ADL measures is that they do not say anything about the underlying cause of the observed disability. It has been suggested, therefore, that performance measures should be added to ADL assessments in order to better understand factors behind ADL disabilities (Guralnick et al., 1989; Lawton, 1990, Guralnick and LeCroix, 1992).

The results of the above-mentioned Evergreen project illustrate the decline in ADL performance with age (Fig.3). Only a small proportion of elderly people under the age of 75 suffer from PADL disabilities whereas in IADL performance more disabilities are observed among the younger age groups.

The findings of the Evergreen project suggest that, in addition to chronic physical illnesses, cognitive capacity and mood problems also affect ADL performance (Laukkanen et al., 1993). Further analyses confirmed a relation between ADL disabilities and reduced sensory and physical performance, and mood and cognitive problems.

These findings support the above mentioned proposal that ADL measurements should be complemented by tests of performance, as this would help to give a clearer picture of the nature of age-related disabilities and help us to understand the pathways of the development of disabilities in different individuals.

Another research line has aimed at describing the changes that take place in physical, psychological, and social capacity with increasing age, in order to identify the factors that

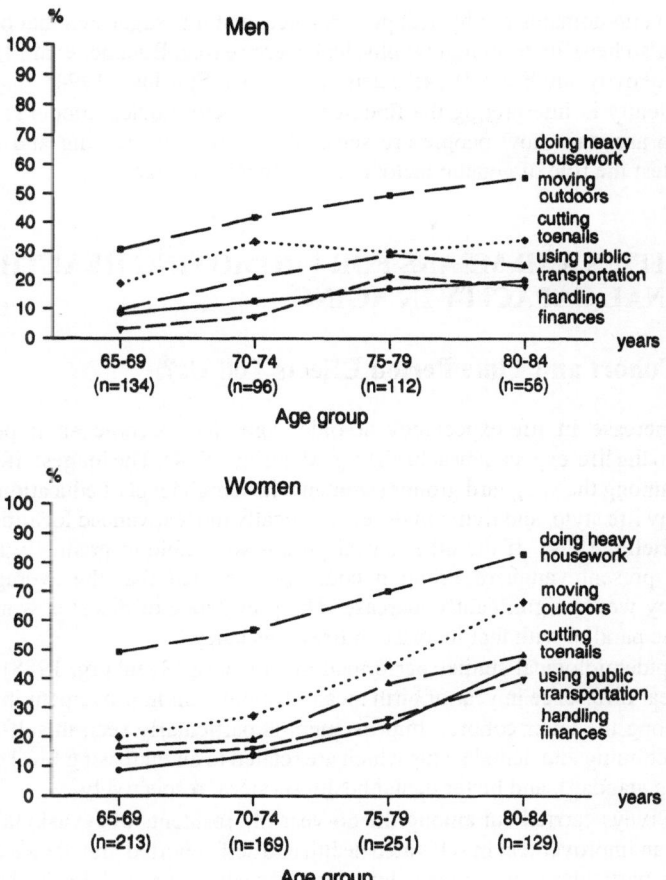

Figure 3. The proportions of 65-84-year-old residents of Jyväskylä who had difficulties or were not able to manage the indicated activities of daily living. The interviewees represent random samples of the total population aged 65 to 84 years. The overall response rate was 80 percent.

modulate those changes, and to look at possible ways of preventing disabilities in the individual's primary roles (e.g. Heikkinen et al., 1984, 1993). This line of research takes a broader interest in attempting to describe generalized functional capacity, which includes different domains and ranges from minimal survival capacity to the fullest possible functional capacity. Several models have been applied in gerontological research to describe generalized functional capacity (e.g. decremental, structural,optimal level, plasticity, biological age and socio-medical models; see Heikkinen et al., 1994; Heikkinen, 1995). The complexity of the concept of generalized functional capacity, its variable definitions, the great number of factors assumed to explain variation in it, and the lack of relevant theoretical tools make it difficult to draw conclusions about the causes of variation in functional capacity.

Both intra- and interindividual variation in generalized functional capacity has been well documented in the gerontological literature (e.g. Era, 1987; Heikkinen et al., 1993). The research findings show that functional capacity is least affected by increasing age among those with high socio-economic status, a high level of education, and healthy life styles. In particular "lifelong" physical activity seems to maintain above average levels of functional capacity in several domains of physical performance, and it is suggested that psychological performance also benefits from regular physical exercise (e.g. Buchner et al., 1992; Wagner et al. 1992; Bokovoy and Blair, 1994; Harris et al., 1994; Spirduso, 1994).

A difficulty in interpreting the findings of the socio-medical model is that it is not easy either to ascertain how people are selected for different training and occupational careers or to test the role of genetic factors in functional capacity.

POSSIBILITIES AND MEANS FOR PROMOTING HEALTH AND FUNCTIONAL CAPACITY IN AGING

What Do Cohort and Time Period Effects Tell Us?

The increase in life-expectancy at older ages has become an important factor contributing to the life-expectancy at birth (e.g. Martelin, 1994). The longest life-expectancy in Finland is among the vanguard groups (women with a high level of education, intellectual work, a healthy life style, and living in the economically most advanced localities within the country. (Martelin, 1994). If the other social groups were able to gradually approach the status of the present vanguard group it could be assumed that the average length of life-expectancy would significantly increase. The prevalence of diseases would then also show a decline parallel with that increase in life expectancy.

The epidemiological studies performed in Göteborg (Svanborg, 1988) suggest that even a five-year difference in year of birth means a significant improvement in health at the age of 75 among the latter cohorts. Improvement is particularly seen in self-rated health, cognitive functioning and dental status which are related to an increasing level of education, a higher living standard, and better dental health services, respectively.

The surveys carried out among the 66-year-old residents of Jyväskylä in 1972 and 1992 showed an improvement in self-rated health and self-reported prevalence of cardiovascular diseases, particularly in women belonging to the latter cohort (Table 2). No significant differences between the cohorts were observed in the prevalence of musculoskeletal diseases.

In the Tampere longitudinal and age cohort comparative study of 60- to 69 year-old people (Jylhä et al., 1992) the differences between the cohorts which were interviewed in 1979 and 1989 were small, but persons belonging to the latter cohort reported somewhat higher levels of self-rated health. In the older age groups no significant cohort differences have been reported in self-rated health.

The trend may not, however, be only in a positive direction, even in the industrialized societies. Recent information, e.g. from Poland and Kiev, suggest that the health situation of older people is worsening as a result of the current social and economic turbulence, leading among other things to decreased levels of self-rated health (Bezrukov et al., in press; Pedich and Bien, in press). Changes in life styles may also involve changes which have negative consequences on health. The increased prevalence of smoking and drinking among women may in the future decrease the present differences between the sexes in life-expectancy and health (Svanborg, 1988).

The question regarding the development of disability-free life expectancy (DFLE) in relation to the development of overall life expectancy (LE) is currently under debate in the gerontological literature (e.g. Guralnick and Schneider, 1987; Verbrugge, 1990; Manton et al., 1991; Robine et al., in press). If all levels of disabilities are taken into account, it seems that DFLE is not increasing at the same rate as LE, and consequently that the prevalence of light disabilities seems to be growing. On the other hand length of life without severe disabilities is increasing at the same rate as LE (Robine et al., in press). No major differences

Table 2. Se.f-rated health and prevalence of self-reported cardiovascular diseases among 66-year-old residents of Jyvaskyla in 1972 and 1992 In 1972 90% of the age group participated in the interviews In 1992 the corresponding figure was 84%

		Year of interviews		
Health variable and sex		1972 (n=387)	1992 (n=490)	p-value
Good self-rated health (%)				
	Men	49 3	56 0	0 228
	Women	45 4	54 2	0 041
Number of cardio-vascular diseases ($\overline{X}$ ± SD)				
	Men	1 0±1 3	1 0±1 1	0 744
	Women	1 1±1 2	0 8±1 0	0 001

in ADL performance were, however, observed in the Finnish cohort comparison studies (Jylha et al 1992, Heikkinen and Helin, unpublished data)

The present information about the prevalence of disability may underestimate the real situation As has been pointed out by Verbrugge (1990) in his review "Iceberg of Disability" only a part of the disability prevalent among elderly people becomes known because there is as yet no systematic registration procedure or generally accepted classification of disabilities The WHO ICIDH document (1980) has been applied in clinical trials but is not systematically used in any country Fried et al (1991) have suggested that nondisabled persons use compensatory strategies to minimize functional restrictions resulting from impairment, thus keeping functional decline at a preclinical level

The forecast about the compression of morbidity (Fries, 1980, 1989) assumes that both morbidity and disability can be postponed or avoided by appropriate preventive activities According to Fries' paradigm, this goal can be achieved through a policy directed toward the modification of the risk factors of chronic diseases, which are main causes of infirmity in aging The development of chronic diseases can be decelerated by improvements in life style and environmental exposure Empirical evidence to support or refute the compression of morbidity hypothesis is not yet available (see Kane et al , 1990)

Many other researchers believe that by setting and maintaining an optimal risk factor profile (cigarette smoking, obesity, lack of exercise, saturated fat intake, hypertension, a diet low in natural fibre, alcohol abuse, and inadequate social support and harmful environmental exposures) in younger ages (before age 30) such that the profile does not change with age, the life-expectancies at birth would be considerably over 85 years (e g Welin et al , 1985, Manton et al , 1991)

The activity theory of aging suggests that people actively involved in various spheres of social life age more successfully than their less active age peers (see Passuth and Bengtson, 1988) Although opinions diverge about the validity of this theory, it can be assumed that there is a reciprocal causal relationship with active people being more healthy and healthy people being more active The age cohort comparisons suggest that the members of the new generations of retired people are more actively involved in social and cultural life than those who retired, say 20-30 years ago, and that gender differences are at the same time diminishing (Svanborg, 1988, McPherson, 1994, Pohjolainen, see in this Volume) Above average social activity also seems to be associated with better life satisfaction (Heikkinen, 1989) If this trend continues it may be assumed in the future elderly people may be socially more active and have better life satisfaction than the present generation of retired people

The interpretation of the findings focusing the health and functional capacity of different birth cohorts is rendered difficult by differences between the cohorts in mortality and the number of survivors. Robine et al. (1992) have recently tackled this question by stating that the evolution of cross-sectionally healthy life-expectancy in survivors as a whole should be studied in relation to the evolution of the proportion of survivors from the initial cohort. This question is exemplified by data from Finland, which show that the proportion of survivors among 66-year-old people increased among women from 53 percent in 1972 to 70 percent in 1992. The corresponding figures for men were 35 percent and 52 percent. If the level of self-rated health or functional capacity is the same in the two cohorts of 66-year-olds, should this not be interpreted as an improvement in health status among this age group if a greater proportion of the later birth cohort reaches the age of 66 years?

The predictors described above in relation to health and functional capacity of elderly people suggest that the rapid social and environmental changes taking place everywhere constitute a time period effect which may appear to be even more important than the cohort effect in relation to the development of health and functional capacity among elderly people in the future.

Time period effects may change attitudes toward both disability and the criteria used in self-ratings of health. The development of medical technology may increase life expectancy but not necessarily DFLE. The proportion of healthy, intact elderly people may be on the increase, but at the same time the proportion of impaired persons will also grow (Kane et al., 1990).

The Future of Healthy Aging

Information about the modulators of self-rated health and functional capacity in aging suggest that the prospects for health promotion and prevention are in principle good. Healthy aging can be regarded as both a utopia and a realistic target. Development in a positive direction depends on many questions related both to the contents and effectiveness of public policies and the focus and to appropriateness of scientific research and the application of its results in social practice. Probably the most important vantage point is the strategy of applying a healthy public policy (see McKinley, in this volume).

Once chronic illnesses and disabilities have developed it is difficult to cure them completely and substantial resources, therefore, need to be set aside for caring. In this paper, however, the main focus is on the possibilities for improving the health and functional capacity of elderly people, on one hand through community-based intervention programmes and on the other hand by applying individual-based interventions.

Effective prevention would require a life-course perspective to include the identification of critical periods of life which could then be used to construct good health and functional capacity as the possibility of doing this in later years is very limited. The formation of bone mass serves here as an example; sufficient physical exercise and an appropriate diet when young help to develop a high amount of bone. Differences in bone mass at the age of 30 will be significant, while during aging bone mass starts to decline, increasing the risk of fractures, particularly among those who already had an under average amount of bone at earlier ages (see Cheng, 1994).

On the other hand muscle mass and strength can be increased at any age, even among the frail residents of nursing homes (e.g. Fiatarone et al., 1990). Life styles are also formed relatively early in life, and it is not easy to change them and adopt new ways of behaviour latest, if it is not a question of attempting to prevent the further progress of a chronic condition.

One possible way of developing prevention and health promotion is to arrange systematic screening procedures for older adults. The methods and contents of the screening procedures that have been applied show great variation and their effectiveness is not sufficiently known (Freer, 1985; Bulpitt et al., 1990; Pathy et al., 1992). The methods used include systematic screening, e.g. in health centres, in geriatric evaluation and management units, and during home visit (Pathy et al., 1992; Stuck et al., 1993; Van Rossum et al., 1993). The use of postal screening questionnaire with selective follow-up has been reported to have a favourable influence on the outcome and use of health care resources by elderly persons living at home. This method is more cost-effective than e.g. multiphasic screening in general populations (Pathy et al., 1992).

Experience shows that in a considerable proportion (30-40%) of elderly people new findings requiring further examination or treatment are usually found. In about a half of cases some kind of a treatment is required. The number of new findings will be even greater if performance assessments are included in the program. A problem related to performance assessments is that there are no generally accepted reference values with cut-off points for use in making decisions focusing on the need for rehabilitation. Screening procedures can also be used for proband counselling, and an additional advantage is that it enables the health professional to acquire a better understanding of the process of aging at the level of the whole population. Communication and contacts with families and care givers can also be improved, and environmental problems related to difficulties in managing with the activities of daily living can be better identified and treated, particularly if those who report ADL problems are accorded a home visit (see Gray et al., 1985; Pathy et al., 1992).

The community-based intervention programme carried out among the elderly residents of Jyväskylä suggests that the response of people in this cultural context is very positive. More than 90% of them, for example, have so far participated in the examinations. The effects of the programme will be studied in the near future. There is an urgent need to collect additional information about the feasibility and effectiveness of different community-based intervention programmes among elderly people to improve their quality of life as well as to delineate, appropriate forms for services from the financial point of view.

Individual-based intervention strategies and programmes may also be applied in order to enhance health in aging. Baker and Martin (1994) have classified intervention strategies according to level of effect and by type of strategy as follows: accommodative (devices that can enhance residual physiological capacity with advancing age), lifestyle alterations (appropriate diet, regular exercise, and cessation of smoking or other deleterious habits), pharmacological interventions, and biological interventions (enhancement of normal age-adjusted physiological function by replacement therapies, hormones and gene therapy). It is highly plausible that many of these intervention strategies will be widely applied in the future in relation both to prevention and health promotion as well as to rehabilitation.

To conclude my presentation I would like to consider some of the theoretical and methodological aspects of health promotion in aging. In as much as health and aging are social as well as cultural and medical concepts, there can be no doubt that these concepts will undergo change in the future. The concept of healthy aging is a cultural construct. People's views as to the causes of illness and aging are based on the one hand on the knowledge and information that is provided by scientific research and how they assimilate that information; but on the other hand notions are also involved that have to do with guilt, shame, the will of God, coincidence, etc. (Sachs, 1994). The difficulties that the academic community is having in piecing together a consistent picture are quite clearly productive of a sense of uncertainty and hesitation in people. It is hardly surprising, then, that alongside so-called official medicine we also have a thriving sector of alternative medicine, which in many cases has different views on the causes of diseases, and which, accordingly, recommends different methods of treatment.

Cultural values include autonomy, toughness, competition, youth and self-control. Loss of health and accelerated aging are no longer an accident but a result of the individual failing to keep him or herself healthy, of eating and living and exercising correctly. This is why preventive health discourse gives priority to these questions; and this is also why the social causes and relations of diseases and aging have become medicalized and individualized rather than politicized and collectivized (see Sachs, 1994).

It is for this reason that there is a present need for research where theory-building starts from the premise that the causes of diseases and unhealthy aging lie in complex systems rather than in the individual's personal qualities, and where the aim is to generate not only models for explanation but also solid grounds for action.

Aging and health should be made visible as cultural constructs. The most important thing is to gain insights, using qualitative research methods, into the ways in which the parties involved understand health and aging. Studies of people's more or less conscious ways of adopting and using preventive programmes could open up explanations for illnesses and aging and help us to understand why those programmes succeed or fail (see Sachs, 1994).

We should also focus our attention on the physical and social environment. Age-associated phenomena should also be studied in relation to global environmental change, including natural hazards and risk assessment in an era period of urbanization (Maylath, 1989; Verhasselt 1993) in which overcrowding, poverty, pollution, and high population density together with growing numbers of slum dwellers affect human health and aging in cities. Furthermore, one way to examine and explain risks and hazards would be to study the relationships between aging, health and functional capacity in the context of cross-national and cross-cultural studies despite the fact that these are extremely laborious, time-consuming and expensive projects which rarely produce truly comparable results (e.g. Liang and Jay, 1992).

It may be that it is only through these studies that we can hope to understand the social processes that are involved in producing the differences observed in aging, health and functional capacity.

REFERENCES

Baker, C T III, and Martin, G R , 1994, Biological aging and longevity Underlying mechanisms and potential intervention strategies, *JAPA*, 2(4). 304-328

Bezrukov, V , Sachuk, N N , Pakin, Y V, and Minaeva, V P , in press, Socio-Demographic Factors of Health, Adaptation, Quality of Life and in Survivorship (A Longitudinal Study of Kiev Aging Residents), in *"Swingin Pendulum Health Related Quality of Life in Elderly Europeans,"* L Ferrucci, E W Waters, E Hekkinen, and A Baroni, eds , INCRA/World Health Organization, Florence

Birren, J E , and Birren, B A , 1990, The Concepts, Models and History of thePsychology of Aging, in *"Handbook of the psychology of aging",* J E Birren, and K W Schaie, eds , Academic Press, New York, pp 3-20

Bokovoy, J L , Blair, S N , 1994, Aging and exercise A health perspective, *JAPA*, 2(3) 243-260

Brook, R H , Ware, J E , Davies-Avery, A , Stewart, A L , Donald, C A , Rogers, W H , Williams, K N , and Johnston, S A , 1979, Overview of adult health status measures fielded in rand's health insurance study, *Med Care*, 17 1-55

Buchner, D M , 1992, Effects of physical activity on health status in older adults II Intervention studies, *Annu Rev Publ Health*, 13 469-488

Bulpitt, C J , Benos, A S , Nicholl, C G , and Fletcher, A E , 1990, Should medical screening of the elderly population be promoted, *Gerontology*, 36 230-245

Campbell, A J , Diep, C , Reinken, J , McCosh, L , 1985, Factors predicting mortality in a total population sample of the elderly, *J Epidemiol Community Health*, 39 337-342

Cheng, S , 1994, *"Bone Mineral Density and Quality in Older People",* Studies in Sport, Physical Education and Health, 34, University of Jyvaskyla, Jyvaskyla, (doctoral thesis)

Era, P , 1987, Sensory, psychomotor, and motor functions in men of different ages, *Scand J Soc Med* , Suppl 39

Fiatarone, M A , Marks, E C , Ryan, N D , Meredith, C N , Lipsitz, L A , and Evans, W J , 1990, High intensity strength training in nonagenarians Effects on skeletal muscle, *JAMA,* 263 3029-3034

Freer, C B , 1985, Screening the elderly, *BMJ,*300 1447-1448

Fried, L P , Herdman, S J , Kuhn, K E , Rubin, G , and Turano, K , 1991, Preclinical disability Hypotheses about the bottom of hte iceberg, *Journal of Aging and Health,* 3(2) 285-300

Fries, J , 1980, Aging, natural death, and the compression of morbidity, *New Eng J Med* , 303 130-135

Fries, J , 1989, The compression of mordidity near or far? *Millbank Q* 67(2) 208-232

Fylkenes, K , and Førde, O H , 1991, The Tromsø Study Predictors of self-evaluated health – has society adopted the expanded health concept? *Soc Sci Med* , 32(2) 141-146

Fylkenes, K , and Førde, O H , 1992, Determinants and dimensions involved in self-evaluation of health, *Soc Sci Med* , 35(2) 271-279

Gray, J A M , Almind, G , Freer, C , Warshaw, G , 1985, Screening and Case Finding, in ' *Prevention of Disease in the Elderly",* J A Muir Gray, ed , Churchill Livingstone, Edinburgh, pp 51-63

Grimley Evans, J , 1988, Ageing and Disease, in *"Research and the Ageing Population",* Ciba Foundation Symposium 134, John Wiley & Sons, Chichester, pp 38-57

Guralnik, J M , Branch, L G , Cummings, S R , Curb, J D , 1989, Physical performance measures in aging research, *J Gerontol* , 44(5) M141-146

Guralnick, J M , and LaCroix, A Z , 1992, Assessing Physical Function in Older Populations, in *"The Epidemiologic Study of the Elderly",* R B Wallace, and R F Woolson, eds , Oxford University Press, New York, pp 159-181

Guralnick, J M , and Schneider, E L , 1987, The compression of morbidity a dream which may come true someday, *Gerontologica Perspecta* 1 8-14

Harris, S , Suominen, H , Era, P , eds , 1994, *"Toward Healthy Aging – International Perspectives, Part 1 Physiological and Biomedical Aspects",* Vol III, Physical Activity, Aging and Sports, Center for the Study of Aging, Albany, New York

Heikkinen, E , 1995, Epidemiologic-ecological models of aging, *Can J Aging* 14(1) 82-99

Heikkinen, E , 1989, Lifestyles and Life Satisfaction, in *"Health, Lifestyles and Services for the Elderly",* W E Waters, E Heikkinen, and A S Dontas, eds , World Health Organization, Publid Health in Europe 29, Copenhagen, pp 39-74

Heikkinen, E , Arajarvi, R-L , Era, P , Jylha, M , Kinnunen, V , Leskinen, A-L , Masseli, E , Pohjolainen, P , Rahkila, P , Suominen, H , Turpeinen, P , Vaisanen, M , and Osterback, L , 1984, Functional capacity of men born in 1906-1910, 1926-30 and 1946-50 A basic report, *Scand J Soc Med* , Suppl 33

Heikkinen, E , Era, P , Jokela, J , Jylha, M , Lyyra, A -L , Pohjolainen, P , 1993, Socioeconomic and Life-Style Factors as Modulatros of Health and Functional Capacity with Age, in ' *Aging, Health and Competence",* J J F Schroots, ed , Elsevier Science Publishers B V , The Netherlands, pp 65-86

Heikkinen, E , Heikkinen, R -L , Kauppinen, M , Laukkanen, P , Ruoppila, I , Suutama, T , eds , 1990, *"Iakkaiden henkiloiden toimintakvky IKIVIHREAT-projekti Osa I'* , [*Functional Capacity in the Elderly Evergreen-project Part I*] Sosiaali- ja terveysministerio, Suunnitteluosasto, 1990 1, Helsinki, English abstract

Heikkinen, E , Suominen, H , Era, P , Lyyra, A -L , 1994, Variations in Aging Parameters, Their Sources, and Possibilities of Predicting Physiological Age, in ' *Practical Handbook of Human Biologic Age Determination'* , A K Balin, ed , CRC Press, Boca Raton, Ann Arbor, London, Tokyo, pp 71-91

Hirdes, J P , and Forbes, W F , 1993, Factors associated with the maintenance of good self-rated health, *Journal of Aging and Health,* 5(1) 101-122

Idler, E L , and Angel, R J , 1990, Self-rated health and mortality in the NHANES-I epidemiologic follow-up study, *Am J Publ Health* 80 446-452

Idler, K , and Kasl, S , 1991, Health perceptions and survival, do global evaluations of health status really predict mortality? *J Gerontol* , *Social Sciences,* 46 55-65

Idler, E L , Kasl , S V , and Lemke, J H , 1990, Self-evaluated health and mortality among the elderly in New Haven, Connecticut, and Iowa and Washington counties, Iowa, 1982-1986, *Am J Epidemiol* , 131(1) 91-103

Jylha, M , 1994, Perceived Health in Old Age – Two Methodological Approaches, in *Experiencing Ageing – Kokemuksellinen Vanheneminen – Att Uppleva Åldrandet,* P Oberg, P Pohjolainen, and I Ruoppila, eds , Svenska social- och kommunalhogskolan vid Helsingfors universitet, Helsinki, pp 171-184

Jylha, M , Jokela, J , Tolvanen, E , Heikkinen, E , Heikkinen, R -L , Koskinen, S , Leskinen, E , Lyyra, A -L , Pohjolainen, P , 1992, The Tampere longitudinal study on ageing, *Scand J Soc Med* Suppl 47

Jylha, M , Leskinen, E , Alanen, E , Leskinen, A -L , and Heikkinen, E , 1986, Self-rated health and associated factors among men of different ages, *J Gerontol* 41(6) 710-717

Kane, R L , Radosevich, D M , Vaupel, J W, 1990, Compression of morbidity issues and irrelevancies, in *"Improving the Health of Older People A World View,"* R L Kane, J G Evans, D Macfadyen, eds , Oxford University Press, Oxford, pp 30-49

Kaplan, G A , Barell, V, and Lusky, A , 1988, Subjective state of health and survival in elderly adults, *J Gerontol , Social Sciences,* 43 114-120

Kaplan, G A , Camacho, T , 1983, Perceived health and mortality A nine-year follow-up of the Human Population Laboratory Cohort, *Am J Epidemiol ,* 117 292-304

Katz, S , 1983, Assessing self-maintenance Acitivities of daily living, mobility and instrumental activities of daily living, *J Am Geriatr Soc* 31 721-726

Katz, S , Ford, A B , Moskowitz, R W , Jackson, B A , and Jaffe, M W , 1963, Studies of illness in the aged The index of ADL A standardized measure of biological and psychological function, *JAMA,* 185 914-919

Laukkanen, P, Kauppinen, M , Era, P , Heikkinen, E , 1993, Factors related to coping with physical and instrumental activities of daily living among people born in 1904-1923, *Int J Ger P,* 8 287-296

Lawton, M P , 1970, The functional assessment of elderly people, *J Am Geriatr Soc ,* XIX 465-481

Lawton, M P , 1990, Aging and performance of home tasks, *Human Factors,* 32 527-536

Liang, J , and Jay, G M , 1992, Cross-cultural Research on Aging and Health, in *"The Epidemiologic Study of the Elderly",* R B Wallace, and R F Woolson, eds , Oxford University Press, New York, pp 301-312

Liang, J , Whitelaw, A , 1990, Assessing the Physical and Mental Health of the Elderly, in *"The Legacy of Longevity Health and Health Care in Later Life",* S M Stahl, ed , Sage Publications, Newbury Park, pp 35-54

Manton, K G , Stallard, E , and Tolley, H D , 1991, Limits to human life exptancy, *Pop Dev Rev ,* 17 603-637

Martelin, T , 1994, *"Differential mortality at older ages Sociodemographic Mortality Differences Among the Finnish Elderly' ,* Publications of the Finnish Demographic Society, 16, (doctoral thesis), Helsinki

Maylath, E , Weyerer, S , and Hafner, H , 1989, Spatial concentration of theincidence of treated psychiatric disorders in Mannheim *Acta Psychiatr Scand ,* 80 650-656

McPherson, B D , 1994, Sociocultural perspectives on aging and physical activity, *JAPA,* 2(4) 329-353

Medvedev, Z A , 1990, An attempt at a rational classification of theories of aging, *Biological Review,* 65 375-398

Mellstrom, D , Nilsson, A , Oden, B , Rundgren, A , and Svanborg, A , 1982, Mortality among the widowed in Sweden, *Scand J Soc Med ,* 10 33-41

Parker, M G , Thorslund, M , Nordstrom, J -L , 1992, Predictors of mortality for the oldest old A 4-year follow-up of community based elderly in Sweden, *Arch Grontol Geriatr,* 14 227-237

Parkerson, G , Broadhead, W , et al , 1992, Quality of life and functional health of primary care patient, *J Clin Epidemiol* 45 1303-1313

Passuth, P M , Bengtson, V L , 1988, Sociological Theories of Aging Current Perspectives and Future Directions, in *"Emergent Theories of Aging",* J E Birren, and V L Bengtson, eds , Springer Publishing Company, New York, pp 333-355

Pathy, M S J , Bayer, A , Harding, K , and Dibble, A , 1992, Randomised trial of case finding and surveillance of elderly people at home, *Lancet,* 340 890-893

Pedich, W , and Bien, B , in press, Quality of Life of the Elderly During the Economic and Political Transformation in Poland, in *"Swingin Pendulum Health Related Quality of Life in Elderly Europeans,"* L Ferrucci, E W Waters, E Hekkinen, and A Baroni, eds , INCRA/World Health Organization, Florence

Pijls, L T J , Feskens, E J M , and Kronhout, D , 1993, Self-rated health, mortality, and chronic diseases in elderly men The Zutphen Study, 1985-1990, *Am J Epidemiol ,* 138(10) 840-848

Popper, K R , 1965, *'Conjectures and Refutations",* 2nd edition, Routledge and Kegan Paul, London, p 46

Robine, J -M , Mathers, C , and Brouard, N , in press, Trens and Differentials in Disability-Free Life Expectancy, in *Congress Proceedings of the Congress on ' Health and Mortality Trends Among elderly Populations Determinants and Implications,"* City of Sendai, Japan

Robine, J-M , Michel, J-P , and Branch, L G , 1992, Measurement and utilization of healthy life expectancy conceptual issues, *Bulletin of the World Health Organization,* 70 791-800

Rose, G , 1992, *The Strategy of Preventive Medicine' ,* Oxford University Press, Oxford

Sachs, L ,1994, Har begreppet stress något gemensamt med "onda ogat", in *"Kampen for folkhalsan '* G Carlsson, and O Arvidsson, eds , Natur och Kultur, Borås

Schipper, H , Clinch, J , and Powell, V , 1990, Definitions and Conceptual Issues, in *Quality of Life Assessments in Clinical Trials",* B Spilker, ed , Raven Press, New York, pp 11-24

Spirduso, W W , 1994, Physical activity and aging Retrospections and visions for the future, *JAPA*, 2(3) 233-242

Stuck, A E , Siu, A L , Wieland, G D , Adams, J , Rubenstein, L Z , 1993, Comprehensive geriatric assessment a meta-analysis of controlled trials, *Lancet*, 342 1032-1033

Svanborg, A , 1988, The Health of the Elderly Population Results from Longitudinal Studies with Age-Cohort Compariso, in *"Research and the Ageing Population"*, Ciba Foundation Symposium 134, John Wiley & Sons, Chichester, pp- 3-16

Twaddle, A C , 1974, The concept of health status, *Soc Sci Med* , 8 29-38

van Rossum, E , Frederiks, C M a , Philipsen, H , Portengen, K , Wiskerke, J , Knipschild, P , 1993, Effects of preventive home visits to elderly people, *BMJ*, 307 27-32

Verbrugge, L M , 1990, The Iceberg of Disability, in *"The Legacy of Longevity"*, M Stahl, ed , Sage Publications, Newbury Park, pp 55-75

Verbrugge, L M , Lepkowski, J M , and Imanaka, Y , 1989, Comorbidity and its impact on disability, *Millbank Q* , 67(3-4) 450-484

Verhasselt, Y , 1993, Geography of health some trends and perspectives, *Soc Sci Med* 36(2) 119-123

Wallace, R B , 1992, Aging and Disease From Laboratory to Community, in *"The Epidemiologic Study of the Elderly"*, R B Wallace, and R F Woolson, eds , Oxford University Press, New York, pp 3-9

Wagner, E H , and Larson, E B , 1992, Effects of physical activity on health status in older adults I Observational studies, *Annu Rev Publ Health*, 13 451-468

Welin, L , Tibblin, G , Svardsudd, K , et al , 1985, Prospective study of social influences on mortality, *Lancet*, 1 915-918

Williams, T F , 1990, Geriatrics A Perspective on Quality of Life and Care for Older People, in   *Quality of Life Assessment in Clinical Trials'* , B Spilker, ed , Raven Press, New York, pp 217-223

World Health Organization, 1980, *"International Classification of Impairments Disabilities and Handicaps'* , Geneva

12

ADDING LIFE TO YEARS! PROMOTING QUALITY OF LIFE IN AN AGING EUROPE

Chris Todd

Health Services Research Group
Department of Community Medicine
Institute of Public Health
University of Cambridge
Cambridge CB2 2SR
England

AGING EUROPE: THE DEMOGRAPHIC TRANSITION

In 1990 some 13% of Europeans, excluding the former USSR, were aged in excess of 65 years (United Nations, 1992a). An European female born in 1950-55 could have expected to live to age 68.0 years, her brother to age 63.6; in 1990-95 she could expect 78.5 years and her brother 72.0 years; by 2020-2025 she could expect 82.2 years and her brother 76.3 years (United Nations 1991; Table 44). European societies therefore are aging. The demographic transition, which is leading to increased absolute numbers of elderly people making up a greater proportion of the population, is to be seen world-wide (Table 1) and results from three factors: decline in mortality, decline in fertility and increase in life expectancy (United Nations, 1992a).

The Challenge of Aging

Of all the challenges facing policy makers throughout Europe today perhaps one of the most important, but least overt, is this demographic transition. The european population in 1995 (excluding the former USSR) is estimated to be 504,247,000 and it is projected that in 2025 it will be 515,212,000 although it may be as high as 549,760,000 (United Nations, 1992b). Life expectancy in Europe differs between countries and between males and females quite considerably. Figure 1 shows the life expectancy for european men and women by country using the most recent data available, normally data based on index years between 1987 and 1990 (United Nations, 1992b). Life expectancy at birth has increased dramatically over the last century, in the UK for example from 44 years for boys and 47 for girls to the present levels of 73 and 78 years respectively (Department of Health, 1991). Such a pattern has been seen across northern and western Europe. Not only is average life expectancy increasing, but also larger numbers of people are living into advanced old age. Table 2 shows data on the age sex structure of the population of England in 1991 and projections for 2031 (Office of Population Censuses and Surveys, 1993). Clearly not only is the over 65 year age

Preparation for Aging, Edited by E. Heikkinen *et al.*
Plenum Press, New York, 1995

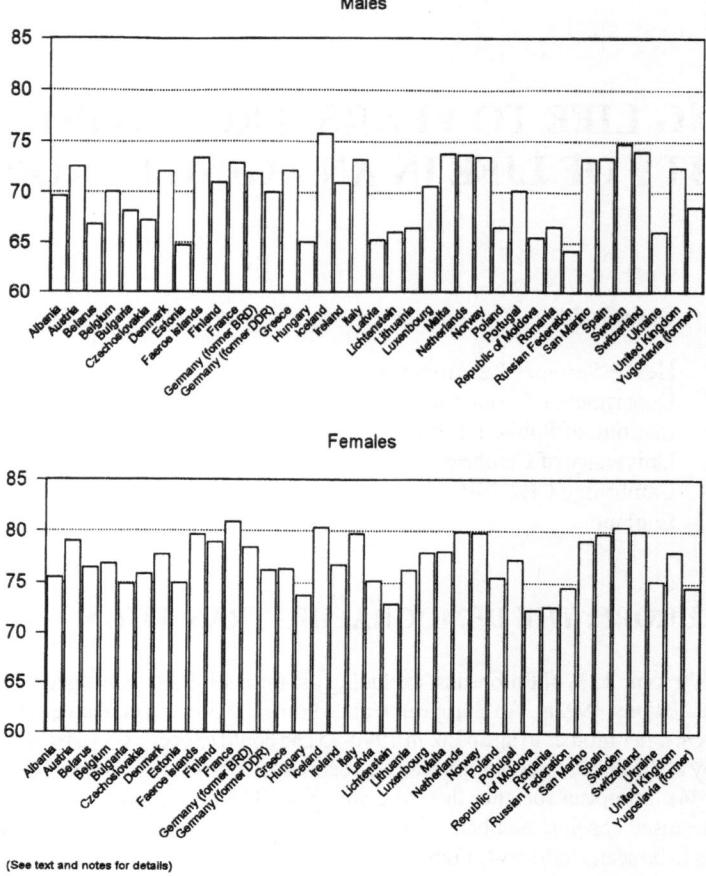

Figure 1. Life expectancy at birth in Europe

group comprising a larger percentage of the total population but the structure within old age is changing with greater numbers living into very old age. Such changes will have effects on the health care systems of our societies and it is time now to plan for these changes (Tinker, 1992; Hollingworth et al., 1995).

Disability and Old Age

Repeated surveys of the population have found that disability and reduced function increase with age (Martin et al., 1988; Office of Population Censuses and Surveys, 1990;

---

* Figure 1 is based on data from United Nations (1992b) which presents life expectancy for the latest years for which data are available. The differing index years are as shown below:Albania 1988-89; Austria 1990; Belarus 1989; Belgium 1979-82; Bulgaria 1988-90; Czechoslovakia 1990; Denmark 1989-90; Estonia 1990; Faeroe Islands 1981-85; Finland 1989; France 1990; Germany (BRD) 1985-87; Germany (former DDR) 1988-89; Greece 1980; Hungary 1990; Iceland 1989-90; Ireland 1985-87; Italy 1988; Latvia 1989; Liechtenstein 1980-84; Lithuania 1990; Luxembourg 1985-87; Malta 1989; Netherlands 1989-90; Norway 1990; Poland 1990; Portugal 1990; Republic of Moldova 1989; Romania 1988-89; Russian Federation 1989; San Marino 1977-86; Spain 1985-86; Sweden 1990; Switzerland 1989-90; Ukraine 1989; United Kingdom 1987-89; Yugoslavia 1988-90.

Table 1. Estimates and projections of percentage of population aged 65 and over, by region 1950-2025 (Source United Nations, 1992a, Table 7, p 17)

	Percentage over 65 years of age			
	1950	1970	1990	2025
World	5 1	5 4	6 2	9 7
Africa	3 2	3 1	3 0	4 1
Latin America	3 3	3 9	4 8	8 6
Northern America	8 1	9 6	12 5	19 9
Asia	4 0	4 0	5 0	9 6
Oceania	7 5	7 3	9 0	13 9
Former USSR	6 1	7 4	9 6	14 8
Europe	8 7	11 4	13 4	20 1
Eastern	7 0	10 4	11 3	17 6
Norhern	10 3	12 7	15 4	19 8
Southern	7 4	9 9	12 8	20 0
Western	10 1	12 8	14 5	22 3

United Nations, 1993) Thus one possibility is that with the increasing age of populations we will have populations with increasing amounts of disability Healthy life expectancy is an important concept which has been introduced over recent years (see e g Robine and Ritchie, 1991) Healthy life expectancy can be used as an index of the population's state of health and is derived from estimates of mortality and disability in the population It essentially addresses the question of whether observed increases in life expectancy are accompanied by decreases in morbidity Put simply, are the extra years spent in good health, or is it a prolonged state of illness and dependency that is associated with longer life? The demographic transition observed over recent decades has resulted in demographers, epidemiologists and public health specialists asking the question, " Is this change prejudicial to quality of life?" Two views have been expressed, one optimistic (Fries, 1980, 1989), one pessimistic (Gruenberg, 1977, Kramer, 1980) The Fries proposal is referred to as a compression of morbidity and it basically posits that the decrease in incidence of disease due to improvements in health care has the effect of delaying onset of disability People

Table 2. Age sex structure of population of England in 1991 and projected for 2031 (rounded to nearest 0 5%)

		Age			
	<65	65-74	75-84	85+	Total
1991					
Population in millions	40 45	4 23	2 62	0 77	48 08
males	87 0%	8 0%	4 0%	1 0%	
females	81 5%	9 5%	6 5%	2 5%	
Total	84 0%	9 0%	5 5%	1 5%	
2031					
Population in millions	40 45	6 11	3 89	1 81	52 27
males	79 5%	11 5%	6 5%	2 5%	
females	75 0%	12 0%	8 5%	4 5%	
Total	77 5%	11 5%	7 5%	3 5%	

Table 3. Life expectancy and disability free life expectancy at age
65 in 3 European countries (Source Robin and Ritchie, 1991)

	Life expectancy	Disability free life expectancy
Females		
England and Wales	17 5	8 9
France	18 5	9 9
Netherlands	18 6	8 9
Males		
England and Wales	13 4	7 7
France	14 3	9 1
Netherlands	14 0	7 9

therefore will live longer in a good state of health Both Kramer and Gruenberg suggest the opposite, that delayed mortality is not accompanied by a decrease in morbidity and that we will see a "pandemic of mental disorders and associated chronic diseases" (Kramer, 1980) A third possibility is that both processes are at work resulting in an increase in numbers in poor health, but a decrease in the prevalence of severe illness The work of Robine and Ritchie (1991) suggests that some 79-89% of life is disability-free for men and some 76-86% of life is disability-free for women, although, of course, there are regional differences (Table 3) The gap between sexes in disability-free life expectancy seems to diminish with age It is very important to recognise that there are considerable differences in disability-free life expectancy between the richest and the poorest parts of populations which have been studied to date According to Robine and Ritchie, based on data from Canada, the poorest fifth of the population differs from the richest fifth of the population by 14 7 years in disability-free life expectancy for men and 7 6 years for women Such differences are considerably greater than the regional differences reported within Europe, which are themselves quite considerable Changes over time of disability-free life expectancy are hard to estimate and not enough is yet known about this (Medical Research Council, 1994) Where they have been calculated, the pattern is not clear, and whether the compression of morbidity hypothesis or the expansion of morbidity hypothesis is correct is still in debate More work in this area is required if we are to be able to use this approach in assessing the quality of life of populations

Age, Sex and Disadvantage

Numerous papers have reported that older women's health is likely to be compromised (Restrepo and Rozental, 1994) This feminisation of ill-health is a result of exacerbated risk for women across the life course and is especially important in older age as they tend to live longer than men, and will often find themselves widowed and entering institutions Furthermore women often have less access to resources than men and are thus further disadvantaged Older women's issues are underrepresented in our research agendas, a lacking that urgently requires attention As Victor (1991) has demonstrated there are profound age, gender and class differences in later life Social inequality can be an issue which is geographically affected as well as being determined by social status (e g Department of Health and Social Security, 1980) The implication of such a finding is quite simply that not only may elderly people, especially women in lower socio-economic groups, be increasingly at risk of disadvantage, but that in some regions of Europe they will be even more at risk than others Given that current thinking tends to see health and especially functional capacity as underpinning quality of life such scenarios suggest need to address quality of life of our aging population with some urgency. especially in those parts of Europe where disadvantage is rife.

QUALITY OF LIFE

The Good Life

Other chapters in this volume address what is meant by quality of life more fully and we will only touch upon it in this chapter. Quality of life has been used to mean many different things. It is often used to refer to those attributes of "the good life", material affluence that were the goals of a world reconstructing after the ravages of World War 2 and which are still the goals of many regardless of the effect this may have on the global system and environment. The US National Commission on Goals set up by President Eisenhower in 1960 widened the remit of "quality of life" to include "education, health and welfare, economic and industrial growth and the defence of the 'free' world" (Fallowfield, 1990). Such an approach is still to be found in the USA in study of quality of life and race. Here, as with age, the challenge continues to be one of how to achieve equity (Farley and Allen, 1987). However more recently the notion of "health related quality of life" has been developed and it is this notion of quality of life that is perhaps most pertinent for the present purposes.

Health Related Quality of Life

Current thinking about health related quality of life is that it has as a basis physical function, but that it reflects the psychological and social functioning of the individual as affected by their health state. Thus, such concepts as subjective well-being, subjective health status, functional status and fulfilment of social roles are all encompassed by health related quality of life. The measurement of these dimensions is now being accepted as augmenting the measures such as mortality in assessing health care (measures which relate more directly to quantity of life). Any measure of health related quality of life must have some basic psychometric characteristics. It must be reliable, valid, and sensitive to change as well as being acceptable to individuals to be measured and suitable for them to complete (Todd and Bradley, 1994). The degree to which scientists have been successful in developing quality of life measures will not be discussed here, but is taken up elsewhere in this volume.

QUALITY OF LIFE OF ELDERLY EUROPEANS

What do we know of the quality of life of elderly Europeans? The European Longitudinal Study of Aging (ELSA) which has been carried out across Europe between 1979 and 1989, collected a number of variables which can cast light on to the quality of life issue in Europe. The results of these studies are more fully reported elsewhere (Heikkinen et al., 1983; Waters et al., 1989; Jylhä et al. 1992), and a volume on quality of life will appear in 1995 (Ferrucci et al., in press). In overview, there is no evidence that the physical function of elderly people across Europe has improved over the ten year period between 1979 and 1989. Nor was there evidence that the health of elderly people had improved over the same period, at least not uniformly. In eastern Europe health of elderly people was clearly compromised by rapid changes in socio-economic and political structures during this period. The standard of living (the "good life" approach to quality of life) of elderly people has also not improved uniformly or greatly. It does appear that elderly people are more likely to have a telephone, washing machine and refrigerator now than they were during the 1970s. But even this standard of living index is not straightforward and elderly people are clearly put at high risk in times of economic crisis.

Measures of functional status, self rated health, life satisfaction, mental health and other psychological components of "quality of life" are reported in the Ferrucci et al volume. Again the results are by no means uniform across Europe. What is clear from the Ferrucci *et al* volume is that the "quality of life" of elderly people is not improving spontaneously in conjunction with increasing life expectancy. At its very best quality of life is being augmented by careful social planning and input of resources, both human and financial, at worst it is deteriorating as society turns its back on the elderly to concentrate on more "pressing" problems of economic destabilisation and change or (more depressingly) to wage war on former neighbours. What is clear is that we cannot expect quality of life to improve of its own accord, it is more likely to drop, unless nurtured.

The Drift to Primary Care

It was also clear that in western Europe there is a move from secondary (hospital) to primary care (see e.g. Dall, 1994). But whether the additional resources implied by these changes are being allocated is considerably less clear. We do not know enough about the implications of changes in health care provision patterns and the implications of the move from secondary to primary care in particular to judge whether they are advantageous or disadvantageous to elderly people. For example, rehabilitation after hip fracture in the community rather than in hospital may be advantageous because it occurs in an older person's familiar environment. But it may be disadvantageous as it does not permit the same nursing and therapy input as in an institution. What is fairly clear is that it "on the face of it" it appears cheaper, at least to a health care purchaser, and in particular permits opportunity benefits to use hospital resources in a different way (Hollingworth et al, 1993). Perhaps, however, one may suspect that involved in the move from secondary to primary care is also a restriction of the resources and it is an issue that has caused considerable debate (Tinker, 1992). In eastern Europe the health care changes that have been observed have been from state provision to market economy. This change has clearly put medicines and health care in general out of the reach of some sectors of the society (Bezrukov et al., in press; Pedich and Bien, in press). On the basis of the available mortality and morbidity data elderly people in eastern Europe are amongst the most needful in Europe for health care provision, but the rapid changes towards a market economy system in the provision of health care has put these basic needs out of reach for many elderly people. The effects on the "quality of life" of the elderly of the change in care provision must be evaluated, but it seems unlikely that the changes being observed are enhancing quality of life. Systems should not be adopted just because they are cheap, or because they conform to the ideological dogma of the moment, but because they facilitate good provision and promote both quality and longevity of life. Unless these issues are addressed we will be in danger of exacerbating the differences between eastern and western Europe at this crucial moment in history when the "Iron Curtain" has fallen. Positive steps are needed because otherwise inaction is likely to reinforce a less tangible but highly discriminating "longitude line" not dissimilar to the racial "colour line" of disadvantage to be found in the USA (Farley and Allen, 1987).

PROMOTING QUALITY OF LIFE

Both good health and good quality of life need to be promoted, but we must not forget that in much of the world (Europe included) health *per se* is still the primary issue. In the same way that preventive strategies have been developed to promote health, it would be possible to promote good quality of life and prevent poor quality of life by a series of strategies, primary prevention, secondary prevention and tertiary prevention Tertiary prevention, that is, cure and care after the event, seldom seems to result in return to previous levels of quality of life or physical function in elderly people This is clearly evident in work conducted on elderly women with hip fractures amongst whom many do not return to their previous levels of physical function and often they do not regain their social integrity and confidence (Todd et al , 1994) Screening for osteoporosis and treatment for osteoporosis using hormone replacement therapy, are probably not the most effective approaches to the reduction of hip fracture (Law et al , 1991) The most effective preventive strategies appear to be more generalist and include changes in lifestyle, diet etc Such generalist lifestyle approaches will probably enhance the quality of life of Europeans in years to come Such health and quality of life promotion can be accomplished now, and will probably prove to be cost effective

In conclusion then, life expectancy has been increasing across Europe this century, but it is less clear if active life expectancy is improving or whether the aging of the population is going to be associated with increasing periods of disability and dependency at the end of life Health and physical function underlie quality of life and almost certainly the quality of life of elderly Europeans has not been improving dramatically over the last decade or so Health and physical function are amenable to promotion Improving function via for example exercise promotion should also result in improved quality of life in an aging Europe, if it has the expected result of increasing disability free life expectancy Dependence on curative routes to improve health and quality of life is probably misguided as seldom is cure complete Recent approaches to quality of life promotion are discussed by Greengross (see in this Volume) and Deeg (see in this Volume) in the pages which follow

ACKNOWLEDGEMENTS

I would like to thank Professor Eino Heikkinen, Dr H Hermanova and the World Health Organisation, Regional Office for Europe for inviting me to organise the plenary session which spawned this paper However, the ideas expressed herein do not reflect any agency views Many thanks to Sue Jukes for typing the manuscript and to colleagues who have discussed the quality of life issue with me I am funded by the Anglia and Oxford Regional Health Authority

REFERENCES

Bezrukov V, Sachuk N N , Pakin, Y V and Minaeva V P in press Socio Demographic Factors of Health Adaptation, Quality of Life and in Survivorship (A Longitudinal Study of Kiev Aging Residents), in *Swingin Pendulum Health Related Quality of Life in Elderly Europeans* L Ferrucci, E W Waters, E Heikkinen, and A Baroni, eds , INCRA/World Health Organization, Florence
Dall, J L C , 1994, The greying of Europe *Br Med J* , 309 1282-85
Department of Health (UK), 1991, *The Health of the Nation A consultation document for health in England* HMSO, London
Department of Health and Social Security (UK), 1980, Inequalities in helath report of a research working group, (Chair, Sir D Black), DHSS, London
Fallowfield, L , 1990   *The Quality of Life The Missing Measurement in Health Care* , Souvenir Press, London
Farley, R , and Allen, W R , 1987, *The Color Line and the Quality of Life in America* , Russell Sage Foundation, New York
Ferrucci, L Waters, E W , Heikkinen, E , in press, *Swingin Pendulum Health Related Quality of Life in Elderly Europeans* INCRA/World Helath Organization, Florence

Fries, J F , 1980, Aging, natural death and the compression of morbidity *New Eng J Med* , 303 130-135

Fries, J F , 1989, Compression of morbidity near or far? *Millbank Q* , 67 208-232

Gruenberg, E M , 1977, The failures of success *Millbank Mem Fund Q Health and Society*, 55 3-24

Heikkinen, E , Waters, W E , and Brzezinski, Z J , eds , 1983, *"The Elderly in Eleven Countries"* Public Health in Europe 21, World Health Organisation, Copenhagen

Hollingworth, W , Todd, C , and Parker, M , 1995, Hospital costs of hip fracture in the 21st century Submitted to *Journal of Epidemiology and Public Health*

Hollingworth, W, Todd ,C , Parker, M , Roberts, J , and Williams, D R R , 1993, Cost analysis of early discharge after hip fracture, *Br Med J* , 307 903-906

Jylha, M , Jokela, J , Tolvanen, E , Heikkinen E, Heikkinen, R -L , Koskinen, S , Keskinen, E , Lyyra, A -L , and Pohjolainen, P , 1992, The Tampere longitudinal study of ageing *Scand J Soc Med* , Supplement 47

Kramer. M , 1980, The rising pandemic of mental disorders and associated chronic diseases and disabilities *Acta Psychiatr Scand*, 62, 282-297

Law, M R , Wald, N J , Meade T W , 1991, Strategies for prevention of osteoporosis and hip fracture, *BMJ* 303 453-459

Martin, J , Meltzer. H , and Elliot, D , 1988, *"The Prevalence of Disability among Adults'* , OPCS Surveys of disability in Great Britain, No 1, HMSO, London

Medical Research Council (UK), 1994, *"The Health of the UK's Elderly People"*, MRC, London

Office of Population Censuses and Surveys, 1990, *"General Household Survey, 1988"*, HMSO, London

Office of Population Censuses and Surveys, 1993, *"National Population Projections 1991 - based"*, HMSO, London

Pedich, W, and Bien. B , in press Quality of Life of the Elderly During the Economic and Political Transformation in Poland, in *"Swingin Pendulum Helath Related Quality of Life in Elderly Europeans* L Ferrucci, E W Waters, E Heikkinen, and A Baroni, eds , INCRA/World Health Organization, Florence

Robine, J M , and Ritchie, K , 1991, Healthy life expectancy Evaluation of global indicator of change in population health, *Br Med J* , 302 457-460

Restrepo, H E , and Rozental, M , 1994, The social impact of aging populations some major issues, *Soc Sci Med* , 39 1323-38

Tinker A , 1992, *Elderly People in Modern Society '*, (Third Edition) Longman, London

Todd, C , and Bradley, C , 1994, The Design and Development of Psychological Scales, in *"Handbook of Psychology and Diabetes A Guide to Psychological Measurement in Diabetes Research and Practice* C Bradley, ed , Harwood Academic, Chur, Switzerland

Todd, C , Mackay-Meakin, S , Llewelyn, J , Pryor, G , Parker, M , and Williams, D R R , 1994, Effects on quality of life of early discharge after hip fracture, *Quality of Life Res* , 3 1, 96

United Nations, 1991, *World Population Prospects 1990"*, Population Studies No 120, Department of International Economic and Social Affairs, UN, New York

United Nations, 1992a, *World population monitoring 1991 with special emphasis on age structure"*, Population Studies No 126, Department of International Economic and Social Affairs, UN, New York

United Nations, 1992b, *Demographic Yearbook 1991"* Issue 43, Department of Economic and Social Development Statistics Division, UN, New York

United Nations, 1993, *Demographic Yearbook Special Issue, Population Ageing and the Situation of Elderly Persons* Department of Economic and Social Development, Statistics Division, UN, New York

Victor, C R , 1991, Continuity or change inequalities in health in later life *Ageing and Society*, 11 23-39

Waters. W E , Heikkinen, E , and Dontas, A S , eds , 1989, *"Health, Lifestyles and Services for the Elderly"*, Public Health in Europe 29, World Health Organisation, Copenhagen

13

HEALTH RELATED QUALITY OF LIFE AS AN OUTCOME MEASURE FOR HEALTH CARE OF ELDERLY PEOPLE

The Emperor's New Clothes

Chris Todd

Health Services Research Group
Department of Community Medicine
Institute of Public Health
University of Cambridge
Cambridge CB2 2SR
England

QUALITY OF LIFE AS OUTCOME

That quality of life should be used as an outcome measure of the health care of elderly people is at first sight an attractive and plausible notion. It is also an idea that has received increasing support over recent years. However, like so many ideas that are attractive at first sight, it is probably a notion upon which we should reflect rather than taking it up uncritically. This paper will look at the way that "quality of life" has been used, in order to open up debate as to its utility.

Outcome

First however, let us consider something which may be slightly less problematic - at least conceptually - "outcome". Donabedian (1966) is usually attributed with introducing the distinction between structure, process and outcome. *The Dictionary of Epidemiology* gives the definition of outcomes as: "All the possible results that may stem from exposure to a causal factor, or from preventive or therapeutic interventions; all identified changes in health status arising as a consequence of the handling of a health problem" (Last, 1988). Outcome, therefore, is often considered to be the result of the treatment or prevention of disease. To give an example, probably the most objective outcome of health care is mortality. Thus in this case we would ask whether people die or live after being treated? Process and outcome are sometimes confused by authors. Numbers of consultant episodes, numbers of beds occupied, numbers of drugs prescribed are clearly process measures of health care. Numbers of re-admissions or re-operations are sometimes used as outcome measures of health care since re-operation is taken to reflect a failure of the first attempt. However,

Preparation for Aging, Edited by E. Heikkinen *et al.*
Plenum Press, New York, 1995

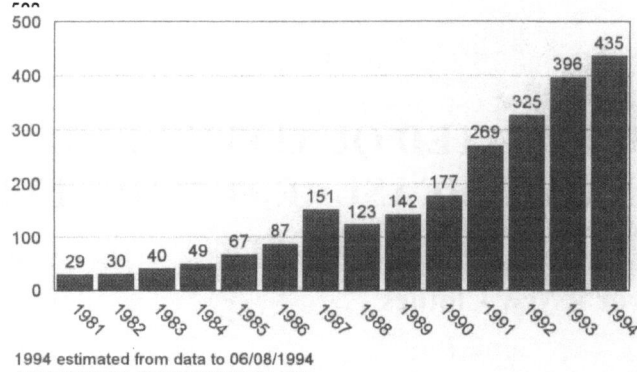

Figure 1. Number of Quality of Life pepers in scientific journals 1981-1994.

re-admission rates and re-operation rates depend on both policy and resources and, as such, are often aspects of the process of care rather than just an outcome of the care offered. (Chambers and Clarke, 1990; Clarke, 1990).

Outcome measurement has become increasingly important over recent years, and, as the focus has moved on to how to practice effective medicine, the questions asked by scientists, clinicians and health care managers have been more and more likely to be in terms of, "Is one treatment better in terms of its outcome than another treatment". Thus, pressure to practice effective medicine, whether this pressure is driven by economic considerations, political considerations, scientific considerations or simple professional motives, has resulted in the need to measure the outcome of care. Whilst the notion of outcome is relatively straightforward, the ways in which outcomes can be measured are often less straightforward. The example of mortality given above is perhaps deceptively simple. Death certification in a country such as the UK may approximate 100% but in many parts of the world this is not true (Alderson, 1988). Identifying death as an outcome may also have its own problems and require information on cause of death alongside event information to make it more meaningful. Such information may be less reliable or valid (Ashley and Devis, 1992) and could result in spurious conclusions (Todd, 1992a). Outcome measures which are less terminal are likely to suffer from these problems even more and care must always be taken to ensure that the measure chosen is in fact causally influenced by the disease condition and /or treatment being assessed.

The Rise of Quality of Life

Over recent years quality of life has been increasingly recognised as an important outcome measure for health care and the concept taken up by scientists and clinicians. This is well illustrated by the numbers of scientific papers which use the concept "quality of life". I conducted an electronic literature search using the on-line BIDS system. [*] I searched the science citation database both by "key words" and by "words in title" using the phrase "Quality of Life" for the years 1981 to 1994 and found a total of 2,030 papers. Figures 1 and 2 present these data graphically by year of publication. Between 1981 and 1984 there has been a fifteen-fold increase in the annual number of papers published in the scientific

---

[*] For more information e-mail BIDSHELP@BATH.UK.AC or ISIHELP@BATH.UK.AC

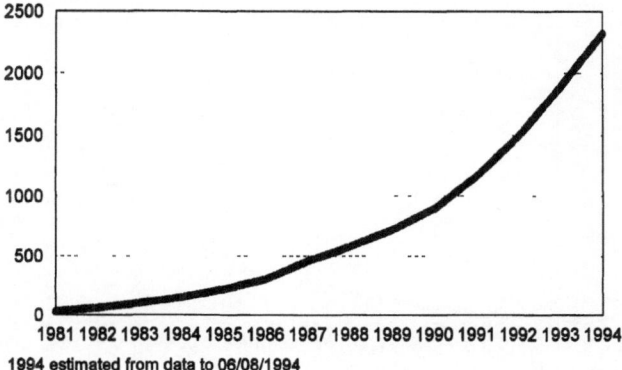

Figure 2. Cumulative number of Quality of Life papers in scientific journals 1981-1994.

literature using the concept "quality of life". I then repeated the search, adding the extra keyword 'elderly'. Over the same time period there has been a twenty-fold increase in the number of papers which are reporting on both "quality of life" and "elderly" people (Figures 3 and 4). On this basis, by the end of 1994 there should be some 85 papers in scientific journals explicitly reporting quality of life of elderly people. This is clearly only a crude estimate of the increase in importance of quality of life measurement. Such a finding depends greatly on the abstracting method used by the electronic database, the completeness and representativeness of the electronic database and, of course, the correct use of keywords. Nonetheless, however crude it may be, this goes some way to show how "quality of life" has become of increasing importance over the last decade. Given that "quality of life" has become an important concept in the literature (in terms of quantity of output at least) it is of considerable importance that the concept is coherently used. The rest of this paper will consider the use of the concept.

Why Measure Quality of Life

There are a number of reasons why quality of life has been taken up by science and medicine, and below I consider some of these.

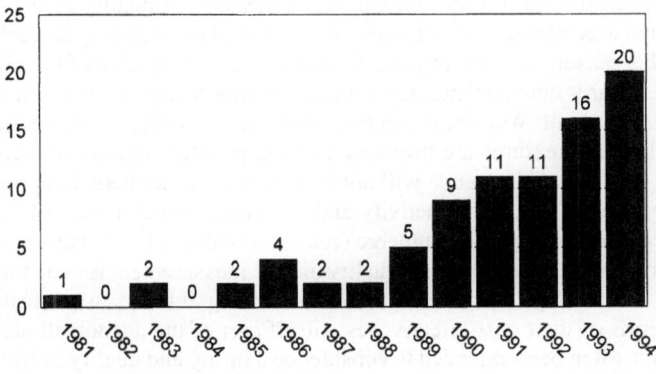

1994 estimated from data to 06/08/1994

Figure 3. Number of papers about Quality of Life and elderly people in scientific journals 1981-1994.

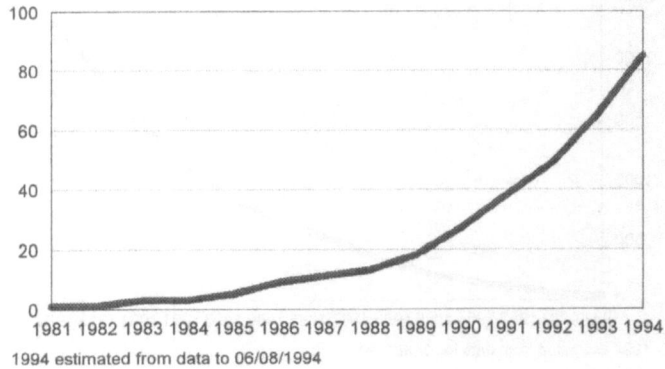

1994 estimated from data to 06/08/1994

Figure 4. Cumulative number of papers about Quality of Life and elderly people in scientific journals 1981-1994.

1. One of the reasons why quality of life has become more important may be because of recognition that in medicine the patient has slipped out of view. As medicine increasingly depends on the use of technology, the technology, be it genetic engineering of new drugs or magnetic resonance imaging has, rather than improving the view of the patient, obscured the view of the patient (see e.g. Foucault, 1973, who made this point more than 20 years ago). Thus it has been recognised that there is a need for a more holistic view. This view should not see the patient as a collection of organs nor be purely focused on clinical outcomes such as length of life, but should be one that also recognises the effects that treatment has on the patient *qua* person. Perhaps the classic example here is to be found in treatment of cancer. It is widely acknowledged that whilst chemotherapy might extend the period of life of a patient, the side effects of treatment are quite substantial and often appear to result in considerably poorer quality of life for the patient. Thus, the clinician and the patient, are left with a dilemma. Is it better to live for longer but suffer the effects of chemotherapy, or would it be preferable to accept a shorter length of life without the nausea, vomiting, alopecia etc. that are the common side effects of chemotherapeutic agents? It is exactly this issue that underlies the development of the first "quality of life measure". Karnofsky and Burchenal (1949) proposed a non-physiological measure of cancer in McClead's book *Evaluation of Chemotherapeutic Agents*. The Karnofsky index is often to be found even today in the literature. However, the Karnofsky index is less a measure of quality of life, and more a measure of global assessment of functioning and it should be used only as such.

2. Another reason over recent years for the increase in "quality of life" is the already identified trend towards outcome measurement. Promoting "health gain" is one view of what medicine should be about. Ways of measuring outcomes of medical intervention which go beyond physiological measures are probably to be applauded, but the underlying agenda may need to be clarified. This agenda will not be explored further here, but it must never be forgotten that medicine is a social activity and as such as much a part of contemporary political and economic change as commerce (see e.g. Freidson, 1970; Turner, 1987).

3. Another driving force behind "quality of life" measurement is to be found in health economics. Over recent years there has been increasing interest in evaluating specific treatments in terms of their cost-effectiveness (itself part of the agenda alluded to above). Such studies have often been extended to consider cost utility and quality of life has become an integral part of the measurement of the utility of a treatment by inclusion into the calculation of the so-called "quality adjusted life year" – (QALY) (Williams, 1985). The

underlying aim of the QALY is to be able to compare treatments not only for the same disease, but also to compare across diseases in a uniform way so as to identify what would be the "best buy" for health care. Thus, for example, Williams (1985) calculated QALYs for a number of diseases and their treatments. On this basis, a health care purchaser driven by purely financial objectives, would find that hip replacement at £750 per QALY, would be a better buy than hospital haemodialysis which costs £14,000 per QALY. However, the QALY has been debated on an number of occasions (see e.g. Drummond, 1988; Buxton, 1992; Jenkins, 1993) and the basic premise may be untenable. However that may be, QALY calculation requires a single index of quality of life; that is a figure which represents the individual's quality of life on a single but generic dimension. Last (1988) in *The Dictionary of Epidemiology* definition of quality of life focuses on this issue:

> Quality of life: In a general sense, that which makes life worth living In a more "quantitative" sense, an estimate of remaining life free of impairment, disability and handicap, as used in the estimation of "quality adjusted life years;" somewhere between these is an estimate of the utility of life - for instance, in clinical decision analysis, the utility of life that is impaired by a disabling degree of angina pectoris may be compared with that of a life that may be shorter in duration but free from disabling pain, as a result of applying therapeutic procedures Such trade offs are part of clinical decision analysis. (Last, 1988, p 108)

Given that the QALY notion has been so widely embraced, the need for a quality of life index for use in cost utility analysis has been promoted. For the rest of this paper I wish to look at the way quality of life as a concept has been used over recent years. Rather than reviewing every journal article, I will look at four collections which specifically set out to review "quality of life" and its measurement. In particular it is important to look at collections because such textbooks are often the first port of call of anybody entering the literature. A number of these books give overviews of various quality of life measures. Whilst the authors themselves point out that they are not "cook books" for quality of life measurement, they may well be used as such by the busy clinician looking for an accepted measure of quality of life. Thus, the use of quality of life by these authors is important as it is likely that the way they use it will persist in the literature. It should be made clear immediately that the publications referred to below are by no means an exhaustive review of quality of life texts. They were selected because of their widespread circulation and ease of access.

Four Collections and a Measure

Spilker's (1990), *"Quality of Life Assessment in Clinical Trials"*, sets itself the task of bringing together quality of life measurement issues for clinical trial design. Spilker points out at the beginning of the book that different authors are "often speaking about totally different topics that emerge from different perspectives .... Quality of Life must be viewed on a number of levels. Although the exact number and definition of levels may vary among authors ..."

(Spilker: 1990, p. 3). Spilker goes on to say that whilst this diffuse use may often be the case in the literature, in his book he will attempt to unify the concept quality of life, at least as a hierarchy. Spilker then asserts that "The major domains of quality of life ... include the following categories.

1. Physical status and functional abilities
2. Psychological status and well-being
3. Social interactions
4. Economic status and factors" (Spilker, 1990, p. 5).

Having thus defined quality of life in the introduction, other authors go on to discuss the notion of quality of life. In the second chapter Schipper, Clinch and Powell write,

"Quality of Life: Patients' perception of performance in 4 areas: physical and occupational function, psychological state, social interaction, somatic sensation" (Schipper et al., 1990, p 11). In the following chapter Fries and Spitz write, "Quality of Life: We do not mean happiness, satisfaction, living standard, climate or environment. Rather, we are speaking of health-oriented quality of life: those aspects that might be affected positively or negatively in clinical studies and the clinical situation ... Essentially it is the World Health Organization definition of health which encompasses our restricted definition of 'quality of life'." (Fries and Spitz: 1990, p 25). Wilson and Goetz (1990) in their chapter on neurologic illness, set themselves the task of discussing, "four dimensions of quality of life: ... physical, financial, social and psychological". (p 347). It is worth noting that in his chapter on rehabilitation, Turner (1990) barely mentions quality of life, but focuses on functional assessment tools and "health assessment scales that are appropriate for rehabilitation", (p 363). Such an approach has much to commend it. Thus, even within one volume there is little consensus on "quality of life" and some author steer clear of it all together. The book, like so many others, is devoted to reviewing specific areas of medical research in which psychological measures, or interviews have been conducted with different patient groups. Whilst offering some insight into the quality of life concept, the volume probably fails to provide a unitary overview, but fragments the concept dependent on the disease being considered.

Also published in 1990 was Fallowfield's book, *Quality of Life: The Missing Measurement in Healthcare*. Fallowfield writes of quality of life, "Conceptually quality of life is a somewhat vague term ... quality of life is not a unitary concept, but rather a complex amalgam of satisfactory functioning in essentially four core or primary domains ...

1. Psychological
    depression
    anxiety
    adjustment to illness
2. Social
    personal and sexual relationships
    engagement in social and leisure activities
3. Occupational
    ability and desire to carry out paid employment
    ability to cope with household duties
4. Physical
    pain
    mobility
    sleep
    appetite and nausea
    sexual functioning" (Fallowfield, 1990, pp. 17-20)

Fallowfield then goes on to review a number of "quality of life" instruments. These include such diverse measures as Katz's ADL scale (Katz et al., 1963), the Nottingham Health Profile Health Status Measure (Hunt et al., 1986) and the Profile of Mood States (McNair et al., 1981). Whilst these instruments are measures of aspects of what may go to make up quality of life, it is not clear that they are "quality of life measures". At the end of the book one is left with the feeling that a clear idea of the nature of "quality of life" is still missing.

Bowling (1991) in her book *Measuring Health: A Review of Quality of Life Measurement Scales* writes, "There is no consensus over a definition of quality of life ... It is becoming fashionable to equate all non-clinical data with 'quality of life' which is likely to be a source of conceptual confusion. Health and functional status are just two dimensions of health-related quality of life". (Bowling: 1991, pp. 8-9). Bowling here identifies a central issue, that "quality of life" is being used to refer to any measure used in medical research which is not clinical. Bowling, like Fallowfield, gives an overview of a variety of measures of health and functional status under the general umbrella of "quality of life". Again, whether these are really quality of life measures is an unresolved issue, but Bowling quite clearly acknowledges the way in which the term is being misused.

Hopkins organised a conference at the Royal College of Physicians, published as *Measures of Quality of Life and the Uses to Which Such Measures May Be Put* (Hopkins, 1992). In his introduction Hopkins writes, "Quality of Life embraces many dimensions, ranging from physical well-being and cognitive competence to the establishment of satisfactory interrelationships, the occupation of housing which is enjoyed, and the possession of a sufficient income to explore the world in ways beyond those necessary to ensure basic biological survival". (Hopkins, 1992, p. 1). An important thing to note from this quotation is the degree to which this definition of quality of life differs from the others we have reviewed. The body of the book consists of contributions by a number of authors on aspects of quality of life and much of the discussion revolves around the QALY and its use. Again the definition of quality of life differs widely between authors, from the generalist who focuses on QALYs as "a common language for expressing health status" (Walker, 1992) to the psychometric approach used to validate the Index of health-related quality of life questionnaires against other instruments (Rosser et al., 1992). Grimley Evans has a chapter in this collection on *Quality of Life: Assessment of Elderly People*, in which he points to the problems of ageism implicit in quality of life assessment as well as the dangers of the fashionable use of the phrase "quality of life" (Grimley Evans, 1992).

What then, have we learnt from looking at these various definitions of quality of life? The first thing is to point out that there is little consensus in terms of the domains of importance for quality of life, other than at the most general level. In overview it seems that quality of life is something to do with the physical, the social and the psychological. This could perhaps be said of all human endeavour and thus does not clarify much. Some authors have written of quality of life as a patient's perception of their performance, others include economic status. There is a general feeling that the concept includes such notions as psychological functions, social function, occupational utility and physical function, and I myself have written that it is generally accepted that quality of life measures relate to three inter-related domains of life: physical, psychological and social (Todd, 1992b). But what does this mean beyond the self evident? Is it acceptable to use a concept which is so ill-defined? If a physiological measure was so poorly defined we would certainly reject it.

Quality of Life and Ageism

Quality of life may appear to be the sort of concept that would be useful for measuring the outcome of health care in elderly people, but, as Grimley Evans (1992) points out, that quality of life is suitable as a measure for older people is based on the assumption that they will be more interested in quality than in quantity. Such an assumption may be partly an inappropriate projection of young people's ideas about being old. Such an "ageist" conception of quality of life is exemplified by Smart and Yates (1987), who write, "Quality of life, in general decreases, with age, with disease and with decreasing socio-economic status", (p. 620).

Quality of Life, Science and Ritual

This brief overview of literature may suggest that the quality of life concept is no longer a useful rubric. It has come to be a fashionable term used to cover all manner of non-clinical measures in health research. It is a poorly defined concept and causes conceptual confusion in the reader. It often disguises conceptual confusion in the author. Quality of life is a notion which has been used to disguise lack of clarity in thinking by many researchers. Very often the quality of life notion also hides vested interests, assumptions which are not made explicit and world views which take on the mantle of science, thereby imbuing themselves with claims of being value-free. This is perhaps most clearly seen in the QALY notion, as is pointed out by Buxton (1992). It is however clearly not understood by many health care purchasers as is exemplified by Walker (1992) in the same volume. Very often quality of life assessment gives the appearance of being scientific, but is scientific ritual only. Claims of scientific respectability must be based upon more than just dogmatic and ritualistic behaviour. Finally, there is a danger that there will be generalised disenchantment with non-clinical measurement if the quality of life research endeavour fails resoundingly to live up to its promise. In its present form it seems to the author to be well on the way to failure and it will perhaps not be long before grant givers, purchasers of health care, providers of health care and clinicians become disenchanted with the quality of life enterprise. There is a real danger that once quality of life does not provide the good that it promises, a way of representing on a single scale, a single index score which would permit us to identify the qualitative effects of our health care systems, then all such attempts will be rejected, including sound psychometric approaches and less generic and less fashionable assessments of function.

QUALITY OF LIFE; THE EMPEROR'S NEW CLOTHES?

What then, can we do instead? First, I would recommend that the use of the phrase "quality of life" be discontinued except for attempts to develop single index scales such as EuroQol (EuroQol Group, 1990). We should therefore refer instead to psycho-social outcomes of health care when considering multidimensional measurement. This is particularly important as it does not imply a single, global measure permitting simplistic comparison of the qualitative component of different diseases, or different age groups, or different people or different cultural experiences of health and illness. Second, concepts should be clearly operationalised. Definitions are useful and operationalisation of concepts is imperative. Third, in comparing health state or outcomes the measures must be chosen so that they are both valid and reliable for the particular sample, that they fit the population of interest and that they measure components of the health state of interest, (see eg Todd and Bradley, 1994). Thus, for our population of interest, elderly people, measures which include such domains as current occupation and career may be of less use, than a measure which can take into account "leisure" activities such as gardening. Obsession with measuring occupation may reflect the protestant work ethic, but may be of less interest to other social groups or cultures including elderly people (Fenwell et al., 1988). We need to continue to access subjective accounts but this will need to be done using qualitative approaches, rather than trying to just come up with a score between 1 and 10. In conclusion then, the present use of the phrase Quality of Life is reminiscent of the Emperor's new clothes in the Grimm Brothers' Fairy Tale.

ACKNOWLEDGMENTS

I would like to thank Dr H Hermanova and the World Health Organisation, Regional Office for Europe, and Professor E Heikkinen and colleagues on the ELSA project for inviting me to take part in the symposium which spawned this paper. However, the ideas expressed herein are my own and do not reflect any agency views. Many thanks to Sue Jukes for typing the manuscript and to colleagues who have discussed the quality of life issue with me. I am funded by the Anglia and Oxford Regional Health Authority.

REFERENCES

Alderson, M R , 1988, *"Mortality, Morbidity and Health Statistics"*, Macmillan, Basingstoke

Ashley, J , and Devis, T , 1992, Death certification from the point of view of the epidemiologist, *Pop Trends*, 67 22-28

Bowling, A , 1991, *"Measuring Health A Review of Quality of Life Measurement Scales"*, Open University Press, Milton Keynes

Buxton, M , 1992, Are We Satisfied with QALYS? What Are the Conceptual and Empirical Uncertainties and What Must We Do to Make Them More Generally Useful? in *"Measures of Quality of Life and the Uses to Which Such Measures May Be Put"*, A Hopkins, ed Royal College of Physicians, London

Chambers, M , and Clarke, A , 1990, Measuring admission rates, *Br Med J* 301 1134-1136

Clarke, A , 1990, Are readmissions avoidable? *Br Med J* , 301 1136-1138

Donabedian, A , 1966, Evaluating the quality of medical care, *Millbank Mem Fund Q Health and Society,* 44 (July) 166 - 206

Drummond, M F , 1988, Resource Allocation Decisions in Health Care A Role for Quality of Life Assessments? in *"Professional Judgement A Reader in Clinical Decision Making"*, J Dowie, and A Elstein, eds , Cambridge University Press, Cambridge

EuroQol Group, 1990, Euroqol – A new facility for the measurement of health related quality of life, *Health Policy,* 16 199-208

Fallowfield, L , 1990, *"The Quality of Life The Missing Measurement in Health Care"*, Souvenir Press, London

Fenell, G , Phillipson, C , and Evers, H , 1988, *"The Sociology of Old Age"*, Open University Press, Milton Keynes

Foucault, M , 1973, *"The Birth of the Clinic"*, Tavistock, London

Freidson, E , 1970, *"Profession of Medicine"*, Dodd, Mead & Co, New York

Fries, J F , and Spitz, P W , 1990, The Hierarchy of Patient Outcomes, in *"Quality of Life Assessments in Clinical Trials"*, B Spilker, ed Raven Press, New York

Grimley Evans, J , 1992, Quality of Life Assessments and Elderly People, in *"Measures of Quality of Life and the Uses to Which such Measures May Be Put"*, A Hopkins, ed, Royal College of Physicians, London

Hopkins, A , ed, 1992, *"Measures of the Quality of Life and the Uses to Which such Measures May Ee Put'*, Royal College of Physicians, London

Hunt, S M , McEwen, J , and McKenna, S P , 1986, *"Measuring Health Status"*, Croom Helm, London

Jenkins, J , 1993, The ethical QALY, *Quality of Life Newsletter,* no 7-8 (June 1-2

Karnofsky, D A , and Burchenal, J H , 1949, The Clinical Education of Chemotherapeutic Agents in Cancer, in *"Evaluation of Chemotherapeutic Agents'* , C M MacLeod, ed , Columbian University Press, New York

Katz, S , Ford, A B , Moskowitz, R W , Jackson, B A , and Jaffe, M W , 1963, Studies of illness in the aged the index of ADL – a standardized measure of biological and psychosocial function, *JAMA,* 185 914-9

Last, J M , 1988, *"A Dictionary of Epidemiology IEA and WHO Handbooks"*, Oxford University Press, Oxford

McNair, D M , Lorr, M , and Doppleman, L F , 1981, *"Manual for the Profile of Mood States'*, Educational and Industrial Testing Service, San Diego

Rosser, R , Cottee, M , Rabin, R , and Selai, C , 1992, Index of Health-Related Quality of Life, in *"Measures of Quality of Life and the Uses to Which such Measures May Be Put"*, A Hopkins, ed , Royal College of Physicians, London

Schipper, H , Clinch, J , Powell, V , 1990, Definitions and Conceptual Issues, in *"Quality of Life Assessments in Clinical Trials'*, B Spilker, ed , Raven Press, New York

Smart, C R , Yates, J W , 1987, Quality of life, *Cancer,* 60 620-622

Spilker, B , ed, 1990, *"Quality of Life Assessments in Clinical Trials"*, Raven Press, New York

Todd, C J , 1992a, Reduction in the incidence of suicide a health gain objective for the NHS, *J Psychopharmaco ,* 6(2) 30-36

Todd, C J , 1992b, Quality of life and diabetes audit, *Health Psychology Update,* 11 9-14

Todd, C J , and Bradley, C , 1994, The Design and Development of Psychological Scales, in *"Handbook of Psychology and Diabetes A Guide to Psychological Measurement in Diabetes Research and Practice"*, C Bradley, ed , Harwood Academic, Chur, Switzerland

Turner, B S , 1987, *"Medical Power and Social Knowledge"*, Sage, London

Turner, R R , 1990, "Rehabilitation, in *"Quality of Life Assessments in Clinical Trials"*, B Spilker, ed , Raven Press, New York

Walker, P , 1992, Are QALYs Going to Be Useful to Me as a Purchaser of Health Services? in *"Measures of Quality of Life and the Uses to Which such Measures May Be Put"*, A Hopkins, ed, Royal College of Physicians, London

Williams, A , 1985, Economics of coronary artery bypass grafting, *Br Med J*, 291 326-329

Wilson, R S , and Goetz, C G , 1990, Neurologic Illness, in *"Quality of Life Assessments in Clinical Trials"*, B Spilker, ed , Raven Press, New York

14

HEALTH-RELATED QUALITY OF LIFE IN OLD AGE: HOW TO DEFINE IT, HOW TO STUDY?

Marja Jylhä

The Finnish Centre for Interdisciplinary Gerontology
University of Jyväskylä
P.O. Box 35
FIN-40351 Jyväskylä, Finland

INTRODUCTION

Quality of life, in the words of Alvan R. Feinstein is something like intelligence. "Everyone knows it exists and thinks they can identify it in various ways, but we may not be able to evoke universal agreement on what it is" (Feinstein, 1987, p. 639). This is also true of quality of life when we consider it from a health point of view. The concept itself seems to be perfectly useful, but at the same time there is something very confusing and even irritating about it. What are we talking about when we talk about health-related quality of life? Do we really need this concept if we think of all the countless number of terms and concepts that refer to different dimensions of health-related phenomena? And if we do need it, for what purposes?

The discussion that follows does not contain a full review that covers the whole field; instead I would like to refer here to Avis and Smith (1994), Faden and German (1994). Schipper et al. (1990) and Walker and Rossner (1992). In this paper, which is a personal comment rather than a systematic piece of research, my main purpose is to briefly discuss the background of the quality- of-life theme in gerontological research and to look at a few theoretical and methodological problems concerning the use of the concept of health-related quality of life as well as at some of the difficulties involved in empirical research.

THE ROOTS

To get some idea of the extent of quality of life research, we ran through the MEDLINE index. In 1973-74 there were 124 titles with the keyword "quality of life"; for 1992-93 the number had climbed to 1848 titles. Restricting the search to those titles that also featured "health policy" as a keyword, the numbers were 4 and 1067, respectively. Clearly the quality of life has made a very forceful and impressive entry into the field of health

Preparation for Aging, Edited by E. Heikkinen *et al.*
Plenum Press, New York, 1995

research. However, the arrival of a new concept does not always mean new themes of research. Investigators who used to talk about health status or subjective health have now started to talk about health-related quality of life—but many of them are using exactly same measurements as before. As Spitzer (1987) has observed, it often seems that the terms health, health-related quality of life and functional status are used interchangeably. It also seems that health-related quality of life can be taken to refer either to health status as viewed from the life quality perspective; or to the consequences of different health conditions to quality of life.

The roots of such a perspective as that on health-related quality of life are certainly easy to trace. This particular approach is grounded in the conviction that traditional medical data on survival, diseases and laboratory values alone can not provide a sufficient picture of the individual's health status or the results of medical care; we are interested not only in the course of disease, but we also need to know how and what happens to the patient, how the condition affects his or her everyday life, how the patient perceives and evaluates his or her health status. In this approach the patient can appear not only as a biological organism but also as an active subject (see Weckroth, 1988). With the advances that are being made in medical knowledge and medical technology, it is important to recognize that the "interests" of the organism are not always the same as the "interests" of the active subject: this is evident in cases where the length of life is weighted against the side-effects of treatment and in numerous everyday decisions in long-term hospitals and nursing homes.

QUALITY OF LIFE IN GERONTOLOGICAL RESEARCH

There are many reasons why the health of elderly people in particular should be looked at from the point of view of quality of life. We know that there is a general trend in most countries towards decreasing mortality in old age, but it is questionable whether this leads to improving health and functional ability for those who are alive (e.g. Verbrugge, 1989; Guralnik, 1991; Robine et al., 1993). The vast majority of persons 65 and over have at least one and usually several medical diagnoses. For the most part these conditions are chronic degenerative diseases which cannot be cured; people just have to learn to live with them. In this situation it is essential to try to contain the adverse consequences of the disease, to support the individual's independence and autonomy in everyday life. In gerontological research, this has highlighted the importance of approaches concerned with functional ability, multi-dimensional assessment and self-rated health.

Traditionally, gerontologists have regarded functioning as a crucial dimension of health status and reasonable functioning as an essential precondition for a good, satisfying life . Lay people seem to share this view: in the recent study of Parkerson et al. (1992), for instance, the most important predictor of poor quality of life was disability in terms of confinement to the home.

As early as 1959 a WHO Advisory Group (WHO 1959) stated that: "health in the elderly is best measured in terms on functioning;...degree of fitness rather than extent of pathology may be used as a measure of the amount of services the aged will require from the community." An important landmark was the publication in 1963 of Stanley Katz's Activities of Daily Living Scale (Katz et al., 1963). Since then several dozen, possibly several hundreds of scales have been published that measure self-rated functional ability in everyday life.

Self-rated health is another health indicator that is traditionally used in gerontological research and that represents a quality of life sensitive approach to health before the big boom. The thing that makes self-rated health particularly interesting is its mediating role between psychology and biology: global self-ratings of health have been found to be a strong predictor of mortality, and researchers are now debating whether self-ratings really provide better summaries of the organism's biological status than ordinary clinical measures, or whether general attitudes or states of mind themselves have an impact on the length of life. (See Idler and Kasl, 1991; Jylhä, 1994b).

THE CONCEPT AND THE MEASURES

In recent years countless papers have been published in health research and gerontology concerning the concept of (health-related) quality of life and its measures. A reading of a few of those papers is sufficient to convince anyone of the importance and the extreme complexity of the theme. Writers seem to agree on the necessity of having and applying the life quality perspective in health research and in evaluating the results of care, as well as on the importance of being able to measure health-related quality of life in a way that is valid, reliable and sensitive to change. There is, by contrast, no agreement on what exactly goes into quality of life itself. What should we be measuring in measurements of health-related quality of life? Some writers stress the importance of functioning, others emphasize sensations, feelings and satisfaction, others still emphasize such dimensions as social contacts or economic security. In spite of these differences in emphasis, most researchers are agreed on the need to measure quality of life as a multidimensional phenomenon. But does this mean we should construct a sum score to indicate a "global" quality of life, or should we have separate indicators for separate dimensions? Should we prefer a "generic" measure of quality of life, or should we construct disease-specific measures?

Indicators of quality of life are needed and used in practical health care and health policy planning, where they should complement or in some cases compete with traditional clinical and economic indicators. This, together with the conventions of empirical mainstream research, help to explain why most researchers take it for granted that whatever health-related quality actually consists in, it must first of all be something that lends itself to quantitative measurement. Indeed in many cases it seems that the requirement of measurability determines the definition and understanding of the phenomenon itself.

In other cases the conceptual understanding seems to conflict with the practical opportunities for measurement. There are several authors who emphasize that quality of life should be understood as a multidimensional, variable and subjective phenomenon:

> Quality of life is multidimensional, each dimension changes over time, and it is a patient-perceived entity (Schipper et al , 1990, p 15)

> Subset components of overall quality of life vary according to time and circumstance The ability to ride a bicycle may be very important to a younger person, but with increasing age or a change in habitus or social circumstance, the ability to perform that skill may take on a very different meaning (Schipper et al , 1990, p 12)

> But the specific domains and their relative importance may differ from one person to another Changes in sexual functioning or physical activity may affect people's overall quality of life differently The question of how to capture these differences is a most difficult one (Avis and Smith, 1994)

> Most contemporary discussions of quality of life in medical ethics have emphasized the subjective or idiosyncratic dimensions of quality of life. An individual's conception of the good life is constructed in the context of deeply personal values, talents, histories, and life experiences Thus, although individuals usually agree about the basic direction of the impact of general states - such as cognitive capacity or being in pain - on quality of life, individuals frequently disagree about the comparative value or importance they attach to these general states, particularly in contexts of conflict of tradeoff This is because these states can have differential impact on individuals' perceptions of the quality of their lives, depending on their life plans and goals As a consequence,

it would not be surprising to find different people making different choices, for example, between
being in pain and preserving cognitive capacity (Faden and German, 1994, p. 542)

The discussions on the subjective, variable and contextual nature of quality of life
do not, however, have a major impact on the practical conclusions and recommendations
that are put forward. Avis and Smith (1994) have constructed a conceptual model of
health-related quality of life that resembles a structural equation model and that includes as
unobserved latent variables both the facets that constitute quality of life and those that affect
it. They argue: "If the quality of life domains were exhaustive, properly conceptualized, and
properly measured, then only these domains should have a direct effect on quality of life."
(Avis and Smith, 1994, 6). Schipper et al. (1990), when writing of the subjectivity of quality
of life conclude: "Appropriate, rigorously designed and evaluated quality of life instruments
can be used in carefully designed studies to provide objective representations of what we
have until recently viewed as essentially intangible subjective processes."

It is hard to avoid a sense of deep confusion in the face of this discussion. It seems
that the elements in the concept of quality of life that are essential, unique and of heuristic
value cannot be approached by tools that are acceptable according to biometrical or
psychometrical criteria. Why is that? Are there any more appropriate methods available?

As far as I can see the concept of QUALITY of life necessarily implies a subject,
someone who is there to make the evaluation. In this respect it seems fair to say that only
judge of quality of life is the individual his- or herself. But who is to define what they should
judge? Here, it might be useful not to have preset definitions but rather to ask, what people
are talking about when they are talking about the importance of health to their everyday life,
about the good life and quality of life. What is the meaning of these concepts and these
things? How are these concepts constructed and used by people in their talk, in the 'voice
of the lifeworld' (see Mishler, 1984)? These are questions that are typically asked in
methodological approaches such as social constructionism (e.g. Berger and Luckman, 1966),
ethnomethodology (e.g. Garfinkel, 1967), conversation analysis and discourse analysis (e.g.
Mishler, 1984, Potter and Wetherell, 1987; Silverman, 1987).

Analysis of people's discourses on health in different situations seems to indicate
that it is often very difficult for them to talk about health as an abstract and universal concept.
In our own research interviews (Jylhä, 1994a) we saw that the 'health' that the interviewees
were talking about took shape in concrete situations of everyday life and could hardly be
evaluated in isolation of these situations. Billig (1988) has pointed out that the same state
of health is described in different ways depending upon the context of the description.

In my opinion there is good reason to look at what we call 'quality of life' from this
point of view. In principle there are innumerable elements out of which a (good) quality of
life can be constructed, but in specific situations only certain things have real importance;
and the thing is that there is no mechanical standardized way of finding out what those
elements are. To someone who has frequent migraine, this condition very probably under-
mines his or her quality of life; whereas someone else who has never experienced migraine
and who has never come to think about it would be very much surprised if the absence of
migraine were to be defined as one of the components of his or her good quality of life.
Further, as Schipper et al. (1990) and Avis and Smith (1994) (see citations above) have
emphasized, different elements, such as physical activity or sexual activity, may have
different significance to different people and at different stages of life. There are no rules
according to which all the possible constituents of quality of life could be listed, and no
general rules according to which the importance of this or that element in a certain person's
life could be estimated; there are always several 'logics' on the basis of which a person can
take into account his or her diseases and ailments when considering quality of life (Eskola,
1993; Jylhä, 1994 b).

According to this line of thinking, which to me gives a more realistic picture of an individual's activity, quality of life takes shape not an abstract, universal phenomenon, but as a particular, relative and contextual phenomenon. Context here does not refer to the external environment but chiefly to the logic of activity in the framework of which the individual evaluates his or her life.

But is not this a very impractical standpoint? Certainly in health care we need standardized, comparable quantitative indicators. Is there any point in talking about health-related quality of life if we cannot measure it?

MEASURES OF FUNCTION, NARRATIVES OF LIFE

The dimension of quality of life provides an extremely important and human perspective on evaluating the situation of elderly people and patients at any age, on making decisions about alternative treatments and on assessing the impacts of medical care. Where quantitative measures are needed, it might indeed be useful to understand quality of life expressly as a perspective, not as a variable. In general terms we know that such aspects as functioning, pains and other sensations, mood, and possibilities to carry on social relationships can provide an important information on the influences of an illness or treatment on an individual's life. These aspects should be examined in a rigorous, valid and reliable way and they should be included in the regular checklist of every general practitioner and geriatrician. But is there any need to combine them mechanically into arbitrary indexes that can hardly be understood and to say that these measure "quality of life"?

In clinical work, as Faden and German remind us, quality of life measures can be viewed as increasingly popular technical, but not always superior alternatives for the physician's question: "How are you?" If the quality of life perspective is taken seriously in face-to-face situations with our elderly patients, no scale or index can replace the simple question: "What do you want?", or "What would you prefer?" (see Feinstein, 1987, 639).

However, different methods will be needed in the quest for a deeper understanding of the meanings of what is called quality of life itself. The big challenge to health research and gerontology now is to develop qualitative research approaches; how to learn to do research using narratives, discussions and personal accounts. This research should not be less rigorous than that which uses quantitative, statistical methods, but it could be more valid in analysing the social, individual and contextual nature of the 'good life' and the logics that people use in evaluating their health and in choosing their preferences in care. This is what the perspective of health-related quality of life is all about.

REFERENCES

Avis, N.E., and Smith, K W, 1994, Conceptual and methodological issues in selecting and developing quality of life measures, in: *"Advances in Medical Sociology, Volume V Quality of Life in Health Care"*, G Albrecht and R Fitzpatrick, eds, in press

Berger, P, and Luckmann, T, 1966, *"The Social Construction of Reality"*, Doubleday, New York

Billig, M, Condor, S, Edwards, D., et al, 1988, *"Ideological Dilemmas"*, Sage, London

Eskola, A, 1993, On the methodological paradigms of psychological peace research, in *"Conflict and Social Psychology"*, K S Larson, ed, Prio, London, Sage, Oslo, pp 176-187

Faden, R, and German, P.S, 1994, Quality of life, considerations in geriatrics, in. *"Clinics in Geriatric Medicine, Clinical Ethics"*, G A. Sachs, and C.K. Cassel, eds, W B Saunders Company, Philadelphia, pp. 541-551

Feinstein, A R, 1987, Clinimetric perspectives, *J Chron Dis* 40 635-640

Garfinkel, H., 1967, *"Studies in Ethnomethodology"*, Prentice Cliff, Englewood Cliffs NJ

Guralnik, J , 1991, Prospects for the compression of morbidity The challenge posed by increasing disability in the years prior to death, *J Aging Hlth* 3 138-154

Idler, E L , and Kasl, S , 1991, Health perceptions and survival do global evaluations of health status really predict mortality? *J Gerontol S,* 46 55-65

Jylha, M , 1994(a), Self-rated health revisited Exploring survey interview episodes with elderly respondents, *Soc Sci Med* 39 983-990

Jylha, M , 1994(b), Perceived health in old age—two methodological approaches, in *"Experiencing Ageing",* P Oberg, P Pohjolainen and R Ruoppila eds, SSKH Skrifter 4, Svenska social- och kommunalhogskolan vid Helsingfors universitet forskningsinstitut, Helsinki, pp 171-184

Katz, S , Ford, A , et al , 1963, Studies of illness in the aged The index of ADL a standardized measure of biological and psychosocial function, *JAMA* 185 914-919

Mishler, E , 1984, "The Discourse of Medicine Dialectics of Medical Interviews", Ablex, Noorwood NJ

Parkerson, G , Broadhead, W , et al , 1992, Quality of life and functional health of primary care patient, *J Clin Epidemiol* 45 1303-1313

Potter, J , and Wetherell, M, 1987, *"Discourse and Social Psychology",* Sage Publications, London

Robine, J -M , Mathers, C , and Brouard, N , in press, Trends and differentials in disability-free life expectancy, in *"Proceedings of the Conference on Health and Mortality Trends among Elderly Populations Determinants and Implications",* Sendai City, Japan, June 21 to 25, pp 1–43

Schipper, H , Clinch, J , and Powell, V, 1990, Definitions and conceptual issues, in *"Quality of Life Assessments in Clinical Trials",* B Spilker, ed , Ravin Press, Ltd, New York

Silverman, D , 1987, *"Communication and Medical Practice Social Relation in the Clinic",* Sage Publications, London

Spitzer, W , 1987, State of Science 1986 Quality of life and functional status as target variables for research, *J Chron Dis* 40 465-471

Walker, S R , and Rossner, R M , eds, 1993, *"Quality of Life Assessment Key Issues in the 1990s",* Kluwer Academic Publishers, Dortdrecht

Verbrugge, L , 1989, Recent, present, and future health of American adults, *Ann Rev Public Health* 10 333-361

Weckroth, K , 1988, The psychological foundations of social psychology, in ' *Blind Alleys in Social Psychology A Search for Way out",* A Eskola, A Kihlstrom, D Kivinen, K Weckroth and O -H Ylijoki, eds, Elsevier Science Publishers, Amsterdam, pp 117-157

World Health Organization, Regional Office for Europe, 1959, *"The Public Health Aspects of the Aging of the Population Report of an Advisory Group, Oslo 28 July—2 August 1958",* Copenhagen

15

HEALTH RELATED QUALITY OF LIFE IN OLD AGE

International Approach in Developing the LEIPAD Questionnaire

Anna-Mari Aalto[1,2], Arja R. Aro[1], Anu Hämäläinen[1] and Jouko Lönnqvist[1]
for the steering group of the LEIPAD -project

[1] National Public Health Institute
Department of Mental Health
Mannerheimintie 166
FIN-00300 Helsinki, Finland
[2] National Research and Development Centre for Welfare and Health
Health Care Services Research Unit
Helsinki, Finland

INTRODUCTION

The LEIPAD-project is an international validation study of a Quality of life instrument for elderly population. The project is under patronage of WHO, and it is carried out by National Public Health Institute in Finland, University of Leiden in the Netherlands and University of Padua in Italy. The project originated from a need to consider the specific conditions of elderly in studying their health and well-being. Aging process includes physical as well as psychosocial changes that affect quality of life (QOL). Elderly people often have various clinical chronic conditions simultaneously and they may suffer from impairments in cognitive capacity. Also the psychosocial situation of elderly includes characteristics important for QOL. Elderly persons have often lower educational level and decreased income compared to general population. Aging involves also changes in social-networks as a result of retirement and loss of age mates. Therefore, most instruments designed for assessing QOL among general population need modifications before addressing them to elderly.

The aim of the LEIPAD-project was to develop an instrument to measure quality of life specifically among elderly population. The frame of reference was the multidimensional concept of quality of life covering various life domains, such as physical functioning, psychological well-being and social interaction. The instrument was designed to be based on self-report, easy to use in primary health care settings and apprehensible also for people with little education. Furthermore, the measure should be sensitive for detecting changes in health and well-being due different kind of interventions. This paper shortly introduces the LEIPAD-instrument and presents some validation results from the Finnish pilot study.

Preparation for Aging, Edited by E Heikkinen *et al*
Plenum Press, New York, 1995

Table 1. The LEIPAD Quality of Life Scales and item examples

Subscales	Item example
Physical functioning	How much does your physical health problems (if any) stand in the way of doing the things you want to do?
Self-care	Can you shop all by your self?
Depression & anxiety	Taking everything into consideration, how anxious do you feel?
Cognitive functioning	How much do your memory problems (if any) stand in the way of doing things you want?
Social functioning	Do you feel emotionally satisfied in your relationships with other people?
Sexual functioning	Is there someone with whom you engage in sexual contact?
Life satisfaction	How satisfied are you in general with your life at present when compared to the past?

THE METHODS IN QUESTIONNAIRE DEVELOPMENT

The items of the LEIPAD-scale were developed on the basis of literature and "know how" of an expert group. The items covered a range of domains, such as perceived physical health, emotional life, interpersonal relationships, financial situation. The subscales with item examples are shown in Table 1. Items, originally generated in English, were translated into Finnish, Dutch and Italian. The accuracy of translations was assessed with back-translations into English by native English speaker. The response scale was a 4-point scale, lower score indicating better well-being.

Two pilot studies were conducted in participating countries. The first pilot study (n=229) was performed to test the wording of items and preliminary internal structure of the scale. In the second pilot study (n=586) the internal structure of the instrument was studied. Participants in the pilot studies were patients of general practitioners and attenders in a memory training program (the Netherlands).

Structural validity of the measure was examined by second order factor analysis, using the subscales as variables. Convergent validity was studied by examining the associations of the LEIPAD-scales with the Rotterdam Symptom Checklist. The Rotterdam Questionnaire comprises of three subscales: Physical distress, Psychological distress and Activities of Daily Living. Further, the relationships of LEIPAD-scales to psychological moderator scales was examined. This was done to test that LEIPAD-scales measure dimensions of life susceptible for change, instead of reflecting merely underlying dispositional personal styles or characteristics. The moderator scales were developed for the present study and they measured Perceived personality disturbances, Anger, Self-esteem and Trust in God.

THE RESULTS CONCERNING THE FINNISH DATA

The Finnish sample was collected in Vantaa, a town adjacent to Helsinki, in five different health centres. The questionnaire was handed out for 199 elderly patients (65 years or older), 39 of these refused to participate due to reasons such as lack of time, hearing or visual problems, and "too thick a questionnaire". The majority of the respondents filled in the questionnaire while waiting for their appointment. The time of filling varied from 15 to 30 minutes. In the Finnish sample (n=162), 56% of participants were men, the mean age was 72 years, the mean years of education was 9 for both genders. Among men 74% of the participants were married, 3% were not married, 18% widowed and 6% divorced. Among women 31% were married, 6% not married, 52% widowed, 7% divorced.

Table 2. Leipad Quality of Life subscales Means, standard deviations, number of items and Cronbach's α_1 in the Finnish sample (n=162) and Cronbach's α_2 for the aggregate sample

LEIPAD-scale	Mean[1]	Std[1]	Range[1]	Number of items	α_1	α_2
Physical Functioning	5 35	2 87	0 15	5	79	74
Self-Care	2 93	3 14	0 12	7	79	74
Depression & Anxiety	2 22	2 06	0 9	4	83	78
Cognitive Functioning	3 54	2 46	0-12	5	79	79
Social Functioning	2 42	1 64	0-8	3	54	61
Sexual Functioning	4 22	1 92	0-6	2	57*	43*
Life Satisfaction	5 71	2 78	0-18	6	69	61
Total scale	25 34	10 82	6-61	32	88	86

*Pearson correlation between two items of the scale
Note For original results see De Leo et al 1994

Table 2 presents the means and reliabilities of the LEIPAD subscales in the Finnish sample For comparison, the reliability estimates of the aggregate sample (Finnish, Dutch and Italian) are presented The reliabilities were in acceptable range, except for the Social Functioning subscale in the Finnish sample, for which the Cronbach's alpha was only 0 54 The means in all the subscales, except the Sexual Functioning Scale, were rather low indicating good well-being among respondents

The two factor solution in the second order factor analysis (see Figure 1) explained 58% of the variation of the subscales in the Finnish sample The first factor was dominated by the scales for Life Satisfaction and Social Functioning, the second by the Self-Care scale and the Physical Functioning scale The Depression & Anxiety and Cognitive Functioning scales loaded on both factors, however more strongly on the first factor

Table 3 presents the correlations between the LEIPAD-subscales and the Rotterdam-subscales The physical dimensions of both scales correlated highly, likewise the psychoso-

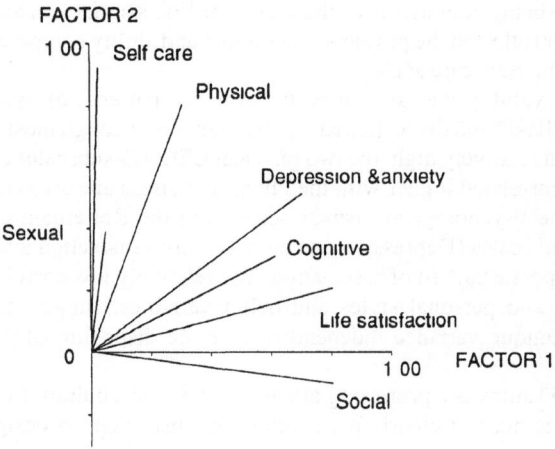

Figure 1. The second order factor analysis of LEIPAD-subscales in the Finnish sample (n=162) Principal component analysis with varimax rotation Note For original results see De Leo et al , 1994

Table 3. Pearson correlations between the LEIPAD Quality of Life
subscales and subscales of the Rotterdam Symptom Checklist
in the Finnish sample (n=162)

	Rotterdam-scales		
LEIPAD-scale	Physical distress	Psychological distress	Activities of daily living
Physical functioning	68***	54***	69***
Self-care	46***	34***	86***
Depression & anxiety	49***	75***	31***
Cognitive functioning	48***	55***	31***
Social functioning	40***	40***	09
Sexual functioning	12	03	27***
Life satisfaction	50***	57***	24**

p*<0 05 **p<0 01 ***p<0 001
Note For original results see De Leo et al , 1994

cial dimensions The scale for sexual functioning had the lowest correlations with the
dimensions of the Rotterdam-scale

The Table 4 presents the associations of four moderator variables and the LEIPAD
Quality of Life dimensions Although most of the correlations were significant they were
generally lower than correlations between the LEIPAD-scales and the Rotterdam-scales
Self-esteem was most consistently related to QOL-dimensions Relatively strong correlations appeared also between the Anger and Perceived Personality Disturbances scales, and
the psychosocial subscales of the LEIPAD Questionnaire

SUMMARY AND CONCLUSIONS

The results indicate that the LEIPAD QOL scales are reasonably reliable, those with
lower reliability were scales with less items In the light of factor analysis, the scales
constructed can be seen as loading on two factors The first factor reflected psychosocial
dimensions of well-being comprising of the scales for Life satisfaction and Social functioning The other factor reflected the physical functioning and ability comprising of the Physical
functioning scale and Self-care scale

Convergent validity was supported by expected patterns of associations between
subscales of the LEIPAD and the Rotterdam questionnaires Though most of the correlations
were significant and relatively high, the two physical LEIPAD-subscales (physical functioning and self-care) correlated higher with the Physical distress and Activities of daily living
scales than with the Psychological distress scale from the Rotterdam questionnaire The
psychosocial Leipad scales (Depression and anxiety, Life satisfaction and Social Functioning), showed an opposite pattern of association The relatively low correlations between the
LEIPAD-subscales and personal styles and belief variables, suggest that the developed
QOL-scales have unique variance independent of more dispositional personality characteristics

The LEIPAD study is a promising attempt to develop a culture free QOL instrument
sensitive for specific need of elderly population The instrument is designed to be easy to
use and applicable in primary care settings It is available in four languages, English, Dutch,
Italian and Finnish, and can therefore be used also for international comparisons Further
studies validating the LEIPAD-questionnaire will concentrate to examine the sensitivity of

Table 4. Pearson correlations between the Leipad Quality of Life subscales and subscales of the Rotterdam Symptom Checklist in the Finnish sample (n=162)

LEIPAD-scale	The moderator scales			
	Perceived personality disturbance	Anger	Self-esteem	Trust in God
Physical functioning	21**	22**	18*	12
Self-care	03	- 02	24**	07
Depression & anxiety	42***	50***	39***	09
Cognitive functioning	31***	34***	27***	09
Social functioning	28***	35***	25***	- 15
Sexual functioning	- 03	- 05	26**	15
Life satisfaction	31***	33***	23**	- 16

*p<0 05, **p<0 01, ***p<0 001
Note For original results see De Leo et al , 1994

the instrument to distinguish between various chronic conditions and to reflect the effectiveness of treatment interventions

REFERENCES

De Leo, D , Diekstra, R F W , Cleiren, M P H , Sampaio Faria, J G , Grigoletto, F , Lonnqvist, J , Trabucchi, M , and Zucchetto, M 1994,   *Quality of Life Assessment – An Instrument to Measure Self Perceived Functioning and Well-Being in the Elderly* , Technical report, EUR/HFA Target 12, Copenhagen, WHO Regional Office for Europe

16

DISABILITY AND QUALITY OF LIFE IN OLD AGE

Luigi Ferrucci, Stefania Bandinelli, Francesca Cecchi, Bernardo Salani, and Alberto Baroni

INRCA Geriatric Department, "I Fraticini"
Via dei Massoni, 21
50139 Florence
Italy

In the past century life expectancy in Europe has increased by more than 20 years. Mortality rates in the elderly have sharply decreased. The oldest-old segment of the population, those age 85 years and older, is currently growing faster than any other age group. Older persons are likely to suffer from one or more diseases or disabilities (Hermanova, 1989; Hermanova et al.,1992). In spite of this, little information is presently available on the dynamic development of chronic diseases, and on their effect on the quality of life in old age.

THE HEALTH SCENARIO

It has been said with some true that illness is the dark side of life. In geriatric research it is often assumed that dynamics in quality of life are the consequences of changes in the health of individuals as they age (Najman and Levine, 1981; Hollandsworth, 1988; Kutner et al., 1992; Anderson et al., 1993; Balducci. 1994). At the turn of the century, mortality and morbidity patterns were dominated by acute, usually infectious, diseases. Nowadays, degenerative diseases play a major role in this health scenario. Since these diseases take long time to develop, they cannot be treated as events "all or nothing", but rather they should be looked as continuous processes. Furthermore, it is generally recognized that, especially in chronic diseases, biological and social factors melt with individual characteristics to create a personal awareness of illness, unique to each human being. This is why most of the traditional epidemiologic measures such as morbidity for single diseases or mortality, cannot capture the broad complexity of health in the older population. Indeed, since medical intervention is focused on prevention, management of symptoms, and treatment of the disabling consequences, improvement and deterioration are regarded as the most important outcomes both in clinical trials as well as in quality control surveys. However, improvement and deterioration are not easy to define. Several interventions may be comparably effective by acting on different aspects of the disease. How effective they are? Which of them results in greater benefits?

Preparation for Aging, Edited by E. Heikkinen *et al*
Plenum Press, New York, 1995

QUALITY OF LIFE

Terms such as comorbidity, active life expectancy and, lately, health related quality of life (HRQL), witness the great deal of conceptual development in new ways of looking at the consequences of diseases. This bio-psychosocial view of the aging process illuminates the illness experience and uncovers consequences of health and functional deterioration. Factors modulating the personal experience of disease are so numerous that a list of them and the description of the nature of their inter-relationships are far from being completed and are still a matter of research. Therefore, it is not surprising that measures and implicit definitions proposed for HRQL range from simple indices of work status and physical functioning, to the most complex multidimensional measures that now dominate the field (Cohen, 1982; Anderson et al., 1993).

A flourishing literature testimonies the widespread but unsupported assumption that all the sensible phenomena in aging can be reduced to a set of variables or to comprehensive global scores. Clinical experience suggests that such an assumption is potentially misleading, because it may obscure the heterogeneity of the aging process. Hence, lacking of a standard definition, the term HRQL has become ambiguous and corrupted by colloquial use.

There is no doubt that the bio-medical and functional approach should be completed by measures of the subjective way in which people handle the experiences of aging and illness. However, switching to the subjective components which may influence HRQL, there is a great uncertainty on what are the essential measures to be considered. the major interest has been focused of the area of research:"coping with stressful events in late life" and "motivation". Indeed, motivation and environment are probably the most important variables conditioning quality of life in healthy elderly people, and coping responses are important elements in modelling the relationship between deterioration of health status and change in quality of life.

Even limiting our analysis to the overmentioned domains, to obtain a global indicator which mix together objective and subjective elements is an intriguing task. First of all, measurable purely objective or purely subjective indicators do not exist when people are asked to report their health experiences in epidemiological studies. It may be easily predicted for example that a depressed subject would be biased in reporting deterioration of health status. Further, in some cases the feasibility of assessing quality of life remains questionable. For example, patients with severe cognitive impairment lack of the minimum cognitive capacity to attach a traditional meaning to life (they may actually experience a special kind of life that we know nothing about).

TOWARD AN OPERATIONAL DEFINITION

In our view, understanding HRQL is an iterative process that begins with a conceptual framework and is followed by the examination of scientific data on hypothesis derived from the framework itself. Each iteration not only gives insight on the nature of the global process that we are trying to capture but also is useful for refining the characteristics of the measurement instrument.

We conceive well-being in the elderly as the ability of an individual to remain independent, that is based on good health, or at least self-evaluation of good health. As several authors have noted, the state of health in more advanced ages tends to result from a number of age- and disease- associated degenerative conditions, that often act interdependently (Guralnik, 1987; Applegate and Curb, 1990; Guralnik et al., 1991). A simple list of pathologic conditions may fail to reflect health status in the aged population because it conveys no information on the severity of diseases, complications related to the diseases, and impacts of comorbidity. In fact, the mutual synergism of a certain pathological condition and its relationships with social, psychological or environmental elements may lead to substantial limitations and a deteriorated quality of life which are not predictable from the severity of the underlying biological impairment.

Given this complexity, medical care cannot be limited to a single disease or symptom. A great deal of effort should be dedicated to understand the dynamic balance between different potential factors which may be targeted for intervention. Most of the literature and the clinical experience of many geriatricians suggest that lowering functional status may represent the final pathway of deteriorating health status in the elderly (Guralnik and Kaplan, 1989). Thus, disability should be the primary target for care and prevention in this population.

To definitively demonstrate that disability in old age is an important component of health related quality of life we must implement intervention studies which demonstrate that prevention of disability is associated with different trajectories in well-being and self-perception of health. We are aware that disability is not an exhaustive measure of health related quality of life. However, we believe that further components should be added to the construct of HRQL, one by one, and tested in specific clinical trials, with the final purpose of finding a reliable and valid comprehensive measure.

The age-associated deterioration of functional status cannot be regarded as irreversible and inevitable. Several lines of research indicate that there are at least two groups of dependent individuals (Fried et al., 1991; Fried, 1992; Colvez and Blanchet, 1994). The first group is comprised of those who, because of a temporary and acute rather than chronic problem, have became dependent. This is a transitory stage from which individuals are likely to recover and return to independence. In contrast, the second group most likely exhibits chronic conditions that are not as well managed, shows untreated and untreatable acute conditions, display slower rates of recovery, and contains individuals who are more likely to stay dependent or to became institutionalized.

In the future, information about the extent and nature of variations in physical functioning and emotional well-being is likely to lead to a new model of treatment intervention. Recent data suggest that a significant amount of incident disability may be prevented and this is likely to improve the average age-specific quality of life.

REFERENCES

Anderson, R T , Aaronson, N K , and Wilkin, D , 1993, Critical review of the international assessments of health-related quality of life, *Qual Life Res* , 2.369-395

Applegate, W B , and Curb, J D , 1990, Designing and executing randomized clinical trials involving elderly persons, *J Am Geriatr Soc* , 38:943-950

Balducci, L , 1994, Perspectives on quality of life of older patients with cancer, *Drugs Aging*, 4 313-324

Cohen, C , 1982, On the quality of life. some philosophical reflections, *Circulation*, 66.29-33

Colvez, A , and Blanchet, M , 1994, Potential gains in life expectancy free of disability· a tool for health planning, *Int J Epidemiol*, 12.224-229.

Fried, L P , 1992, Conference on the physiologic basis of frailty, *Aging*, 4.251-252.

Fried, L P , Herdman, S J., Kuhn, K.E., Rubin, G., and Turano, K., 1991, Preclinical disability. hypothesis about the bottom of the iceberg, *J Aging and Health*, 3.285-300.

Guralnik, J M., 1987, *"Capturing the Full Range of Functioning in Older Populations"*, DHHS, DHHS Pub., Hyattsville, M D., 236 p

Guralnik, J M., and Kaplan, G A , 1989, Predictors of healthy aging· prospective evidence from the Alameda County study, *Am J Public Health*, 79·703-708

Guralnik, J M , LaCroix, A Z , Branch, L G , Kasl, S V, and Wallace, R B , 1991, Morbidity and disability in older persons in the years prior to death, *Am J Public Health*, 81 443-447

Hermanova, H , 1989, Multidisciplinary health assessment of the elderly An international perspective, *Dan Med Bull* , 7(Suppl) 4-6

Hermanova, H , Wertheimer, J , Evans, J G , Michel, J P, and Butler, R N , 1992, Geriatric primary care an European perspective, Part I, *Geriatrics*, 47 31-32, 35, 39-41

Hollandsworth, J C , 1988, Evaluating the impact of medical treatment on the quality of life A five-year update, *Soc Sci Med* , 26 425-434

Kutner, N G , Ory, M G , Baker, D I , Schechtman, K B , Hornbrook, M C , and Mulrow, C D , 1992, Measuring the quality of life of the elderly in health promotion intervention clinical trials, *Public Health Rep* , 107 530-539

Najman, J M , and Levine, S , 1981, Evaluating the impact of medical care and technology on the quality of life a review and critique, *Soc Sci Med* , 15 107-115

17

RESEARCH AND THE PROMOTION OF QUALITY OF LIFE IN OLDER PERSONS IN THE NETHERLANDS

Dorly J.H. Deeg

Vrije Universiteit
Departments of Psychiatry and Social Gerontology
Prins Hendriklaan 29
1075 AZ Amsterdam
The Netherlands

INTRODUCTION

Quality of life is important for humans regardless of their age. Yet, quality of life in older persons deserves special attention. The quality of older persons' lives is more often compromised than that of younger persons, because changes are more likely to occur in their health and in their social world. Although a change may mean an opportunity to improve quality of life, most changes in older age can be expected to constitute losses rather than gains (Heckhausen et al., 1989).

Efforts to promote or maintain quality of life, then, are most needed in older persons who are experiencing a loss. Compensation for the loss is to be found within the older persons' remaining capacity range. The options to find such compensation may be manifold. Consider, for example, a 70-year-old women who has to give up driving her car because her vision has deteriorated. She may opt to start walking, provided her physical condition is good enough. She may also opt to get rides in others' cars, provided her social network includes sufficient people who drive a car. If both her physical fitness and her social network are deficient, this women will have to try to compensate for the loss of transportation ability in non-transportation-related aspects of life.

OBJECTIVES OF PROMOTION OF QUALITY OF LIFE

Turning our attention to health-related quality of life, the broad objective of promotion of quality of life can now be specified in two directions. First, to provide adequate health care, either cure or care or both. Second, to find non-health factors that compensate for a

Preparation for Aging, Edited by E. Heikkinen *et al.*
Plenum Press, New York, 1995

loss of good health. Without diminishing the importance of the first direction, in this contribution we will attempt to substantiate the second direction. Beforehand, we will briefly review possible agents instrumental in promoting and maintaining quality of life.

AGENTS IN THE PROMOTION OF QUALITY OF LIFE

Quality of life is best promoted and maintained by the persons whose life it concerns themselves. That is, provided they have the capacity to mobilize the resources needed. Resources may include personal resilience, or support obtained from social network members such as relatives, neighbors or friends. If such resources are lacking, formal organizations and services will have to fill the gap. On the highest level, government policy should facilitate the promotion and maintenance of quality of life. We describe briefly how a government might contribute to older individuals' quality of life by focusing on the Netherlands.

GOVERNMENT POLICY ON AGING: THE NETHERLANDS

The objective of the Dutch national government's policy on aging (Ministry of Welfare, Health, and Cultural Affairs, 1991a and 1991b) is to increase and maintain autonomy, social integration, and quality of life. It bases its policy on aging on three principles. First, policy on aging is more than health care policy. This principle acknowledges that quality of life is not to be equated with good health only. Second, policy on aging is additional policy. This principle stresses that older persons are not excluded from general policy. Third, policy on aging is predominantly a risk group policy. This principle implies that many older people are doing quite well on their own. Explicit mention is made of the risk group of older women living alone and with a state pension as their only source of income.

The Dutch government indicates three primary policy instruments in the field of aging: experiments, innovative programs, and research. An example of an experiment is the integration of care services for the elderly on the local level. Innovative programs include the promotion of participation of older persons in social organizations or in sports activities. Research should be instrumental in monitoring change, both in older individuals and in society.

RESEARCH AND POLICY ON AGING

Most non-commercial research activities in the Netherlands are indirectly or directly supported by the national government. Currently, a relatively great portion of research funds is allocated to research programs on aging. The Netherlands Organization for Scientific Research (NWO), traditionally focusing on fundamental research, recently stresses societal relevance. It has designated three priorities relevant to older persons (Netherlands Organization for Scientific Research, 1994): quality of care (including home care and care of the chronically ill), chronic diseases (including issues of quality of life), and health and validity in old age. Furthermore, the Netherlands Program for Research on Aging (NESTOR) has been established by the Ministries of Education and Sciences and of Welfare, Health and Cultural Affairs. It has launched six five-year subprograms, on determinants of diseases and disorders, living arrangements and social networks, cognition and compensation, determinants of need of care, economic aspects of aging, and clinical geriatrics, respectively

(Netherlands Programme for Research on Aging, 1989). These programs are not interded to be policy relevant, but should create a firmer place for research on aging in research institutions such as universities. Each of the subprograms takes place within one existing academic discipline. Thus, integration in academic research should be facilitated. The NESTOR program emphasizes the importance of longitudinal research for the understanding of aging processes. This is research in which the same persons are observed with respect to the same characteristics, more often than two times, and during a period long enough to enable the ascertainment of changes in these characteristics (Deeg, 1989).

In addition, the Ministry of Welfare, Health and Cultural Affairs has allocated direct funds for a period of 10 years (1991-2000) to support a comprehensive, longitudinal study which brings together various disciplines and should have policy relevance. This is the Longitudinal Aging Study Amsterdam, a longitudinal research project on predictors and consequences of changes in physical, cognitive, emotional, and social functioning in older adults (Deeg et al., 1993).

The Longitudinal Aging Study Amsterdam (LASA) is based on a sample of 3107 older persons (ages 55-85 years) who were first examined during 1992-1993. The examination consists of a face to face interview and several medical tests, administered by trained interviewers in the homes of the participants. The sample is based in three geographical areas in the Netherlands, which vary in socio-cultural background. Each area includes a larger town and several semi-rural municipalities. The sample includes greater numbers of men and women in the older age years in order to retain sufficient numbers for study by the end of the study period. In 1995-1996, and in 1998-1999 two follow-up cycles are planned.

The interdisciplinary nature of LASA reflects the fact that policy-oriented research questions often extend across disciplines. The study should provide information which may serve to identify policy relevant aspects of aging and to develop new policy aims. Moreover, it may serve to test assumptions from which policy measures are developed. As stated earlier, the promotion or maintenance of quality of life is central to the Dutch government's policy on aging. In the remaining paragraphs of this contribution, we will exemplify this issue using research findings from a pilot study in preparation of LASA.

HEALTH-RELATED QUALITY OF LIFE: SOME FINDINGS

With age, the risk of chronic disease increases, and health is likely to decline. An important research question is: How does a change in health affect quality of life? Since research questions on change require longitudinal data, and LASA has only cross-sectional data so far, this question is reformulated as: How do aspects of quality of life differ for persons without and with chronic illness?

Methods

The data used to answer this question were collected in a random population sample in the town of Sassenheim, which is located in an urbanized rural area in the West of the Netherlands. The sample was drawn such that in each five-year age group (55-59 through 85-89 years) men and women were approximately equally represented. Of the 550 persons sampled, 543 were alive at first contact, and 359 (66.1%) were willing to participate. The reasons for non-participation were refusal (23.8%), incapacity (9.0%) and impossibility to contact (2.4%). Non-participants were slightly older and more often female than participants (Smit, 1993). Data on chronic diseases were collected in a random subsample of 243 persons. The findings reported are based on this subsample.

Self-reports were obtained for six physical chronic diseases that are most prevalent in the older population: respiratory disease or chronic non-specific lung disease (CNSLD), cardiovascular disease, stroke or cerebrovascular disease (CVA), diabetes, cancer, and arthritis. The number of chronic diseases per person is considered to be an index of comorbidity.

Health-related quality of life was considered to have two components: functional disability, and subjective experience of both health and life as a whole. Functional disability was measured by three indicators: 1. whether the need of help was reported for one or more out of seven basic activities of daily living (ADL) including eating and drinking, washing hands and face, bathing, dressing and undressing, sitting down and getting up from a chair, toiletting, and transferring; 2. whether the need of help was reported for one or more of six instrumental activities of daily living (IADL) including preparing warm meals, shopping, cleaning house, taking out garbage, managing money, and filling out forms; 3. whether difficulty was reported doing one or more of six activities (functional limitations, FL) including using transportation, walking for five minutes without stopping, running to catch a bus, carrying an object of 5 kg, stooping and picking up something, cutting one's toenails. Each of these three indicators formed a variable with two categories: needing help or having difficulty vs. not needing help or not having difficulty.

Subjective experience of health was measured by two questions: one asking about health in general (self-perceived health), and one about health in comparison to persons of the same age (comparative health). Subjective experience of life as a whole was measured by a question on satisfaction with life at this moment. Each of these three questions had five answering categories, ranging from very positive to very negative. The first two and the last three answering categories were taken together, resulting in dichotomous variables (good vs. not good).

In addition, subjective experience of health was considered to be determined by the importance an individual attaches to health relative to the importance attached to other domains of life (Flanagan, 1982; Netherlands Central Bureau of Statistics, 1990). The respondents were asked to indicate the three domains of life they considered most important in their present situation out of the following eight domains: good health, a good marriage, a nice family, religion, good housing, a good income, useful time spending, and many friends.

Bivariate associations were tested by chi-square statistics, using a significance level of 0.05. Furthermore, the effect of comorbidity on each of the quality of life indicators was tested while controlling for the effect of age and sex using logistic regression models with comorbidity, age, and sex as the independent variables.

Results

As is shown in Figure 1, all chronic diseases except diabetes and cancer are more prevalent in older age groups ($p < 0.05$). The fact that the prevalence of diabetes and cancer are not greater in the oldest group may be due to underascertainment (Fried and Wallace, 1992). Alternatively, this may be explained by a greater increase with age of mortality caused by these diseases as compared to their incidence (Netherlands Cancer Registry, 1989; Netherlands Central Bureau of Statistics, 1993). Figure 1 also shows that the percentage of persons reporting more than one chronic disease (comorbidity) increases with age: 25% of persons aged 75 years or over report two or more of the six diseases.

The three indicators of functional disability show clear increases across age groups ($p < 0.05$: Figure 2). Above age 75, over 20% need help with ADL, and even over 80% need help with IADL or have functional limitations.

However, subjective health-related quality of life does not show such clear associations with age (Figure 3). Even a greater percentage of 65-74-year-olds report good

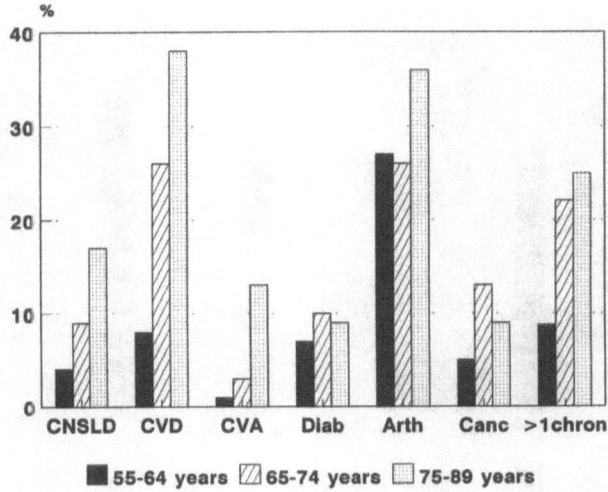

Figure 1. Prevalence by age of chronic diseases, Sassenheim 1992.

self-perceived health than is the case for 55-64-year-olds (not statistically significant). Moreover, increasing percentages of persons in older age groups report to be in better health than their age peers ($p < 0.05$). In all age groups, around 70% of persons report to be satisfied with life as a whole. Interestingly, the percentage of persons who rank good health among their priorities is smaller above age 75 ($p < 0.05$).

In order to determine how health-related quality of life depends on the presence or absence of illness, associations are evaluated between comorbidity and quality of life indicators. As the number of diseases increases, there is a clear increase in help needed with ADL and IADL and in functional limitations ($p < 0.05$: Figure 4).

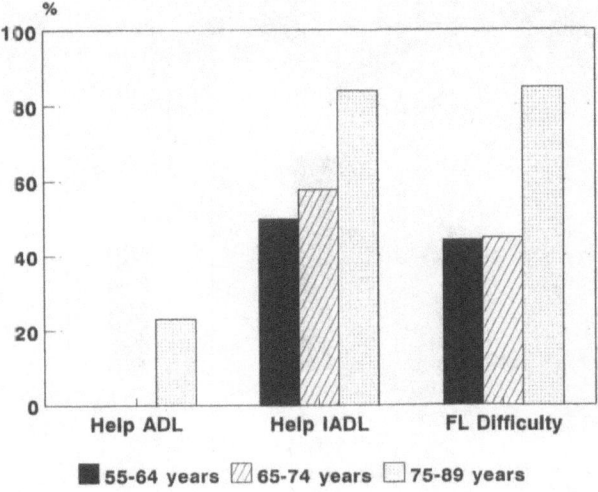

Figure 2. Functional disability by age, Sassenheim 1992.

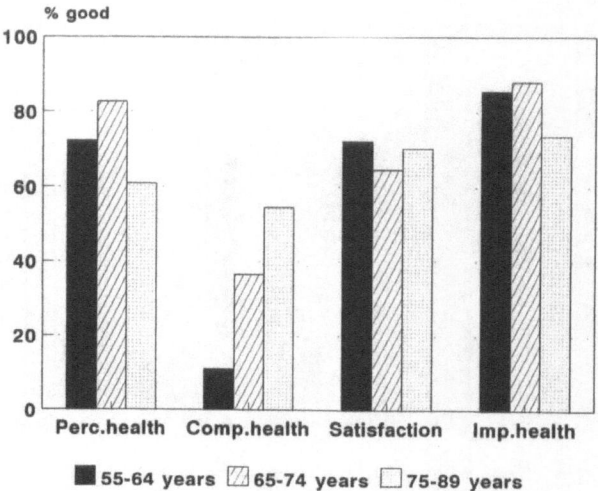

Figure 3. Subjective quality of life by age, Sassenheim 1992.

Also, the percentage of persons who report good self-perceived health steeply decreases with increasing comorbidity ($p < 0.05$: Figure 5). Still, over 20% of persons having more than one chronic disease report their health to be good. There is no significant association between comparative health and number of chronic diseases. Similarly, the association between satisfaction with life and comorbidity does not reach significance. However, a significantly smaller percentage among persons with one or more chronic diseases report good health to be a priority than among persons without a chronic disease ($p < 0.05$).

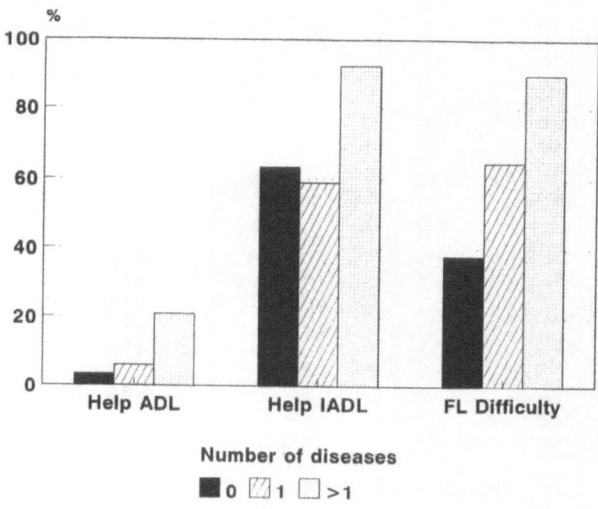

Figure 4. Association of comorbidity with functional disability, Sassenheim 1992.

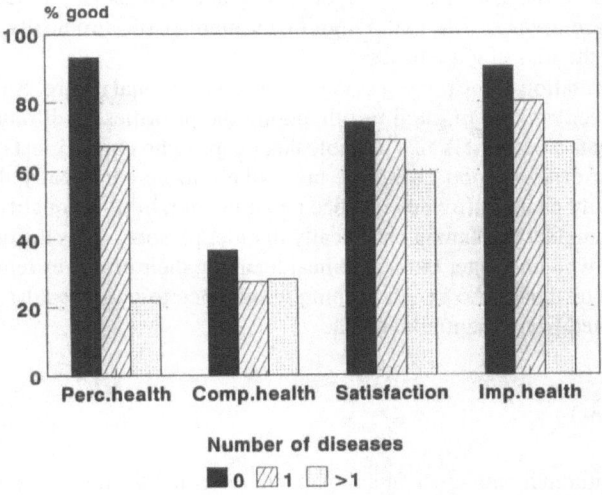

Figure 5. Association of comorbidity with subjective quality of life, Sassenheim 1992.

All statistically significant bivariate associations between comorbidity and quality of life indicators remained significant when controlling for age and sex in logistic regression analyses, except those involving help needed with ADL and IADL.

Discussion

The essence of the findings from this study seems to be that although more objective indicators such as functional disability show a poorer quality of life in those who have one or more chronic diseases, the more subjective indicators do not show convincing differences. In particular, satisfaction with life as a whole does not seem to be greatly affected by the presence of chronic diseases. Apparently, life satisfaction is preserved in spite of poorer health. A possible explanation may be found in the evidence of a shift in priorities: for a significant number of chronically ill persons, the importance attached to good health appears to decrease in the face of adverse health. This evidence suggests that as people are confronted with disease, they employ a cognitive strategy that enables them to maintain their life satisfaction by reconsidering the importance of good health in favor of other domains of life. Similar results have been reported in two studies of physically disabled adults of various ages. One study showed that physically disabled and non-disabled rated their lives as being equally satisfying (Cameron et al., 1973). The other study takes a closer look at the disabled's perception of their disability (Weinberg and Williams, 1978). The subjects in this study viewed their disability as a fact of life, inconvenient and frustrating, but also providing advantages. Seven categories of advantages were mentioned: a challenge to work at, a greater sensitivity and tolerance toward other people, the entitlement to special treatment, a wider range of experiences, a greater appreciation of life, a more realistic view of life, and the ability to give hope to other people. In other words, life seemed no less worthwhile because there were still things to be achieved. It remains to be studied whether these categories also apply to older people who become chronically ill.

When the priority of one domain of life becomes lower, the priority of other domains should become higher. Domains of life that gain in importance for older chronically ill persons proved to be the family and religion (Deeg and Kriegsman, 1993; Braam et al., 1994). It is in these areas of life that a significant number of chronically ill elders seek compensation for the loss of good health.

A major limitation of the present study is its cross-sectional nature. Although it seems likely that the lower ranking of good health among the priorities of ill older persons is a consequence of their disease, it is also possible that people who during most of their lifetime have given low priority to good health are more likely to be chronically ill in older age. Moreover, this study cannot provide evidence about the durability of quality of life, and in particular of life satisfaction, among chronically ill older persons. Does the quality of life of the chronically ill who no longer rank good health among their priorities remain better than the quality of life of those who keep attaching importance to good health? Such evidence can only be obtained from longitudinal data.

CONCLUSION

Until longitudinal data are available, a tentative conclusion may be drawn from the research findings presented. The promotion of quality of life, then, is not only to provide cure and care to the chronically ill, but also to help them to find compensation for their loss of health in non-health factors by shifting their priorities to other domains of life.

REFERENCES

Braam, A W , Beekman, A T F , Deeg, D J H , Van Tilburg, W , 1994, Religiosity and depressive symptoms among elderly persons, a study among elder inhabitants of Sassenheim, *Tijdschr Psychiatr*, 36(7) 509-519 In Dutch, summary in English

Cameron, P , Gnadinger, D T , Kostin, J , Kostin, M , 1973, The life satisfaction of nonnormal persons, *J Consult Clin Psychol* , 41(2) 207-214

Deeg, D J H , 1989, *"Experiences from longitudinal studies of aging Conceptualization, organization, and output"*, NIG Trend studies no 3, Netherlands Institute of Gerontology, Nijmegen, The Netherlands

Deeg, D J H , Knipscheer, C P M , Van Tilburg, W , 1993, *"Autonomy and Well-Being in the Aging Population Concepts and Design of the Longitudinal Aging Study Amsterdam"*, NIG Trend studies, No 7, Netherlands Institute of Gerontology, Bunnik, The Netherlands

Deeg, D J H , Kriegsman, D M W , 1993, Chronic disease, autonomy and well-being The Longitudinal Aging Study Amsterdam XVth Congress of the International Association of Gerontology, Budapest, July 4-9

Flanagan, J F , 1982, Measurement of quality of life Current state of the art, *Arch Phys Med Rehabil* , 63 56-59

Fried, L P , Wallace, R B , 1992, The complexity of chronic illness in the elderly From clinic to community, in *"The Epidemiologic Study of the Elderly"*, R B Wallace, R F Woolson, eds , Oxford University Press, New York/Oxford, pp 10-19

Heckhausen, J , Dixon, R A , Baltes, P B , 1989, Gains and losses in development throughout adulthood as perceived by different adult age groups, *Dev Psychol* , 25 109-121

Ministry of Welfare, Health and Cultural Affairs, 1991a, *"Aging Matters Portrait and Policy Focus on the Elderly 1990-1994"*, Summary, Ministry of Welfare, Health and Cultural Affairs, Directorate of Policy for the Elderly, Rijswijk, The Netherlands

Ministry of Welfare, Health and Cultural Affairs, 1991b, *"Working Together along New Pathways Welfare Policy in the 1990's ,"* Ministry of Welfare, Health and Cultural Affairs, Rijswijk, The Netherlands, in Dutch

Netherlands Cancer Registry, 1989, *"Incidence of Cancer in the Netherlands"*, Coordinating Council of Comprehensive Cancer Centres, Utrecht, The Netherlands

Netherlands Central Bureau of Statistics, 1990, *"Trends from the Life Situation Surveys ,"* Mens en Maatschappij, 65/4, in Dutch

Netherlands Central Bureau of Statistics (1993) *"Vademecum Health Statistics The Netherlands"*, SDU publishers/CBS publications, The Hague, The Netherlands

Netherlands Organization for Scientific Research (NWO), 1994, *"Annual Report 1993, Medical Sciences,"* Netherlands Organization for Scientific Research, The Hague, The Netherlands, in Dutch

Netherlands Programme for Research on Aging (NESTOR), 1989, *"Long term programme"*, Netherlands Institute of Gerontology, Nijmegen, The Netherlands

Smit, J H , 1993, [Field Work and (Non)response in "Autonomy of Older Persons in Sassenheim", *Report of the pilot study of the Longitudinal Aging Study Amsterdam* D J H Deeg, and J H Smit , eds , Vrije Universiteit, Department of Psychiatry and Department of Sociology and Social Gerontology, pp 6-11, Amsterdam, The Netherlands, in Dutch

Weinberg, N , Williams, J , 1978, How the physically disabled perceive their disabilities, *J Rehabil* , July/August/September, 31-33

18

DOES WORK STRESS ENHANCE THE RATE OF AGING?

Willem J. A. Goedhard

Department of Social Medicine
Vrije University
Amsterdam, The Netherlands

INTRODUCTION

In the twentieth century mankind has witnessed an extraordinary increase in life expectancy, at least in the so-called developed countries. Mean life expectancy at birth in my country rose between 1900 and 1985 from 51 years to nearly 74 years (+45%) for men and from 53 to 80 years (+51%) for women (Knook, 1988). Also at higher ages an increase was observed, for example in the same period of time, life expectancy for 65-year olds rose from 11 to 14 years (+27%) for men and from 11 to 18 years (+64%) for women. It is a well-known fact however that the observed quantitative changes were not matched by a proportional increase in qualitative good life expectancy. Despite many studies, conferences and books on gerontology, growing older often implies illnesses, disablements, and suffering. There is a clear discrepancy between the gain in years and the net gain in healthy years. Besides, although the increase in years has been considerable, there is still a big gap between life expectancy and maximal life span, which is about 110 - 115 years.

Now the question is whether we will be able to improve the quality of life for older people. One of the measures of this quality is the period of life we are able to maintain our work ability. The ultimate goal of occupational health care is to protect the health of workers and to prevent work-related diseases. People should be able to continue working as long as they want to, without age-discriminative mandatory retirement rules. However, the opposite can be observed during the last few decades. Although mandatory retirement age is 60-65 years in many countries, many workers are unable to reach this age without problems with their working ability. In the Netherlands labour participation of 55-64 year-old men has fallen from 81% in 1970 to 46% in 1990; above 65 years labour participation is virtually zero (Trommel, 1993). This strong decrease has two main causes: 1. disablement and 2. (voluntary) early retirement. Table 1 gives some data of men in the age range 55-64 years.

Work stress is a major cause of early retirement and disablement in the Netherlands. Disablement due to work stress-related mental disorders increased from about 10% of all causes in 1970 to about 32% in 1990 (GMD, 1990). If we want to be able to successfully intervene we need to know the relationship between work stress and disease and influences on the aging organism. In general it can be stated that advancing age coincides with increased

Preparation for Aging, Edited by E. Heikkinen *et al.*
Plenum Press, New York, 1995

Table 1. Disablement and early retirement of men in the
Netherlands (source Trommel, 1993)

	Disabled (D.I.A.)		Early retirement	
Year	55-59 %	60-64 %	55-59 %	60-64
1980	30	41	—	—
1989	33	39	—	35%

D.I.A. = disablement Insurance Act

vulnerability, but we do not know whether this is also true for perceived psycho-social stress. There are thus three main entities involved: 1. work stress, 2. the aging process, 3. development of diseases.

In this paper I want to concentrate on the possible relation between work stress and aging, in particular the aging process and secondly on the question of a possible relation between rate of aging and disease. For the latter high blood pressure (hypertension) will serve as subject of study.

THE AGING PROCESS

The aging process itself is still a mystery. There are numerous theories of aging, all of which are able to satisfactorily describe some aspects of the aging process but they cannot provide a complete understanding. There are two groups of theories: stochastic and non-stochastic theories (van Bezooijen, 1994). Stochastic theories are of particular interest since herein it is assumed that aging is a random event. It can be speculated that if an event is random, and not strictly programmed, there are opportunities for external interventions. Another subdivision of the aging process assumes the existence of primary aging and secondary aging; the first being genetically based, the latter being based on 'environmental' conditions. This might imply the possibility of intervention, i.e. creation of optimal environmental conditions would induce a prolonged life, whereas adverse conditions would induce harm and eventually cessation of life.

Most body functions are subjected to the aging process and subsequently to a decline of these functions. The rate with which this happens can be called the rate of aging. However what we observe when we measure such functions, either longitudinally in the same person or cross-sectionally in a group of people, is probably a mixture of primary and secondary aging effects. In this respect occupational health care plays a major role, since labour circumstances usually last many years and may have effects long after cessation of work. Since age-associated diseases, like cardiovascular diseases, osteoporosis and cancer, usually have long latency periods, preventive measures for older people can best be started at middle age: around 50 (Berg and Cassels, 1992) This means during our active working period. When we are looking at body functions it can generally be observed that the decline of function usually begins after the age of 30. In particular the reserve capacities appear to decrease with advancing age. These reserve capacities are especially important when we are actively working or during physically demanding leisure activities. If we want to oppose the aging effects, for example a decrease in muscular strength, this can be achieved by training and regular physical exercise. Thus, the rate of aging may strongly differ from person to person leading to differences between chronological and biological ages. Roughly it can be stated that the average decline of many body functions is about one percent per year after the age

Table 2. Comparison of some characteristics of older workers with health and/or work capability problems and geriatric patients

	Geriatric patient	Older worker
Age	above 70	above 45
Multiple pathology	usual	usual
Permanent disability	usual	often
Prone to psychic problems	high	moderate
Social problems	often	often
Diminished musculo-skeletal performance	often	often

of 30. There are however big variations among people in general and among older workers. The inter-person variability in body functions increases with age (Heikkinen, 1992).

WHAT IS AN 'OLDER WORKER'

Many times the question is raised: 'Can we define the concept of older worker?' There is obviously not a strict age-limit. In many cases the words older worker refer to workers over fifty. In Sweden a 45-plus project is running. If we want to define an older worker with health problems we could look at parellels with the concept of a geriatric patient. In table 2 some of the characteristics of older workers and geriatric patients are mentioned. Not every worker over 50 has problems and many are able to continue working to 65 or 70. The same holds for aged people: not every 75-year old is a geriatric patient.

Health promotion programmes and periodic occupational health examinations of older workers may detect early deviations from optimal functioning. Older workers may remain actively working despite a decrease in their functional ability (Ilmarinen et al., 1992). This might induce harm to the organism and possibly enhance the effects of aging.

WORK STRESS: CAN IT BE MEASURED?

The stresses that people may experience can be divided in acute and chronic stresses. Work stress is usually experienced as chronic stress, although acute stressful situations at work are sometimes, and in some occupations often, present. Since work is an important aspect of life for most people, work stress is a relatively important factor in determining the well-being of life. To find out whether work stress can be deleterious to one's health, it is useful or even necessary to measure stress and to present stress factors in a (semi-)quantitative manner. For this purpose questionnaires are used that are filled out either by the employee or by a researcher. In this study a questionnaire was used that was published by Frommer et al. (1986). This questionnaire had been validated in a longitudinal study among civil servants in Australia. It looked especially useful for the occupational health practice because of its concise nature. The questionnaire was translated into Dutch and tested in a group of 50 employees. The acceptance was high and the results of the testing were in good agreement with the presumed levels of stress based on the medical histories of the examined subjects. The major advantage of the questionnaire is that it provides quantitative data of various (six) stress factors, showing also the relative differences between the factors. These factors are presented on a five-point scale (1 to 5; 1= eustress, 5 = extreme dysstress). Using this questionnaire scores of the following stress factors (sf) are obtained:

- sf1: boredom
- sf2: lack of support
- sf3: quantitative overload
- sf4: qualitative overload
- sf5: inadequate salary or prospects
- sf6: unsatisfactory physical work conditions

The results of the questionnaire can then be used to study possible relations with various body functions for example blood pressure. It should be realized, however, that the obtained data represent subjectively perceived stress.

Perceived Stress and Absence from Work

It can be reasoned that subjects under too much work stress have a tendency to escape from their job. However, studies sofar do not strongly endorse the hypothesis that work stress will lead to increased absence from work (McKee et al., 1992). In a study on 162 subjects using the questionnaire that was described above, positive associations were found between frequency of absenteeism and four separate stress factors (sf1, sf2, sf4 and sf5) (Goedhard, 1993). There was no association of duration of absence and any of the stress factors. However, combinations of two stress factors showed much stronger correlations with absenteeism. Table 3 shows some results. As can be seen from this table the so called 'active' group experiences a higher than average work load but shows the lowest figures of absenteeism. Obviously the influence of a relative good work support may be a compensation for the amount of work. If however the second factor becomes unfavourably high too, then the figures for absenteeism become high.

BLOOD PRESSURE

When we talk about 'our blood pressure' we usually refer to the readings on the blood pressure (BP) manometer, made by our physician. Actually, by this method only the highest (systolic) and lowest (diastolic) blood pressure values are measured in the arterial system of the systemic circulation. The real BP waveform is much more complicated than the two values. (Figure 1). This BP waveform can not be measured by the 'old-fashioned' device that is available in everyday medical practice. Until some ten years ago the full waveform could only be recorded by means of invasive techniques i.e. by putting a needle into an artery and connect this to a suitable pressure manometer. Since approximately 1980 a new

Table 3. Combinations of sf2 and sf3 and their relation with absenteeism

Lack of support (sf 2)	Quantitative overload (sf 3)	Absence duration (%)	Absence frequency (per year)
low	low ('low stress')	4.7	1.5
low	high ('active)	0.5	0.7
high	low ('passive')	1.9	1.2
high	high ('stress')	4.6	2.4

('high' and 'low' refer to higher and lower than mean values respectively)

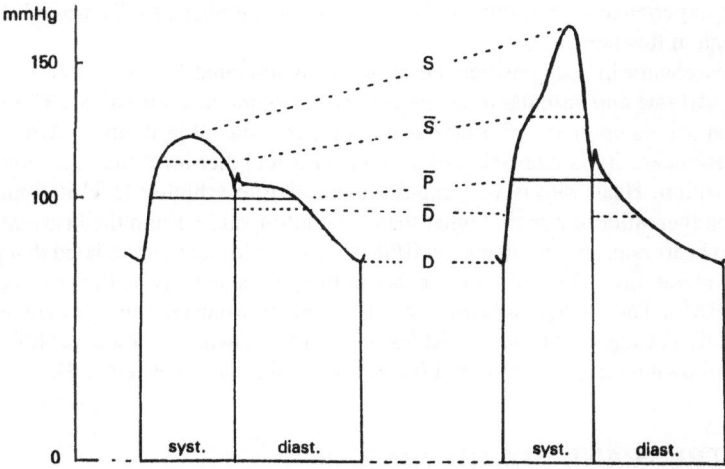

Figure 1. Two schematic drawings of an arterial blood pressure curve. left: at young adult age; right: at old age; due to arterial stiffness at higher age the pressure curve changes; S = systolic pressure; P = mean arterial pressure; D= diastolic pressure; Psyst. shows the biggest change, in this example from about 120 mm Hg to about 165 mm Hg. (source: Knoop, 1984).

noninvasive technique has become available, the Finapres (Settels and Wesseling, 1986). With this method, using a small manchet wrapped around a finger, the BP waveform can be recorded continuously on a beat-to-beat basis. A major advantage of this method is that it provides information about the mean arterial blood pressure (BPmean). BPmean can be regarded as the systemic pressure that determines the perfusion of body organs. BPmean and systolic BP differ in this respect since systolic BP is dependent on the elastic properties of the arterial wall. Arterial stiffness usually increases with advancing age and therefore systolic BP is often increased in older people.

When BP becomes too high we speak of hypertension. The limits of 'normal' and 'abnormal' BP have rather artificially been set. Since systolic BP (BPsys) and diastolic BP (BPdias) have been measured millions of times the BP limits for hypertension are given in reference values of BPsys and BPdias. In the future reference values of BP mean may possibly be used as limits of hypertension. There are several limits for hypertension varying from 140/90 (BPsys/BPdias) to 160/100. For older people it is usually assumed that the higher limits are more appropriate to use in medical practice. For the very old it is even doubtful whether treatment of hypertension is beneficial. The risk of lowering BPmean too much may lead to an ill perfusion of vital organs (brain, kidneys, heart) and may therefore be harmful.

BLOOD PRESSURE REGULATION: THE BAROREFLEX

An advantage of the Finapres method is that BP can easily be measured in various body positions and also during a change in body position. An example is standing up from supine position. It will be clear that this change in body position is quite a challenge to the circulation. To provide adequate blood perfusion to the brain in standing position, BP must be rapidly adapted to the initial decline due to hydrostatic changes. Sometimes we may experience some dizziness when we stand up too quickly from supine position. Elderly people often experience such spells of dizziness and the number of falls around the bed is relatively high in this age group.

Upon a change in body position the baroreflex is activated. This reflex system induces changes in heart rate and vascular diameter in order to maintain a normal BP. The sequence of effects that occurs upon an orthostatic manoeuvre is depicted in figure 2. After an initial decrease in BPmean, BP is increasing usually towards a higher level than the control level in supine position. Heart rate is very rapidly increasing to a higher level and may either decrease soon thereafter or remain higher than the control value. From the heart rate at rest, the change in heart rate, and the change in BPmean a variable can be calculated that provides information about the effectivity of the baroreflex; it is usually called the baroreflex sensitivity (BRS). The change in heart rate (dF) upon an orthostatic manoeuvre as well as BRS appeared to be age-associated variables. In a study on healthy volunteers (20-97 years of age) the following results were found (Goedhard et al., 1984) (see table 4).

WORK STRESS AND BLOOD PRESSURE

A study on the possible influence of work stress on blood pressure was done in a group of 40 workers, aged 45-62, mean age: 51.8 years). All subjects filled out the stress questionnaire and blood pressure was measured in standing position. Some of the subjects were known as hypertensives and were medically treated accordingly. The following mean values of blood pressure were observed:

- BPsyst: 159 +/- 25 (sd) mm Hg;
- BPmean: 112 +/- 19 (sd) mm Hg;
- BPdias: 89 +/- 15 (sd) mm Hg.

Subjects were then divided in groups of 'high' and 'low' blood pressure' for pressures higher or lower than 160 (BP syst) and/or 115 (BP mean) and or 95 (BPdias) mmHg. Stress scores were divided in an analogous manner. Subsequently 4 groups were obtained:

- A low pressure, low stress (N=6)
- B low pressure, high stress (N=13)
- C high pressure, low stress (N=4)
- D high pressure, high stress (N=17)

Comparing the groups B and D (the groups A and C are too small to allow meaningful conclusions) for baroreflex functioning the following results were obtained (see table 5).

Table 4. Average changes in blood pressure (dP, mmHg) and heart rate (df, beats per minute, bpm) in some age groups and the subsequent average values of BRS (ms/mm Hg)

Age group (years)	Number	dP (mm Hg)	dF (bpm)	BRS (ms/mm Hg)
20-24	17	18	27	15
30-34	36	15	27	13
40-44	13	15	22	11
50-54	18	12	19	10
60-64	11	10	14	9
70-74	7	12	13	7
80-84	8	12	8	6

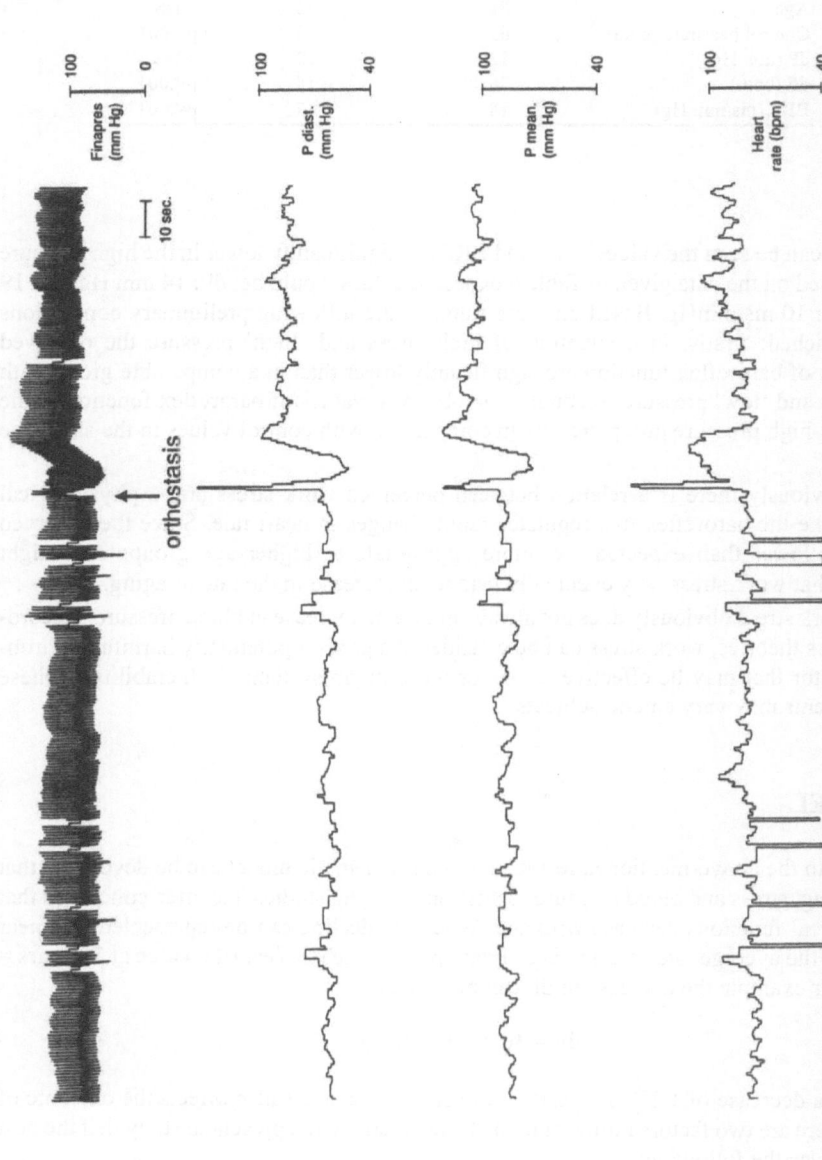

Figure 2. Blood pressure changes during an orthostatic manoeuvre (from supine position to standing position) From top to bottom: 1) Finapres tracing; 2) diastolic pressure; 3) mean arterial pressure; 4) heart rate After standing up (arrow) we observe a drop in pressure, reaching its minimum in approximately 10 seconds; then pressure increases again to a level which is higher than the control level; heart rate shows an immediate increase upon the manoeuvre, lowering to a new level after about 30 seconds.

Table 5. Mean values of parameters of baroreflex function in two
groups of workers divided according to their levels of perceived
work stress and arterial blood pressure

	BP low/stress-high	BP high/stress-high	Diff.
Number	13	17	
Age	51	52	NS
Control heartrate (bpm)	63	75	p<.001
dP (mm Hg)	12	12	NS
dF (bpm)	26	15	p<.001
BRS (ms/mm Hg)	18	7	p<.001

As can be seen the values of dF and BRS are significantly lower in the high pressure group. Based on the data given in Table 4 expected values would be: dP: 14 mm Hg, dF: 19 bpm, BRS: 10 ms/mmHg. Based on these findings the following preliminary conclusions can be reached: firstly, in a situation of high stress and 'high' pressure the observed parameters of baroreflex function are significantly lower than in a comparable group with high stress and 'low' pressure; secondly: the observed values of baroreflex function in the high stress-high pressure group are low in comparison with control values in the same age group.

Obviously there is a relation between perceived work stress and a physiological function like the baroreflex that regulates rapid changes in heart rate. Since the observed values are lower than expected, i.e. more appropriate to higher age groups one might conclude that work stress may eventually lead to an increase in the rate of aging.

Work stress obviously does not always induce an increase in blood pressure. According to stress theories, work stress can be considered a general potentially harmful environmental factor that may be effective in one or more organ systems. Vulnerability of these organ systems may vary among subjects.

A MODEL

With the above-mentioned results in mind a schematic model can be developed that relates aging, stress and blood pressure regulation. In aging studies it is often concluded that physiological functions decrease with age. When the decline can be represented by linear regression the average rate of aging is a certain percentage per year (the value at 30 years = 100%). For example the decrease in dF can be given as:

$$dF = 36.2 - 0.33 \times age;$$

this gives a decrease of 1.1% per year. However, if work stress also affects the outcome of dF then there are two factors influencing dF. It was found in the presented study that the best model fit was the following:

$$dF = 54 - 0.19 \times age - 3.2 \times sf1 - 5.4 \times sf3 \text{ (bpm)}$$

It can easily be appreciated that if stress is low, e.g. sf1=sf3=1 the this equation becomes:

$$dF = 45.4 - 0.19 \times age \text{ (bpm)}.$$

This model would reduce the age effect on dF to about 0.5% per year. If this is true then there may be practical consequences. Should we be able to limit environmental stress then the rate of aging might be lowered, with a subsequent decrease in vulnerability to disease.

CONCLUSION

The preliminary conclusion can be reached that primary and secondary age-associated changes in fast blood pressure regulation through the baroreflex are of the same order of magnitude. When this conclusion might be applied to age-associated changes in general, this offers good opportunities for preventive measures or programmes for people over fifty years of age.

A second conclusion is that too high work stress can be a health hazard and may influence important physiological functions in a negative manner.

REFERENCES

Berg, R L , and Cassels, J S , eds, 1992, *"The Second Fifty Years,"* National Academic Press, Washington

Frommer, S , Edye, B V , Mandryk, J A , Berry, G , and Ferguson, D A , 1986, Systolic blood pressure in relation to occupation and perceived work stress", *Scand J Environm Health*, 12 476-485

GMD, Netherlands Joint Medical Service, 1990, *Year report*, Gemeenschappelijke Medische Dienst, Amsterdam

Goedhard, W J A , 1993, "The Relation Between Psychosocial Stress and Age in a Working Population", in *"Aging and Work"* Ilmarinen, J , ed, Helsinki, Finnish Institute of Occupational Health, p 25-32

Goedhard, W J A , Settels, J J and Wesseling, K H , 1984, "Aspects of Blood Pressure Regulation in Relation to the Aging Process" (in Dutch) in *"De bloeddruk op oudere leeftijd"* W J A Goedhard, and D L Knook DL, eds, Samson-Stafleu, Alphen ald Rijn/Brussel, p 37-53

Heikkinen, E , 1992, "Aging and Functional Capacity, Theoretical and Research Models", in *"Aging and Work"*, W J A Goedhard, ed, Pasmans, The Hague, p 62-77

Ilmarinen, J , Louhevaara, V, Korhonen, O , Nygård C-H , and Hakola, T , 1992, "Aging and Physical Capacity to Work A Four-Year Follow-up of 51-55 Year-Old Municipal Employees", in *"Aging and Work"*, W J A Goedhard, ed, Pasmans, The Hague, p 78-98

Knook, D L , 1988, "Biomedical Gerontology Research for the Health of the Elderly A National Perspective" in *"Health and Aging"*, J J F Schroots, J E Birren, A Svanborg, eds, Springer, New York

Knoop, A A 1984, Physiological Backgrounds of Blood Pressure at Old Age", in *"De bloeddruk op oudere leeftijd"*, W J A Goedhard, and D L Knook, eds, Samson/Stafleu, Alphen a/d Rijn/Brussels, p 26-36 (in Dutch)

McKee, G H , Markham, S E and Dow Scott, K , 1992, "Job Stress and Employee Withdrawal from Work", in *"Stress and Well-being at work"*, J Campbell Quick, L R Murphy, and J J Hurrell, eds, American Psychological Association, Washington, p 153-163

Settels, J J , and Wesseling, K H , 1986, "Finapres, Noninvasive Finger Arterial Pressure Waveform Registration", in *"Psychophysiology of Cardiovascular Control"*, J F Orlebeke, et al , eds, NATO conference series III, Plenum Press, New York, vol 26 267-283

Trommel, W A , 1993, *"Eigentijds met pensioen"*, VUGA, The Hague, (in Dutch)

van Bezooijen, C F A , 1994 "Theories of Ageing, Age-Associated Disease and Their Prevention", in ICOH Workshop *"Add Life to Your Working Years"*, Book of Abstracts, p 4

19

INTERRELATIONS BETWEEN GENERATIONS IN HISTORICAL PERSPECTIVE

Birgitta Odén

Department of History
University of Lund
Lund, Sweden

Relations between the generations—between the old and their grown-up children—are determined by several factors:

- *biologically* through the genetic *likenesses* between individuals in the same family but different generations.
- *emotionally* through the *feelings* of affection—or hatred – which arise during the human child's long dependence on his or her parents.
- *socially* through the *cultural codes* which every society develops for its collective survival

In my paper I will describe and analyze the last of these factors: the cultural codes in the Scandinavian countries and the changes they have undergone up to the present day.

The perspective is *historical*. Since cultural codes are highly tenacious and are passed on from one generation to the next in the form of ethical and moral commandments, a historical perspective can shed light on today's *ties* to tradition and the *changes* which have nevertheless occurred as a consequence of profound economic, social, and political processes of change.

THE MIDDLE AGES

The Middle Ages is the oldest period that has left sufficient source material to allow historians to create a picture of relations between generations.

The medieval Nordic societies were peasant societies. Roughly half of the peasants farmed their own freeholds, while the rest were tenants of the king, the nobility, or the church. The proportion of freehold tenure was greatest in the north of Scandinavia and least in Denmark. The land was not farmed collectively but through a form of cooperation which demanded that every farm in a village did certain jobs at established times. The form of production required that the head of the peasant household was active and fit to work (Odén, 1994).

Preparation for Aging, Edited by E. Heikkinen *et al.*
Plenum Press, New York, 1995

This form of production gave rise to intensive and varied legislation in the thirteenth-century laws of the various provinces, which regulated what happened when the aging peasant was no longer fit to work. The rules varied from one region to another and from one period to another, but they had one thing in common: the aging peasant had to suffer a loss of status in relation to the new head of the farm. He had to give up the high seat and accept an inferior position in the household.

Medieval relations between generations were not just determined by production but also by the prevailing religious ideology, that of the Catholic church.

The church's ethical foundation for relations between the generations was the fifth commandment given to Moses: "Honour thy father and thy mother". The book of Exodus also has rules which prescribe the death penalty for those who strike or curse their parents.

The Catholic theologians had intense discussions about whether Mosaic law should really be applicable in Christianity. A treatise that was translated into Swedish in the early fifteenth century asserted that the fifth commandment was applicable to Christians, not just because it was God's law but because it was a part of *natural law* to love one's parents. On the other hand, the death penalty was not accepted for physical and verbal abuse of parents. In the life to come, God would exact punishment for these sins. There was therefore no reason for society to penalize them.

Paintings on the vaults of Swedish churches and in books of popular devotion conveyed the instructional message to the congregation. The fifth commandment was interpreted as the obligation to provide one's parents with food and drink when they were old and infirm. God's commandment was associated with the same ethical rules as the secular laws: the maintenance of the elderly was more important than their status in the family.

One may wonder whether there are any reminiscences of pre-Christian ethics in the rules for relations between generations which shaped medieval society in Scandinavia. In my view, there are such reminiscences. Most of the medieval Scandinavian laws have extremely harsh penalties for those who assaulted or killed anyone within the walls of the house. Such a crime was known as a "dastardly deed" (*nidingsdåd*), and the offender could be killed with impunity by anyone. Since the elderly lived within the walls of the household in the Middle Ages, these strict rules for peace in the home were a kind of guarantee that violence against old members of the family would be curbed.

The relations between generations which I have hitherto concentrated on concern relations between biological or quasi-biological generations. The fact that the biological family was seen as the ideal model for the care and maintenance of the elderly in the Middle Ages is interestingly illustrated by the contracts for caring that have been preserved from Stockholm in the late Middle Ages. In these contracts old men or women left their property to someone who was not their own child—often a niece—in return for care until their death. The forms for the contracts contained biological metaphors: an old woman would ask to be treated like a true mother (Odén, 1987). The same forms of contractual agreements using biological metaphors have been found in eastern Finland. The party contracting to provide help is called "caring son" or "caring boy" (Jutikkala, 1963).

But far from all old people were in the situation that they had grown-up children to care for them or property which allowed them to plan for their old age. They were dependent on collective measures.

In the Middle Ages, the collective care of old people without biological care systems was organized by the charitable institutions of the church. It would be beyond today's theme to show how the charitable institutions worked. Only one thing needs to be stressed: in the Nordic countries the church to a large extent left the care of the needy to the individual households. The form this took was that the peasants were excused from paying one-third of the tithe in return for looking after the poor themselves. The church could thereby continue a pre-Christian form of organization with a firm ethical base: hospitality.

FROM THE SIXTEENTH TO THE EIGHTEENTH CENTURY

In the early sixteenth century the Nordic countries converted to Lutheranism. At the same time, a new view of the state emerged—the nation state. Feudal and patriarchal patterns of thought gained a foothold in the highest circles in society. The church became an instrument of royal power. The idea was propagated that the king is God's representative on earth. It was then logical that the king saw it as his duty to further the application of God's commandments within his realm. This was obvious in Sweden, as well as in Denmark. In 1608 a number of "Mosaic" laws were introduced in Sweden, including the rule from Exodus that those who struck or cursed their parents should be put to death. In accordance with this law, grown-up children were sentenced to death during a period of 260 years (Odén, 1991).

There is clear evidence, however, that the country's legal experts found this law in conflict with an old Swedish perception of justice. Even if the local courts passed the death sentence, it was generally annulled in the court of appeal. Yet if the case reached the king's supreme court, he personally refused to grant a pardon. There could be no exceptions to God's law, according to God's representative on earth.

In the first half of the nineteenth century there was a dramatic rise in the number of cases of violence against parents which were brought to court. Since crime statistics began to be kept at this time, the authorities became aware of the problem, which was a topic of lively political discussion.

The authorities thought that the violence between the generations was primarily due to a new form for the maintenance of the elderly which was gradually introduced in the eighteenth century and which became the most common form in the nineteenth century: the pensioner's allotment—in Sweden called *undantaget*, literally "the displacement". This meant that old, infirm parents no longer stayed in the household with the son or son-in-law who took over the farm. Instead the old couple set up their own household with the aid of a fixed quantity of products from the farm. The change was occasioned by new forms of production and changed ownership structures in agriculture. The change coincided with a lower retirement age for the old farmer, which fell from 60 towards 50. Fulfilling the terms of the contract could be a heavy burden for the young household and was thus a seed-bed of conflicts which had to be resolved at the local court, but which also triggered violence and harsh words between the generations (Gaunt, 1983).

The opinion of the authorities did not capture the whole truth, however. The assaulters and the victims often belonged to the proletarian and pauperized groups in society. The early nineteenth century was a time of dissolving norms. The influence of the church was weakened, there was increased mobility in the countryside, patriarchal relations were broken, alcohol abuse increased, and there was a general rise in violence. This society also saw an increase in violence against parents, while the state simultaneously heightened its efforts to discipline the population.

Relations between generations on the collective side long continued to be based on the medieval tradition. The church and the crown agreed that responsibility for the old and the poor should rest with the individual households, while it was the duty of the parishes to ensure that the care system worked.

The institutions which the state took over from the church—the asylums (known as hospitals)—were primarily intended for the care of the mentally ill and the handicapped. Orphans and old people could also be cared for there, but the state's constantly repeated intention was that the old should be removed from the state-run asylums and put in the care of their own parish. The state-run medical care provided in hospitals (lazarets), which developed in the eighteenth century, avoided taking in the elderly. Old age was not a curable disease. The hospitals had been set up to restore sick manpower to health. The mean age of the patients was 30. Old sick people were cared for in their homes—by relatives or by neighbours (Odén, 1993).

LIBERALISM

The early nineteenth century in the Nordic countries was dominated by liberal ideas in the field of social policy as well. In the discussions about new poor-law legislation there were frequently heard liberal demands that the ties between the elderly and their grown-up children should be dissolved. Children should no longer be obliged to pay the costs their poor parents caused the municipalities. Conservative politicians saw a threat in this: the family ethic could be broken up. The political result was a compromise.

In Sweden grown-up children continued to be responsible for the economic mainte-nance of their parents until 1956 in social legislation and until 1970 in family legislation. Long before this, however, adult children had in practice ceased to be responsible for their aged parents. This has to do with the development of collective care. In this period a new model for collective caring expanded: the voluntary, philanthropic associations, where religious ladies in the new bourgeoisie found a mission in helping the lonely and the poor.

Municipal administration was reorganized in the mid-nineteenth century. Effective institutions were created for discharging the municipality's care duties. Medical care was also institutionalized at county council level. With the building of municipal poor-houses, which were later transformed into old people's homes, the responsibility for providing care in most parts of Sweden was transferred from the households to institutions. The number of boarders (*inhyses*) in the households fell. The number of old people living with their grown-up children also fell. After the introduction of the old-age pension in 1913, old people also had better opportunities to go on living in their homes, especially if they had extra income or could produce their own food so as to eke out the meagre pension.

A decisive step towards change was that *the state* assumed responsibility for the pensioners through the old-age pension system. The alternative of municipal pensions was rejected, although every municipality acquired a board which determined the need for supplementary pensions. Behind this stance there was a new view of the state which developed in the latter half of the nineteenth century. The responsibility of the state increased in pace with the increase in tax revenues through the growth of industrial production.

We learn from ethnologists—through the material created by oral history—that the turn of the century was a period when relations between old people and their children were good. Crime statistics no longer included data on violence against parents, so there was nothing to contradict this idyllic picture. The reason for this—in my view—is that the close economic ties between the generations were dissolved. There was less potential for conflict, and the positive emotional ties could be developed into relations which benefited both sides, allowing them to exchange care and spend leisure time together (Odén, 1993).

THE WELFARE STATE

In the middle of the twentieth century the elderly once again became a weighty political issue. There was criticism of the compulsory character of the old people's homes, their poor quality and their mixed clientele. The right of the elderly to homes of their own was stressed. Pensioners' homes were built and home help for the aged was taken over from the church's welfare workers by municipal service agencies. Medical care for the elderly was expanded, and geriatrics became a discipline of medical study. Old-age pensions were changed into occupational pensions for all wage-earners, which guaranteed that they could maintain their standard of living after retirement. Behind these changes lay anxiety about the growing proportion of old people among the electorate and the need for a mobile workforce so that "the Swedish model" could work.

In the 1960s and 1970s there was further expansion of the old peoples' portion of the welfare state: service houses, transport service, long-stay clinics, food distribution, technological security alarms, and home medical care satisfied a variety of needs in the everyday life of the elderly.

We learn from sociologists that relations between the elderly and their adult children were very happy during these years. The so-called loneliness of the elderly was more of a theoretical construction of scholars and a stereotyped view held by young people than a widespread reality. Statisticians were able to show, for example, with the aid of studies of the living standards of different groups, that most old people had a satisfactory standard of living and a good network. Only the group of elderly single women—unfortunately a growing group in society—were threatened in their security and had no political or social resources to rely on.

The costs of eldercare and pensions rose to become huge sums, which were borne by the state in the first hand, by the municipalities in the second hand. Taxes rose to become the highest in the world. But it is important to note that, until the middle of the 1980s, the majority of the Swedish people backed this support for the elderly. The political parties' election programmes and the statisticians' opinion polls agreed in showing that the Swedish people were willing to pay high taxes for the elderly and for good medical care. It is only in the past ten years that there has been a noticeable shift in public opinion.

Today's debaters in newspapers and books start with the assumption that the increased costs of the public sector have reduced the gross domestic product of the Swedish state and the potential for growth in the economy. They look back to the past, when relatives and voluntary organizations shouldered the responsibility for the care of the elderly. "Civil society" is assigned the role of the ideal model, where people take responsibility for themselves, and where state control is reduced to a minimum.

Perhaps "the Swedish model" has outlived its role as a future Utopia. But the new Utopias which will help us to "choose the future" should not be sought among the rejected alternatives of a distant past. They must instead be designed with a view to renewal and creative initiatives, which should not come about at the cost of increased insecurity for old people or by excluding young women from the labour market and tying them to care within the framework of the household, which once was the only solution to the problems of care when Sweden was a land of poverty. This should *not* be the solution for the future.

REFERENCES

Gaunt, D , 1983, *"Familjeliv i Norden,"* [*"Family Life in the Nordic Countries"*], Gidlund, Stockholm, in Swedish

Jutikkala, E , 1963, *"Bonden i Finland genom tiderna,"* [*"The Peasant in Finland through the Ages"*] (Swedish translation of "Suomen talonpojan historia," Helsinki, 1958), LT, Stockholm, in Swedish and in Finnish

Oden, B , 1987, Planering for ålderdomen i senmedeltidens Stockholm", [Planning for old age in late medieval Sweden] in. *"Manliga strukturer och kvinnliga strategier En bok till Gunhild Kyle, december 1987,"* [*"Male Structures and Female Strategies A Book for Gunhild Kyle, December 1987"*], B Sawyer, and A Goransson, eds , Historiska institutionen, Goteborgs universitet, Goteborg, in Swedish

Oden, B , 1991, Relationer mellan generationerna. Rattslaget 1300–1900, [Relations between the generations The juridical position 1300-1900] in· *"Maktpolitik och husfrid, Studier i internationell och svensk historia tillagnade Goran Rystad"* [*"Power Politics and Domestic Peace Studies in International and Swedish History Dedicated to Goran Rystad"*], B Ankarloo, et al , eds , Lund University Press, Lund, in Swedish

Oden, B , 1993, *"Att åldras i Sverige"* [*"Aging in Sweden"*], (with Alvar Svanborg and Lars Tornstam), Natur och kultur, Stockholm, in Swedish

Oden, B , 1994, Skyddet for de gamla i medeltidens lagstiftning, [Protection for the aged in medieval legislation] in " *och framtrader landsbygdens manniskor Studier i nordisk och smålandsk historia tillagnade Lars-Olof Larsson på 60-årsdagen den 15 november 1994"*, [" *and the People of the Countryside Come Forth " Studies in the History of Scandinavia and Småland Dedicated to Lars-Olof Larsson on His 60th Birthday, 15 November, 1994"*], Hogskolan i Vaxjo, Vaxjo, in Swedish

20

GENDER AND AGING

Christine Castelain Meunier

CADIS CNRS
54, boulevard Raspail
75006 Paris
France

In this article, we will study the differences between aging men and women and their evolution.

QUERYING THE TERNARY PATTERN OF THE AGE CLASSIFICATION

First, we will analyze how the ternary pattern of the age classification as defined by A.M. Guillemard et al. (1991) can be queried based on the study of middle class men and women who are involved in community life.

Our aim is to question and criticise the classification of ages by cycle. Industrial society has established a pattern that divides life into three phases. The first one is youth or learning period, the second, adulthood or working period, and the third retirement or period of leisure and rest.

We will show, through the analysis of qualitative interviews, how aging men and women refuse to comply with this three-phase division with their representations, practices and values. We will base our analysis on a large study (Guillemard et al., 1991) carried out on early retired, pre-retired and retired people who are actively involved in community life[*].

The three-phase age division is rejected because people refuse to be socially and culturally marginalised after the loss of social identity or of active citizenship following retirement, and to be classified in a category that often becomes an "age ghetto". The division of a lifetime into age brackets does not encourage the development of personality. The phases of learning, working and resting appear to have to be separated, permanently alternated—and not spread over long periods of time—a condition that renders them exclusive and incompatible. This explains why these people often try to redefine the concept of work through voluntary work. They enjoy the absence of hierarchy, attempt to blend their social and private lives, and develop a new notion of work. The characteristics include the fact that there is no

---

[*] Associations working in the fields of training or unemployment, inter-generation solidarity, psychological aid, local or regional movements, aid for domestic or international companies, etc.

financial reward, a larger scope and choice of tasks, a certain degree of self-sufficiency, and wider social interactions. These people are motivated by a need to assert their personality and not by financial compensation. They seek pleasure rather than promotion. Their way of dealing with time and space is different: they tend to live in the present (Bidou, 1984) and are more closely involved in the local community. This is the result of a deep-seated opposition to the inappropriate division of the age cycles. Retirees refuse to be classified as a member of an age segment that is contrary to their needs and expectancies, especially if they are young and in early retirement.

SOCIAL FORCE AND HANDICAPS

What can be said then about the differences between men and women? The refusal to retire, rest and be inactive is very strong and applies to both sexes. Men are more present in economic associations and are generally more active in the community, whereas women are involved in social associations that favour the development of relationships and personality.

Men undertake more responsibilities than women. However—and it is an interesting paradox—the values, ways of life and practices these community networks convey are more "feminine". Women are nevertheless minority members, and the active social force that older women represent does not weigh as much as it could. There are well-known phenomena that explain why older women are less active social protagonists:

- women live longer than men and so suffer more from loneliness and widowhood,
- women's pensions are smaller and there are often problems surrounding benefits for surviving spouses, they have shorter careers and suffer more from unemployment.

Family life takes up most of a woman's social, emotional and affective reserves in the home. When women grow older, they pass through major three stages (Vannemaa, 1994).

- the first is the transition from employment to retirement, with the issue of the "double" day.
- the second, the transition from marriage to widowhood.
- the third, the transition from a position of love-giving (conjugal, parental) to loneliness.

These three transitions represent a social, affective and purely female handicap for participation in community life. Elderly women are more likely to be placed in institutions against their will than men (Asili, 1994). Their autonomy and capacity to choose their own destiny is taken from them. Their health and degree of anxiety are closely linked to their voluntary or involuntary entry to an institution.

It would appear that the improvement of women's place in modern society, a process that began at the turn of the century and matured internationally and collectively in the 70s, was displaying its contradictions and fragility. Finland should be congratulated for being the first country who granted women the right to vote.

Female handicaps accumulate as women grow older, preventing many from becoming social protagonists. Women find themselves caught between tradition and modernity. Nothing, however, should stop them from becoming social and cultural forces in rebuilding the social links in a society that increasingly lacks centrality and integration networks. Women's lives are dominated by a series of questions and issues: the evolution of their sexuality; motherhood; their ability to run everything at once (see Schuller in this Volume) and combine the social and private aspects of their lives; their psychological curiosity. All these endow them with a particular awareness of present-day problems. They have a social and cultural complicity with young people. The wealth of the feminist movement lies in its two-fold dimension (Touraine et al., 1986) the fight for equality on the one hand and the fight for identity on the other hand. Everything is based on the fact that women are different. In other words, it is a matter of respect for specificity and individuality. Also of the search of the meaning of life, which has become so important to us now, in a society that encourages standardisation and focuses on consumption, preventing people from being individuals, and what's more responsible individuals.

DIFFERENCES AND SIMILARITIES BETWEEN MEN AND WOMEN

What will the future bring? How will the differences between elderly men and women change? Will female handicaps grow in number? In sum, will the gap between elderly men and women widen or narrow?

We have to look into the past to define the future of male and female identities, the relationship between men and women and the place and role of children.

We will study men, women and children through the analysis of three types of families. Louis Roussel (1988) has classified families into three different types: the traditional family of the farming society; the modern family of the industrialised society; and the contemporary family, i.e. of the present-day.

In the traditional family, men and women had to "mould" their personalities into different, complementary roles (Illich, 1973). Men were in charge of production, women of reproduction. Their lives were dominated by the struggle against famines, epidemics and death, the urge to preserve tradition and the need to improve filiation. Children were seen as miniature adults (Aries, 1973) and had no individuality. There was a high infant mortality, and a child who reached adulthood was rare indeed. At that time, there was a solidarity between generations as adults covered their aging parents' needs. Paternal authority had an almost religious aspect within the family. It was the royal right, the incarnation of the authority of the divine right on earth. The father had full rights over his children. Punishment was commonplace. Women had no civil rights.

The modern family is more intimate and romantic. Children are slightly more individualised. Developments in sanitary facilities made for great improvements in hygiene. Paternal authority decreased with the establishment of family courts in 1790. The State had more authority over the children: primary education became compulsory for girls and boys in 1881-1882. Girls started to emancipate themselves through education but a woman's duty remained her husband's career (Blunden, 1982).

The contemporary family is characterised by women's emancipation. Women now benefit from civil rights, have the opportunity to work and the possibility to assert themselves as social subjects. With contraception, women can plan their families. Living together now means coming to terms with a partner's identity. Parents hand over more authority over their children to the State, as mixed secondary education becomes compulsory. Psychologists are consulted to solve family and personal problems (De Singly, 1994). Children have become a source of happiness for the parents and the issue consists in offering affection while encouraging independence (Roussel, 1988). Parental authority replaced paternal authority in 1970. The battles over the custody of children during conjugal separations have increased. The role played by the grand-parents has become more diversified and complex against such a background of family uncertainties. With the development of the state pension scheme, children no longer have to cover their elderly parents' needs. Women are asserting their independence and more apply for divorce than men. Men are encouraged to get more involved in the family circle, which results in occasional clashes, ambiguities and contra-dictions from both parts. Sharing complementary roles between both sexes has become an item of negotiation rather than a reflection of well-defined cultural models. The roles are badly defined but the feeling of freedom is greater than before. As a result, stress and anxiety appear. Men may have the impression they have lost their prerogatives and privileges, but they are now entitled to express their feelings (Castelain Meunier, 1988, 1989, 1992). There seems to be a standardisation of both sexes' ways of life, an undifferentiated cultural environment that is approached through different perspectives.

Eino Heikkinen (see in this Volume) thinks that, as far as life expectancy is concerned, the gap between men and women will narrow in the future.

Gerald E. McClearn (see in this Volume) has demonstrated that an androgynous tendency develops between men and women as they grow older. Women gain more autonomy and more responsibilities. Men are less infantilised and become "apprentices" in the family sphere. They sacrifice less their family life for their career. They start to be concerned about the flexibility of working hours mentioned by Tom Schuller (see Schuller in this Volume)

CONCLUSION

Will female handicaps lessen with the narrowing of the gap between men and women, and if so, to what extend? Will older women be more capable of taking over responsibilities in the community? Will men become more integrated in the family sphere and still take on responsibilities as a social and cultural force? It is also the cultural and social reactions of a society faced with the aging problems which is at stake. We have returned to our opening issue: the ternary classification of ages and the refusal to be marginalised. It is a necessary but insufficient condition if older men and women are to be equals.

However, this does not mean that staff who deal with the elderly should not receive adequate training. We can avoid sex, age and cultural "ghettos" if we encourage better communication and more appropriate relationships. As Alain Touraine (1992) suggested in his analysis of the contradictions of the contemporary society, everybody should be allowed to play an active part in the historical evolution of the society, by objecting to:

- identity recess, particularly of women
- instrumentalism which dehumanises and desocialises, and which is particularly appealing to men
- objectivisation which concerns both men and women living in a marketing society which prevents people from asserting themselves as subjects.

In sum, if we remain vigilant and try to blend modernity and tradition, the discrimi-nating differences between men and women will disappear. But these differences also have to be respected, all the more so since authenticity (Giddens, 1992) is now playing a larger part in private lives.

REFERENCES

Asılı, N , 1994, Form of entry, institution and gender: Repercussions upon anxiety in institutionalized elderly patients, paper presented at the XVII International IAUTA Congress, August 1994, Jyvaskyla, Finland
Bidou, C , 1984, *"Les Aventuriers du quotidien", ["Everyday Adventures"]*, PUF, Paris, in French.
Blunden, C , 1982, *"Le travail et la vertu", ["Work and Virtue"]*, Payot, Paris, in French

Castelain Meunier, C , 1988, *"Les hommes aujourd'hui, virilite et identite", ["Men Today, Virility and Identity"]*, Acropole, Paris, in French

Castelain Meunier, C , 1989, *"L'amour en moins, l'apprentissage sentimental' , [' Less Love, Sentimental Initiation"]*, Olivier Orban, Paris, in French

Castelain Meunier, C , 1992, *"Cramponnez-vous les peres Les hommes face a leur femme et a leurs enfants", ["Be Careful Fathers Men Facing Their Wife and Their Children"]*, Albin Michel, Paris, in French

De Singly, 1994, *"Sociologie de la famille contemporaine", ["Sociology upon contemporary Family"]*, Nathan, Paris, in French

Giddens, A , 1992, *"The Transformation of Intimacy, Sexuality, Love and Eroticism in Modern Society"*, Cambridge Polity Press, Cambridge

Guillemard, A -M , Castelain Meunier, C , Vercauteren, R , 1991, *"La retraite en Mutation", ["Retirement in Mutation"]*, Recherche FEN, Paris, in French

Illich, I , 1973, *"La convivialite' , ["Convivielity"]*, Seuil, Paris, in French

Roussel, L , 1988, *'La famille incertaine", ["The Incertain Family"]*, Odile Jacob, Paris

Touraine, A , Gillon, C , Goele, N , Jaquin, D , Castelain Meunier, C , 1986, *"Etude du movements des femmes", ["Study upon Feminist Movement"]*, Rapport, Paris, CEE, in French

Touraine, A , 1992, *"Critique de la modernite", ["Critisism of Modernity"]*, Ed Fayard, Paris, in French

Vannemaa, M , 1994, Elderly women in the Nordic countries, paper presented at the Workshop Gender and Ageing, XVII International IAUTA Congress, August 1994, Jyvaskyla, Finland

21

GENDER, AGING, AND QUALITY OF LIFE

Gerald E. McClearn,[1,2] Pamela J. Maxson,[1] and Debra A. Heller[2]

[1] Center for Developmental and Health Genetics
College of Health and Human Development
The Pennsylvania State University
University Park, PA 16802
[2] Program in Biobehavioral Health
College of Health and Human Development
The Pennsylvania State University
University Park, PA 16802

INTRODUCTION

One of the most striking demographic aspects of aging in developed nations is the sex difference in longevity and in age-related disease. Understanding the determinants of this phenomenon might powerfully illuminate basic aspects of aging processes and provide for the design of rational interventions.

Although the basis of this sex differential is imperfectly understood at present, it seems likely that the underlying mechanisms are complex and heterogeneous. In particular, it seems *un*likely that the causal factors are unidirectionally distributed so that all of the longevity-promoting influences are in females and the longevity-restricting ones in males. Based on these considerations, we suggest that there may exist multiple quantitative dimensions of factors affecting sex-differences in age-related processes, and that the distributions of males and females on these dimensions may overlap. If so, the possibility is raised that scores on these dimensions may prove to have greater predictive value than the simple classification of biological sex.

We have begun a program of research with the objective of identifying some of these dimensions, or what might be called "androgyny continua," using data from the Swedish Adoption/Twin Study of Aging (SATSA). This study is a longitudinal examination of Swedish twins with an average age of about 60 years at the time of the first measurements in 1984. Assessments of the domains of personality, cognition, health, activities of daily living, work and home environment, and basic biomedical status are made by mail-out questionnaires and in-person examinations at three-year intervals (Pedersen et al., 1991).

The research objectives of the present initiative are to define, from various domains of age-related variables, multivariate composite variables that yield overlapping sexual distributions, and to associate individual differences on these dimensions to late-life outcome variables related to health and survival. Longer-term objectives include the estimation of relative contributions of genetic and environmental factors to individual differences on these dimensions.

Preparation for Aging, Edited by E. Heikkinen *et al.*
Plenum Press, New York, 1995

This presentation is a progress report on very preliminary stages of this research program.

Some Androgyny Dimensions

The basic logical requirement for an "androgyny dimension" is, of course, that male and female means should differ. However, to exploit the hypothesized quantitative variability in the underlying processes, there should be substantial variability within each sex, and it is entirely appropriate for the female and male distributions to overlap extensively. Many familiar variables meet these requirements; body height is one obvious example. One present goal is to create composite variables satisfying these criteria from gerontologically relevant data sets in different domains where this sexual dimorphism is less obvious.

DISCRIMINANT STRUCTURE

Our first efforts involve the *personality* and *cognitive* panels of measures from the first SATSA questionnaire, which have been described in detail elsewhere (Pedersen et al., 1991). The putative androgyny dimensions were created simply by generating a discriminant function between male (n=273) and female (n=304) individuals, and by calculating, for each individual, a score based on the compositing weights of this discriminant function. In the following abbreviated descriptions of the structure of the discriminant functions, correlations of key variables with the composite are shown parenthetically.

Cognition. The discriminant structure is oriented so that high scores are more "female-like." Based on the coefficients obtained from our discriminant analysis, a high composite score reflects relatively higher performance on measures of certain *memory* functions (.24 to .29) and on *speed and accuracy* measures (.20) than on crystallized intelligence measures of *analogies* (-.23) and *information* (-.46) and the fluid intelligence measures of *rotation* (-.46) and *figure logic* (-.21). Note that the scale refers to pattern of intellectual performance as well as to overall level.

Personality. On this continuum, the higher, more female-like, scores represent a pattern of relatively lower sense of *directedness* to life (-.36), higher *emotionality* (.41), greater *inhibition of anger* (.74), and greater *fearfulness* (.74) as these terms are defined by these particular test instruments.

Distributions. The distributions of the individual values from these scales are shown in Figure 1. For each continuum, essentially the whole range is represented within each sex, but the distributions differ in central tendency. By predicting everyone with a score above mid-scale to be a female and everyone below mid-scale to be a male, one attains 65% and 75% correct assignment for females and males, respectively, from the personality androgyny dimension and 76% correct assignment for each sex from the cognition androgyny dimension.

Relationships of Androgyny Scores to Age-related Outcomes

The outcome variables of greatest interest are, of course, survival and advanced-age disease status, for which data will become available as the study cohort ages. In the meantime, we can obtain some information from examining the relationship of the androgyny dimensions to some contemporary variables. These relationships can be illustrated by

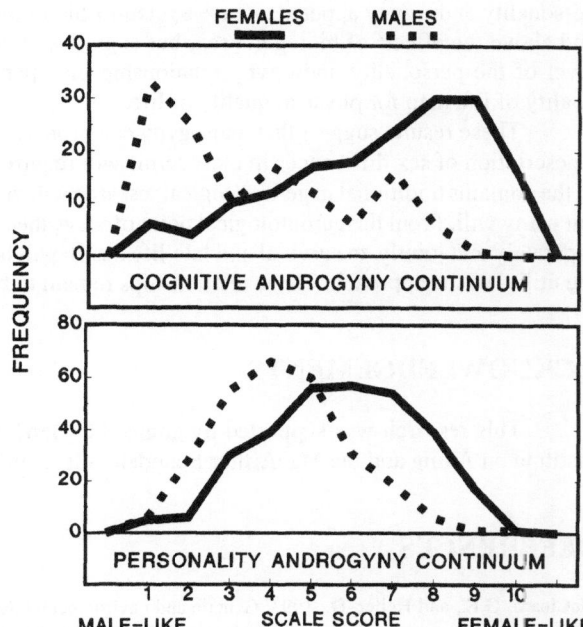

Figure 1. Distributions of biological females and males on the cognitive and the personality androgyny continua.

two composite variables representing, respectively, physical quality of life and psychosocial quality of life (McClearn and Heller, 1993). These composites were derived by factor analysis of a panel of sex-residualized variables including measures of life satisfaction, sum of illnesses, self-rated health, activities of daily living, depressed mood, interpersonal relationships, somaticism, subjective socioeconomic status and perceived social support.

For the whole sample, correlations of .11 and .21 were obtained between the personality androgyny dimension and the factor scores for the Physical (I) and Psychosocial (II) factors, respectively. The relationship could, of course, be different at different ages, so the sample was subdivided into those 50 to 65 years of age and those 65 and older. The correlations for the younger group were non-significant for both factors, whereas that for the older group was .14 and .24 for Factors I and II, respectively. It is of particular interest to examine the relationships of androgyny continua *within* biological sexes. Within biological females, the relationship of the personality androgyny score to Factor I was not significant, but the correlation with Factor II was .22. The corresponding values for biological males were .16 and .25, respectively.

None of these correlations are large, but it should be reiterated that the outcome measures were based on sex-residualized scores, so these values represent relationships of the hypothesized processes underlying the androgyny continua after the effects of biological sex *per se* have been statistically removed.

No significant correlation was obtained between the cognitive androgyny dimension and either of the factors in any of the groupings described above.

SUMMARY

Conclusions from these preliminary results must be guarded, naturally, but there are hints as to the type of results that may be obtained with further analyses. For example, personality androgyny appears to have a relationship to these quality of life variables, over and above the effects of biological sex, but cognitive androgyny does not, and the overall level of the personality androgyny relationship may perhaps be higher for psychosocial quality of life than for physical quality of life.

These results suggest that androgyny can properly be regarded as complex and that a description of sex differences in these terms will require a number of dimensions. Not all of the domains traditional in gerontological research will yield useful androgyny dimensions, but many will. From the gerontological point of view, the relationships of these dimensions, separately and jointly, to survival and late-life health will be key considerations in assessing the utility of the approach. These relationships remain to be elucidated.

ACKNOWLEDGEMENTS

This research was supported by grants AG04563 and AG10175 from the National Institute on Aging and the MacArthur Foundation Research Network on Successful Aging.

REFERENCES

McClearn, G.E., and Heller, D., 1993, Genetic and environmental determinants of quality of life of elderly people, paper presented at WHO Workshop on Quality of Life of Elderly People, Jyväskylä, Finland, 26-28 May 1993.

Pedersen, N.L., McClearn, G.E., Plomin, R., Nesselroade, J.R., Berg, S., and deFaire, U., 1991, The Swedish Adoption Twin Study of Aging: An update, *Acta. Genet. Med. Gemellol.* 40:7-20 .

22

COLLECTIVISM, INDIVIDUALISM AND GRANDCHILD-GRANDPARENT RELATIONS

Helena Hurme

Åbo Akademi University
PB 311
FIN-65101 Vasa
Finland

INTRODUCTION

Since the beginning of the eighties, there has been a steady rise in the number of studies concerned with grandchild-grandparent relations. Most of the studies have been carried out in the US , but there are studies from Europe as well. Direct cross-cultural comparisons have been peculiarly rare, however, with the exception of Apple's (1956) anthropological analysis and McCready's (1985) article which compared Americans of, among others, Scandinavian and Polish origin. Nahemow (1984) compared two traditional societies in Africa.

The hidden assumption seems to have been that in "Western" cultures, these family relations are more or less identical. This was one of the reasons for carrying out the comparative study of grandparent-grandchild relations in Finland and Poland reported below *.

The lack of cross-cultural comparisons is all the more surprising when one takes into account that Hofstede (1984) in his study with 117 000 IBM employees from all around the world showed that also "Western" cultures vary on several dimensions. One of them is collectivism versus individualism. Hofstede showed that Finland was quite high on individualism but not as high as the US. Poland was not included in the study. Of the Slavic countries in Europe which might resemble Poland most, Yugoslavia was quite high on collectivism.

Triandis (1990, p. 42) defines individualism as follows:

> In individualistic cultures most people's social behavior is largely determined by personal goals that overlap only slightly with such as the family, the work group, the tribe, political allies, coreligionists, fellow countrymen and the state When conflict arises between personal and group

---

* The study reported here was performed in collaboration with the late professor Maria Tyszkowa from the Adam Mickiewicz University in Poznan, Poland, who died before its completion The data collection was planned together and supervised by both of us The author has analyzed the material and written this text When the words we or our etc is used in the text, it refers to thoughts which were developed together (The word author refers to Helena Hurme)

goals, it is considered acceptable for the individual to place personal goals ahead of collective goals. (p. 42)

Collectivism is the opposite; if conflict arises, the group goals take precedence. Triandis (1990) adds that one should not think of them as one dimension as they may coexist in a certain culture. Triandis et al. (1988) come to the conclusion that collectivism concerns the in-group and not everyone in society. This is an important factor with reference to the family.

According to Triandis (1990), the most important contrast between individualistic and collectivistic societies is a factor called "Family integration", with items like "Aging parents should live at home with their children", which is typical of a collectivistic orientation. Triandis also mentions results which show that small distance and little emotional detachment from their relatives, broad family concerns and more emphasis on family integrity and cooperation is more typical for collectivists than for individualists.

Socialization toward obedience is associated with collectivism (Triandis, 1990) and the description of the self of collectivists are more often in terms of "We" than "I" (Reykowski and Smolenska, 1993). Van den Heuvel et al. (1992) found that children from collectivistic cultures referred more to social aspects when describing the self than did children from an individualistic culture.

The assumption in this study is that Poles support an extended family model more strongly than Finns and they show more family integration in the form of more contacts and cooperation between the generations and less emotional attachment from their kin. Authority is also assumed to rest more with the older generations. Socialization is supposed to be more directed towards obedience and there is rather a We orientation than an I orientation.

METHODS AND SAMPLE

Two cities, Poznan in Poland and Vasa in Finland were chosen for data collection. 417 Finnish and 300 Polish school children and adolescents between 11 and 20 wrote an essay on their grandparents which were then content analyzed and 79 Finnish and 40 Polish preschool children were interviewed using a semi-structured method. They also drew their family and their grandparents. Later, 105 of the Finnish adolescents were interviewed and 135 Finnish and 53 Polish adolescents completed incomplete sentences concerning the grandparents. A random sample of the parents and grandparents of these children got a questionnaire concerning intergenerational relations. *. 146 Finnish and 58 Polish parents and 102 Finnish and 73 Polish grandparents completed the questionnaire.

RESULTS

Family Interaction and Integration

The Polish and Finnish adults of both generations differed highly significantly on most of the items concerning intergenerational norms. The Poles accord the grandparents more rights in relation to their children and grandchildren but also more duties in relation to them. The same concerns the younger generation. Thus the authority rests with more often with the family and its leader and not with the individual in Poland as compared with Finland. The family drawings of 4-7-year-old Polish children showed up to 23 different person while

---

* (An earlier version has been used in Hurme 1988)

the typical Finnish family drawing concerned the nuclear family with four to five persons. This difference in conceptions of the family is also seen in the Finnish adolescent interviews, where only around 20 per cent include the grandparents in the concept of the family. The representations in the mind of the Finnish and Polish adolescents concerning their grandparents differ highly significantly in many respects: the Polish adolescents describe their grandparents using more personality descriptions, more descriptions of their emotions and skills and more positive expression as well as mentions of more common activities and getting something emotional and/or intellectual from them whereas the Finns use more descriptions of their appearance and their activities. These differences are seen both when using compositions and incomplete sentences. For most dependent variables, the country variable increases prediction significantly when the effect of age, sex and geographical distance is taken into account. Preschool interviews point in the same direction. The crucial variables seems to be contact frequency and common activities: when they were controlled for, the country differences disappeared for emotionally related variables and was smaller for the other variables in the compositions. The Polish adolescents mentioned significantly more contacts than the Finns and this result was corroborated by data from the questionnaires. In families where the grandmother lived less than 15 minutes away, the Poles still had more frequent contact with them.

There were significant differences between the Finnish and Polish subjects as to the patterns of help and exchange between the generations, Polish adult children considering practical help from their parents as more important than the Finns, who again consider emotional help as more important. Polish adult children considered practical help, financial help and help in illness as more important forms of help to their parents than the Finns who consider emotional help/support to be more important than the Poles.

Socialization towards independence versus obedience

The Polish and Finnish parents differ highly significantly as to their conceptions of the goals and methods of child rearing. When 79.2% of the Finns put discipline in the three lowest places, 37% of the Poles did. The Finns again stress self-fulfillment more. When 68.3% of the Finns have placed it in seventh place or higher, only 37.2% of the Poles have done this. The Finns also stress independence significantly more than do the Poles. Polish parents consider physical sanctions as more important than the Finns. 8.5% of the Finns and 20% of the Poles place them in fourth place of importance or higher. The Poles also consider creating rules as more important. 34.6% of them has placed this in first place as opposed to 6.1% of the Finns.

Orientation towards "I" versus "We"

The tendency to refer to social aspects when describing the self was reflected in the sentence completion tasks of the Polish adolescents in this study where the stem "In my opinion my grandmother..." was answered by a reference to the relationship of the two in 48.9% of the Polish cases as compared with 15.7% of the Finnish (p<.001). The Finns again in 27.5% of the cases stressed her appearance whereas none of the Poles did so (p<.001).

CONCLUSIONS

The results show that the concepts of individualism and collectivism are central in explaining differences in grandparent-grandchild relations and intergenerational relations in general also between two European countries. They also show that contact frequency distinguishes individualistic and collectivistic cultures. One can assume that an interactional style is transmitted between generations which in turn is reflected in how the grandparents are depicted. In the more collectivistic, Polish culture, contacts between grandparents and grandchildren are more frequent than in Finland even when distance has been taken into account and the Poles describe their grandparents in more personal and positive terms.These results should be followed up in larger cross-cultural studies with measurement of the individual form of individualism-collectivism, viz. allocentrism vs idiocentrism (c.f. Triandis et al. 1986, Triandis, 1990) as well as standardized measures of the grandparent-grandchild relationship.

REFERENCES

Apple, D , 1956, The social structure of grandparenthood, *Am Anthrop* 58 656-663

Hofstede, G , 1984, *"Culture's Consequences International Differences in Work-Related Values,"* Abridged edition, Sage, Beverly Hills

Hurme, H , 1988, *"Child, Mother and Grandmother Intergenerational Interaction in Finnish Families,"* Jyvaskyla Studies in Education, Psychology and Social Research, 64, University of Jyvaskyla, Jyvaskyla

Nahemow, N R , 1984, Grandparenthood in Transition, in *"Life-Span Developmental Psychology Historical and Developmental Effects,"* K A McCluskey, and H W Reese, eds , Academic Press, New York

McCready, W , 1985, Styles of grandparenting among white ethnics, in *"Grandparenthood,"* V C Bengtson, and J F Robertson, eds , Sage, Beverly Hills

Reykowski, J , and Smolenska, Z , 1993, Collectivism, individualism and interpretation of social change limitations of a simplistic model, *Polish Psychological Bulletin*, 24 89-107

Triandis, H , 1990, Cross-Cultural Studies of Individualism and Collectivism, in *"Cross-Cultural Perspectives,"* J J Berman, ed , Nebraska Symposium of motivation, vol 37, University of Nebraska Press, Lincoln

Triandis, H C , Bontempo, R , Villareal, M , Asai, M , and Lucca, N , 1988, Individualis-collectivism Cross-cultural perspectives on self-ingroup relationships, *Journal of Personality and Social Psychology*, 54 323-338

Triandis, H C , Leung, K , Villareal, M , and Clark, F , 1985, Allocentric versus idiocentric tendencies convergent and discriminant validation, *J Res Pers* , 19 395-415

van den Heuvel, H , Tellegen G , and Koomen, W , 1992, Cultural differences in the use of psychological and social characteristics in children's self-understanding, *Eur J Soc P*, 22 353-362

23

MID-LIFE

Opening Remarks on Prevention

Hélène Reboul

Lumière University Lyon 2
Lyon, France

The topic of preparation for aging is an extension of the concerns and of the researches carried out by members of the Scientific Comittee of I.A.U.T.A. on retirement, addressed by our colleagues *Jeanneret* (Neuchatel) and *Lemieux* (Montreal), inter alia, during the *Hull Congress* (at the University of Quebec, Canada), in 1990.

For looking at the way one cares for one's own body, particularly through exercise and movement, contemplating retraining and rehabilitation in case of physical ailment, was the subject of the *Barcelona* (Spain) Congress in 1992.

The topic now being proposed gives us the possibility of contemplating the time preceding retirement with the awareness of a more or less marked progress of aging. The sequence of these topics shows the consistency of I.A.U.T.A.'s Scientific Council's strategy in organizing its congress. The underlying concern here has to do with the field of gerontology, as we all know that our Third Age Universities are helping each student to achieve a freer, more harmonious aging process.

For me as a Professor of Gerontologic Psychology, the interest of this subject lies in personal development based on my professional practise over a 40-year period:

- early in my career, my concern was directed at the elderly who were either physically and/or psychologically dependent. Wasn't it a way of exorcizing this coming event that upsets all who are no longer young ?

Then the setting up and putting in place of the Third Age University in Lyon (in 1975) enabled me to meet and learn about active pensioned employees, and even younger people who wished to age gracefully, thanks to the cultural achievements and the conviviality prevailing in our classes.

And now, after five years, having added to my knowledge of the aging process and how it contributes to life, I realize that scientific interest should focus on the time immediately preceding retirement, starting at midlife. Thus, I became aware—and convinced—that both medical and social gerontologists seem to want to extend their domain and their "area" of influence. In fact, their only concern is to contribute to a better aging of the population, to encourage a better adjustment to the progress of age, which is dealt with by prevention. But first, let us try and define what is *midlife*.

Preparation for Aging, Edited by E Heikkinen *et al*
Plenum Press, New York, 1995

The first question to be asked is "When does it happen?". The answer will depend on the time one estimates for one's lifetime—80, 90, 100 years or more. Usually, when people are nearing 40, 45 or 50, they wish to push that often critical threshold forward, since it can cause a crisis very aptly described by Jacques Eliot (1974). It must be pointed out that in France, where criticism is very strong, round numbers in decades have a special attraction and are considered as thresholds to reach and cross. In other words, "midlife" is determined by each individual as a function of that individual's personality, his/her original culture and his/her genetic assets, which causes some to live better in certain families while in others, for instance, there are cardiac or circulatory illnesses.

There is, then, an individual perception often conditioned by the family environment. However, in the merciless and uncompromising working environment one's view on aging is more deterministic: 45 year old males are deprived of the possibility of night work or alternative working schedules.

In France, the number of wage-earning trainees is dwindling. Thus, those over 45 account for no more than 5% of training employees, while numerically speaking, forty-fivers stand for a significant category, that of the post-war baby-boomers (II WW). We must stress the impact of these considerations whose ramifications reach the retired for many reasons: learning new things may seem unrealistic when we are dealing with a population of retired people, as we do in our Third Age Universities.

Facts concerning aging in the working environment have been addressed in a Colloquium called *"Santé, travail et vieillissement. Age et entreprise, un enjeu de société."* (Health, Work and Ageing. Age and the Organization, a Society Topic) published in the *Archives des maladies professionnelles* (Archives on Working Diseases) in 1993 - n° 3 under the coordination of sociologist Xavier Gaullier (1993).

The concept thus expressed shed light an all negative bromides and stereotypes against which an aging person should struggle at a certain point in time when he may no longer wish to fight on or fight back.

In other words, that means it is up to each individual to find the means required to advance and develop.

TO PROGRESS AFTER MID-LIFE

Going forward often requires going over one's past life.

This enables one to see where one started off, with what ideals, as the result of a contract with oneself. This is an assessment that takes into account gains and benefits as well as losses and shortfalls. In other words, one takes stock of one's life, and this should allow one to leave one's past behind to be able to use the time to come in a continuum of achievements. This also enables one to compare what is left or becomes important and what is over. One's life would then be guided by this new eventual happening. It would make it easier to find consistency between the view of oneself and the look that one's own society sheds on those who have overstepped the threshold of midlife.

In my view, the perception of this consistency, in harmony with the environment, could make aging more fruitful, knowing however that nothing is achieved forever and that the life project one has chosen for oneself may be changed by social, economic and political circumstances, as well as by unpredictable family developments.

This is why, in our Universities for All Ages—or Interage Universities as we call them—we are asked to provide courses in hygiene, dietetics, biological and philosophical basics: so as to acquire the personal tools needed to better manage daily life, aiming at prolonging it under the best possible health and socialization conditions. However, this progress can occur amid greater or lesser difficulties, and trigger a midlife crisis, like the crisis of adolescence or that of retirement.

THE MID-LIFE CRISIS

It can come up in several ways at different times, with various manifestations. This is what we have been researching in Lyon, and that my colleague Prof. Zilma Cavalcante also investigates at the Ceará State University in Fortaleza, Brazil, and at the University Without Boundaries (either social or age). In Lyon, men and women from 40 to 60 were interviewed ; with women giving more answers as a rule. The interviews were carried out at the respondent home, either in the evening or during the weekend ! and were broken down into three parts :

1. The aging of parents and what considerations are elicited ;
2. The respondents' own health status, drugs taken, insomnia, etc...
3. The respondents' own idea of his/her aging with regard to their mates and children.

Women as a whole were happy to have an interview that helped them be aware of their own progress. Men, on the other hand, became defensive to the point of denial, since, their crises would come up later than menopause, at the age of andropause.

One often connects the emergence of puberty with that of menopause. The former can bring anxiety because sexual intercourse occurs increasingly early and can result in undesired pregnancy among teenagers. In France, menopause is considered freeing, although it means the end of fertility. In Brazil, where the body is experienced differently, these feelings are further heightened.

For both sexes, identification with an elderly parent brings about thoughts about one's aging as compared to that of one's father or mother.

Elements of affection transpire through statements about physical aspects, all the more so in France because the number of divorces is estimated at one for every three marriages. Midlife often corresponds to 15, 20, 25 years of married life. That is, at a time when a certain weariness occurs. Once the children leave home, the couple is left face-to-face with a new life project to build. Within this context, another element can be introduced with the taking in charge of elderly parents requiring help and assistance... which may change the marriage relationship at this sensitive time.

My Brazilian colleague shows in her survey that there are three ways to respond to midlife crisis:

- overactivity everywhere, particularly at work,
- mysticism, that either alienates the individual or favours an inner life. To quote Saint Paul, "the outer man is destroyed, the inner man is renewed".
- the search for a new social identity.

So, the characteristic of this stage, in a number of cases, is the need to reorganize daily realities, which helps to become aware that the crisis - sometimes a painful passage— allows or has allowed an enrichment of life prospects.

To conclude, I would like to quote the authors of a small book, a recent one maybe, called *Les Histoires de Vie* (Life's Own Stories) (Pineau and Le Grand, 1993) "Making one's life has never been an easy thing. Neither is making a living. Even less understanding it. The end of the millenium does not change these vital difficulties. While the course of human life is now enriched by new possibilities and new horizons, it is also going through a bio-ethical revolution where birth and death, organisms and the environment must be adjusted to the biosphere and biogenetics."

So, every aging individual who is part of a social group must lead his own life as best he can, for his own sake but at the same time with a responsibility before the coming generations to which one should convey the desire and the pleasure of aging.

REFERENCES

Eliot, J , 1974, *"Mort et crise du milieu de la vie in Psychoanalyse du genie createur"*, Dunod, in French
Gaullier, X , 1993, Sante, travail et vieillissement Age et entreprise, un enjeu de societe [Health, work and ageing Age and the organization, a society topic] *Arch Mal Prof* 54(3) 1983-208, in French
Pineau, G , and Le Grand, J -L , 1993, "Histoires de Vie" *[Life's stories]*, *Que sais-je* Vol 2760, PUF, 127, Paris, in French

24

LIFE-STYLE AND ITS DETERMINANTS IN TWO COHORTS IN THE ELDERLY

Pertti Pohjolainen

Kuntokallio, Center for Gerontological Training and Research
FIN-01100 Östersundom
Finland

INTRODUCTION

It has been found in earlier research that certain components of life-style are crucial to aging and health status. There are many different approaches to defining the concept of life-style (Berbalk and Hahn, 1980; Taylor and Ford, 1981; Tokarski, 1985). In this study life-style is defined as a theoretical category consisting of models of behaviour and the choices of an individual as a member of society (Pohjolainen, 1991). The different areas of life-style include, for example, living habits, social participation, hobbies and life-satisfaction (Fig. 1). Life-style consists of two main components, the objective and subjective. In the elderly life-style is determined by a number of different factors: socio-economic status, health, aging processes, earlier life history, genetic factors, etc.

The results of the study project "The Elderly in Eleven Countries" tend to show that the differences between age groups can only partly be explained by age changes (Heikkinen et al., 1983; Waters et al., 1989). In many areas the differences are due to the changes in cultural, social or urbanizational factors. These are so-called cohort differences. Cross-sectional studies do not reveal cohort differences. We need a research design in which persons at the same age are compared with each other in different years. This means a cohort comparison study.

The purpose of this study was to describe the differences in life-style variables between two cohorts, and find out how background factors were related to some components of life-style.

MATERIAL AND METHODS

The population of this study, which is a part of a more comprehensive study project "The European Longitudinal Study on Aging (ELSA)", consisted of two cohorts (born 1910-19 and 1920-29). In the cohort comparison the 60-69-year-olds in 1979 (185 men, 179 women) were compared with the 60-69-year-olds in 1989 (208 men, 187 women) (Fig 2).

Preparation for Aging, Edited by E. Heikkinen *et al.*
Plenum Press, New York, 1995

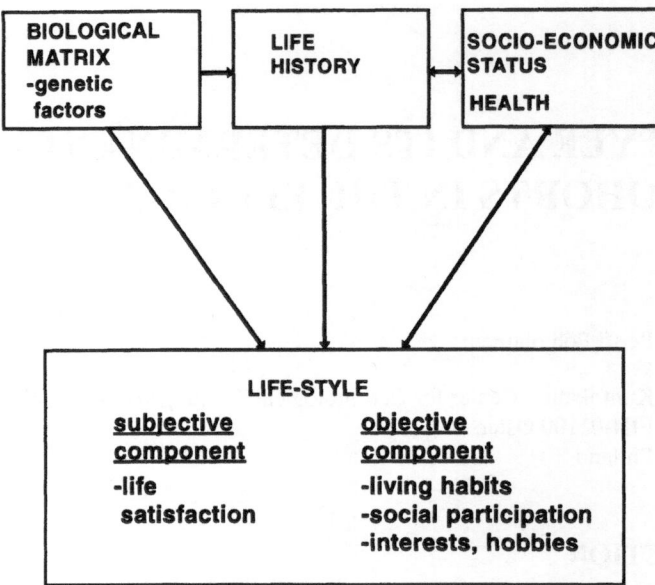

Figure 1. Model of life-style and its determinants in the elderly.

The subjects were residents of the city of Tampere. Interviews were used as a study method. The statistical method used was chi-square analysis.

The life-style variables in this study were living habits (smoking, alcohol consumption and physical exercise), hobbies, social participation and life-satisfaction. Socio-economic status was defined on the basis of education and occupational status. The variables describing health status were functional ability (ADL-index) and self-rated health.

RESULTS

The most distinct differences between the cohorts in living habit variables were in alcohol consumption for both men and women (Fig. 3 and 4). In particular alcohol drinking had increased among women during the previous decade. In smoking or physical exercise there were no significant cohort differences.

Formal social participation (active participation in various societies and organizations) among the 60-69-year-olds was somewhat more extensive than among the same age group ten years earlier. In addition, informal social participation, described in this study by participation in different social occasions, had increased somewhat in the later cohort. The largest increase was seen in visits to public libraries and to foreign travel. In 1979, 30% of the men had visited libraries but in 1989 the figure was 44%. Among the women the respective percentages were 25% and 48%. In 1979, 26% of the men had travelled abroad but ten years later this figure was 56%. Among women the respective percentages were 32% and 53%.

In 1989, 60-69-year-olds were generally more satisfied with their lives than persons of the same age in 1979. Of the men, 70% were satisfied with their lives but ten years later this figure was 80%. Among the women the percentage of satisfied persons had increased from 65% to 77%. Satisfaction with economic status, especially, had increased.

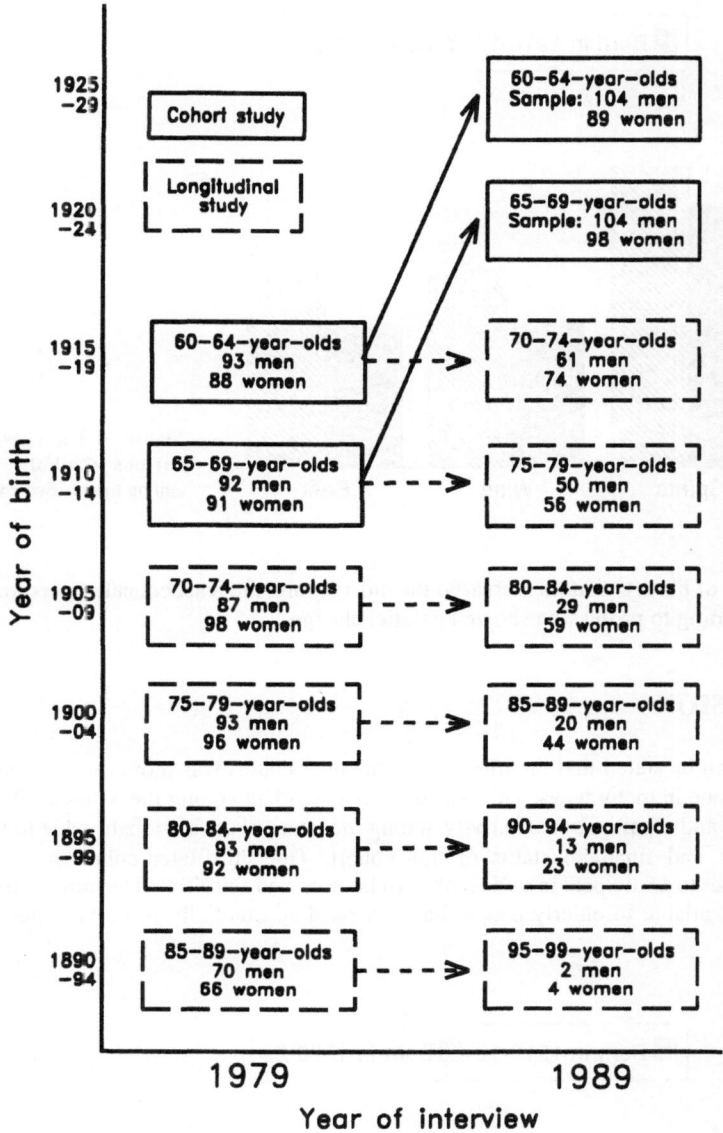

Figure 2. Study design of the Tampere Longitudinal Study on Ageing.

Health variables (self-rated health and functional ability) had a connection with physical exercise in both cohorts among both sexes (Table 1). Those who reported good health exercised the most. Those who had a good functional ability (ADL-index) exercised more than those with a fair or poor functional ability.

Level of education and occupational status correlated with formal and informal social participation in both cohorts among the men and the women. As an example of social participation was "visiting libraries" (Table 2). Among the men with the most advanced education, 47% visited libraries in 1979. In 1989 this figure was 69%. Among the men with least education the respective percentages were 0% and 42%. Also among the women the

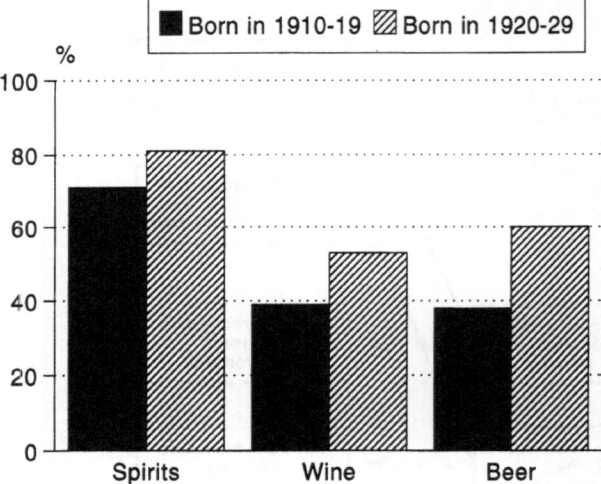

Figure 3. The consumption of various alcoholic beverages among men in two cohorts.

percentage of library visitors increased the most among the least educated persons. Assessment according to profession showed parallel changes.

DISCUSSION

It can be stated that the life-style of the later cohort was more "active" than that of the earlier one in many ways. This was especially evident among the women. The increase in interests and organizational activity among the later cohort is probably due to the higher educational and financial status of this cohort. Thus, the later cohort has had better opportunities to participate in a wide range of hobbies and activities. The number of different activities available to elderly people has increased substantially in Finland during recent decades.

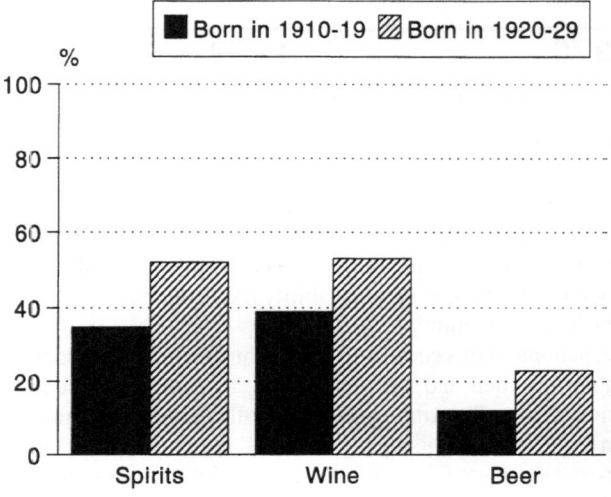

Figure 4. The consumption of various alcoholic beverages among women in two cohorts.

Table 1. The percentage (%) of men and women exercising physical activity in various cohorts according to health variables

	Men		Women	
	Born 1910-19	Born 1920-29	Born 1910-19	Born 1920-29
Self-rated health				
good	81	80	73	85
average	68	81	73	78
poor	42	65	55	38
Functional ability				
good	75	82	80	84
moderate/poor	40	68	53	63

Table 2. The percentage (%) of men and women visiting libraries in various cohorts according to education and occupation

	Men		Women	
	Born 1910-19	Born 1920-29	Born 1910-19	Born 1920-29
Educational level				
less than primary	0	42	8	30
primary	31	39	25	44
more than primary	47	69	60	68
Occupational status				
white-collar	33	55	47	59
blue-collar	29	37	17	44
housewife	---	—	10	29

The differences between the cohorts detected in this study may be due to both cohort and period effects. For example, new age groups reaching pensionable age have already assumed different styles of alcohol consumption from those of the previous generations. Changing social trends (e.g. more general consumption of alcohol) also change the life-styles of the elderly.

The greater life-satisfaction among the later cohort may be explained by increased social activity. A number of earlier studies have shown a clear connection between activity and life-satisfaction. Another explanatory factor may be the improved economic status of the later cohort due to better pension systems.

The significance of socio-economic status as a determinant of life-style was greater in the earlier cohort than the later one. The present study indicated that educational level, occupational status and health status were important determinants of life-style in the elderly. The results of this study show that many of the differences observed between the age groups are due to the so-called cohort and/or period effect, and not caused by actual aging.

REFERENCES

Berbalk, H., and Hahn, K.D., 1980, Lebensstil, psychisch-somatische Anpassung und klinisch-psychologische Intervention [Life-Style, Psycho-Somatic Adjustment and Clinical-Psychological Intervention], in: *"Klinische Psychologie - Trends in Forschung und Praxis" ["A Clinical Psychology—Trends in Research and Practice"]*, Bauman, U. et al., eds, Band 3, Thieme, Stuttgart, pp. 22-71.

Heikkinen, E , Waters, W E , and Brzezinski, Z J , eds, 1983, *"The Elderly in Eleven Countries A Sociomedical Survey"*, Public Health in Europe 21, WHO, Copenhagen

Pohjolainen, P , 1991, Social participation and life-style A longitudinal and cohort study, *JCCG*, 6 109-117

Taylor, R, and Ford, G , 1981, Lifestyle and ageing Three traditions in lifestyle research, *Ageing and Society* 1 329-345

Tokarski, W , 1985, Freizeitstile im Alter uber die Notwendigkeit und Moglichkeiten einer Analyse der Freizeit Alterer [Leisure styles of the elderly necessity and possibilities of an analysis of old people's leisure time], *Z Gerontol* 18 72-75

Waters, W E , Heikkinen, E , and Dontas, A S , eds , 1989, *"Health, Lifestyles and Services for the Elderly"*, Public Health in Europe 29, WHO, Copenhagen

25

PSYCHOLOGICAL ISSUES OF AGING AND WORK

Pekka Huuhtanen

Institute of Occupational Health
Department of Psychology
Topeliuksenk. 41 aA
FIN-00250 Helsinki
Finland

INTRODUCTION

Increased use of modern technology and changes in the boundaries between work and non-work periods of people are common trends in the current working life. Simultaneously, the increasing average age of the population will be a challenge for health and manpower policies in most industrialized countries. The oldest workforces of OECD countries in the year 2000 will be in Finland, Japan, Luxembourg, Sweden and Switzerland, where more than 40% of the working population will be older than 45 years (Aging and work capacity, 1993). Simultaneously, early exit from work and early retirement has been increasing. In Finland, the ratio of people working to those retired is estimated to increase from 5/2 to 2/5 between 1990 and 2030. At the same time, industry will be able to recruit fewer younger workers.

The relationship between aging and work capacity is increasing in importance because of these demographic trends. During middle and late adulthood, significant individual differences exist for life history, work experiences, aspirations, and health conditions. The work capacity of older people is often incompatible with work demands, which can lead to stress and health problems. Age-related changes occur in human physiological and psychological functions, in attitudes and ways of learning, and in the acquisition of new skills with increasing age (McEvoy and Cascio, 1989; Rhodes, 1983). Changes in one's work and in the new technology have both positive and negative impacts on stress and well-being, depending on the way changes and learning possibilities are arranged (e.g. Czaja, 1988; Huuhtanen et al., 1991; Huuhtanen and Leino, 1992). The effects of new technology are mediated by the existing organization of work. The decisions about the division of tasks between computers and people are made in connection with the system design process.

These trends suggest that the aging workers and ways to address them will be an emergent issue, as regards quality of working life and job stress in the future. Too little thought and planning has yet been directed to the later stages of working life. Most studies

Preparation for Aging, Edited by E. Heikkinen *et al.*
Plenum Press, New York, 1995

on physical and mental capabilities are based on the comparison of persons over 60 with 20-year-olds, neglecting middle-aged people (Ageing at work, 1993).

Generation Gap

Structural changes in the labour market, economic recession and the high rate of unemployment in Europe have raised the issue of intergenerational equity, the competition for resources between age groups or generations. On the societal level, retirement has been a cornerstone of moral economy, a right that can be claimed after a long working career (Kohli and Rein, 1991). At the work place level, variability in age and differences between age cohorts can create conflicts due to differences in training, experience, values, and basis of work motivation. The determinants of job satisfaction and organizational commitment may vary, the skills of elderly workers may become obsolete. The financial squeeze of the welfare states can lead to a hardening view of the relationship between age and declining capabilities.

Attitudes toward aging workers are contradictory, characterized by organizational ambivalence. On the one hand, older workers usually earn more than comparable younger ones, but are on the average less well educated, and trained on old technologies. As a consequence, they are less flexible and the costs of requalification are higher (Kohli and Rein, 1991; Laczko and Phillipson, 1991). On the other hand, numerous studies (see e.g. Rhodes, 1983) have consistently shown that age is positively related to overall job satisfaction, job involvement, work motivation, and negatively with turnover.

NEW PATTERNS OF EARLY RETIREMENT

Explanations for the transition from work to early exit and retirement have focused either on 'pull' or 'push' factors (e.g. Beehr, 1986). Changes in the organization of work, work contents and the work environment, high rates of unemployment and rationalization can have a pushing effect toward early retirement.

According to the pull view, increased interest in early retirement is explained as a result of social policies that have created attractive exit possibilities. These include early old-age public pensions, new institutional pathways to retirement, lowering of age boundaries, and increased pension incentives (Kohli and Rein, 1991). The development of statutory pension schemes is related to the country-specific characteristics of political, social and economic structures (Kilbom, 1992; Salminen, 1993).

Prior to the 1970s, most research on retirement focused on financial and health factors as antecedents of retirement decisions (Hanisch and Hulin, 1990). Research conducted in the 1970s and 1980s found that individuals' job satisfaction and the importance of their jobs were related to the age at which they retired. The changing balance between work and retirement and the boundaries between work and non-work have generated new research activities (Floyd et al., 1992). Retirement as a stage of life is a new historical phenomena. Since the 1960s, a new patterning of life course and the lengthening of retirement has prompted social scientists to find new theoretical explanations for aging and life and career stages.

The transition from work to retirement has become less clear-cut, more a matter of personal timing. A better health status and lengthening of pension time have de-standardized the modern tripartition of the life stages: pension is no longer a restricted left-over period of life after 'active' work.

Because the 'third age' can last even one third of the whole human life-span, attention has been paid to the planning for it. This planning of retirement should begin early enough in the pre-retirement phase. In Finland, Tikkanen and Kuusinen found that for their sample of persons born in the year 1927 and 1929 (age 60 or 62), retirement was a wanted and long waited life-event by majority of the participants in the study. There were still persons who would like to continue working. Authors suggest more flexible retirement systems together with retirement planning (Tikkanen and Kuusinen, 1992).

Retirement Process as a Part of Individual's Life Career

On the individual level, retirement is a time-consuming process affected by both personal and environmental factors. It proceeds through the stages of preference to retire, decision to retire, and act of retirement (Beehr, 1986). Life stage and life career development theories (e.g. Super, 1980; Levinson, 1986; Ornstein et al., 1989) emphasize multidirection-ality of personal development in adulthood, consisting of a series of transitions brought about by experiences. Major life events alter the individual's social roles, personal identity, goals, and expectations. Some social roles are beginning early in life, e.g. that of child, and others late in life, e.g. that of pensioner (Super, 1980).

Decision points occur before and at the time of taking on a new role, of giving up an old role, and of making significant changes in the nature of an existing role. One important decision point is to make the decision as to whether, when, how, and where to retire. The actual decision-making process consists of the periods of anticipation, planning, action, and adaptation at each major worker career decision points. Coping with the developmental tasks of one life stage is basic to coping with those of the next life stage. It has been shown a real though imperfect relationship between vocational success and satisfaction in the preretire-ment years and satisfaction with one's life in retirement (Super, 1980).

Studies of adult development for older persons have often neglected the role of the retirement experience or have limited their analysis to a single aspect of retirement such as global satisfaction or economic changes. Understanding the transition from work to retire-ment can be attained by analysing past experiences and feelings surrounding the transition, present satisfaction in retirement, and prospects for future adjustment (Floyd et al., 1992). Too often, theories of retirement are theories about the general process of aging. More elaborated psychological theories of different forms of retirement are needed, because in different phases of the transition from work to retirement, pull and push factors may have different psychological connotations.

AGING AND PSYCHOLOGICAL FUNCTIONS

The relationship between aging and work behaviour is reciprocal. The individual develops during his entire work and life history. Psychological functional capability, abili-ties, needs and attitudes change as people get older. The direction and speed of the changes vary between individuals and age cohorts.

The fundamental problem in analysing the relationship between age and work is, how to make distinction between the role and impacts of individual aging, age cohorts, and periodic factors. Most studies on e.g. intelligence and aptitudes are based on cross-sectional data, comparing test results between age groups at the same time. In these study samples, persons under the age of 30 years have quite different training than those over 60 or 70 years of age. This in turn may affect the test results more than individual aging. Even follow-up studies have some problems which make reliable and valid comparisons difficult. The main problem is the drop-out of persons under the study period, which leads to the selection bias in the study sample.

Aging and Work Behaviour

Most studies regarding aging and work are dealing with the attitudes of workers toward work and organization As was noted earlier, most studies reveal that job satisfaction, work motivation and commitment to organization tend to grow with advancing age (e g Rhodes, 1983) However, selection bias should be remembered those most dissatisfied might have leaved the organization earlier

By psychological functional capacity has been normally referred to individual's capacity to cope with every day tasks demanding intellectual and other mental efforts Psychological work ability is always associated with the demands of the work and working environment This relationship is culturally and socially determined, associated with the values, economic constraints, and technical development People in different age cohorts have to cope with different kind of working life This means that also the definitions of intelligence and corresponding testing methods are not developed in isolation They are changing over time and are connected with the education system and concepts of the nature of the human beings

As regards psycho-motor performance, we are dealing with the human information processing This can be divided into attention, cognitive system like memory, thinking and decision making, and motor performance By aging, typical changes happen in sensory functions (loss, impairment of vision, hear) Fortunately, these can be corrected by helping aids, facilities, and technical tools

In the human work performance, also anticipation and training have impact on the every day work situations Experienced workers can effectively use their former experiences in new situations, on the condition that they have had possibilities to learn and effectively solve problems during their work-life There are, however, marked differences between different occupations in this respect According to the polarization hypothesis, computerization has led to both more and less skilled tasks, depending on the position of the worker in the work process Health and stress complaints have been most often associated with a lack of independence and intense VDT use in office tasks

The most often discussed psychological issue in connection with aging at work is the possible deterioration of memory functions and learning capabilities Researchers have started to make a distinction between fluent vs crystallized intelligence (e g verbal capacity and reasoning, based on the knowledge of culture) People are using different learning and decision making strategies These are based on formal rules among younger people and on experience and wisdom among elderly workers

In addition, metacognition and self-concept play an important role in learning situations Those with less formal training and lack of challenges during their earlier work-life career tend to underestimate their own capabilities, which may add extra difficulties during the changes at work Lower self-confidence of the elderly workers can lead to vicious circle in learning situations, if employees are not encouraged to take part in training

Risk of Information Overflow

Criteria of healthy work" emphasize that workers should be given possibilities to control their work and to develop their own abilities and vocational skills (Lindstrom, 1994) Current changes at work can have both positive and negative impacts on stress and well-being of the workers When working with more flexible technological tools, there is e g a risk that information overflow will exceed the cognitive capacity of people working with computers, under conditions of rising productivity demands Experience from both research and practice witnesses that the updating of skills to fulfil these new requirements at work demands additional mental energy especially from older employees (Huuhtanen,

1993) They might need more encouragement and time, they must have the possibility to proceed in self-determined work and learning pace The new material to be learned should be fitted to the earlier experience or expertise A "brush-up" of learning strategies might be sometimes useful

Not only cognitive demands of work are changing For elderly employees, changes in the work role many times have impacts on the working habits and personal values The creation of occupational identity and original occupational choices might have been based on a quite different quality of work and society than today

Banking tasks are one illustrative example of this type of drastic change People were recruited and tested for banks by quite different criteria some twenty thirty years ago than today Carefulness, submissiveness, kindness, clerical dexterity, and interest in numbers were characteristics emphasized earlier Today, active result-oriented selling, persuading role with the use of changing computers tools is needed This means that older employees have to change their personal style when serving the customers They must have mental energy and motivation to continuous learning, under highly competitive business circumstances (Huuhtanen and Leino, 1992)

In concluding, individual differences are increasing by age The clearest impact of aging on psychological performance can be seen in tasks which demand complex combination of visual and psychomotor performance, under the time pressure These tasks demand simultaneous shift of attention between e g different computer screens and motor activities In most daily tasks at work, however, higher motivation, experience and wisdom of the elderly workers compensate the possible decrease in speed In addition, some psychological and social abilities and performance are developing during the whole work and life career, most of all verbal capabilities and feeling of empathy All what has been said holds for aging individuals without diagnosed disease or deterioration of central nervous systems or motor or sensory functions

What does this all mean as regards aging and work? A life-long training and learning by doing are the key elements in ensuring the healthy and productive aging Too often, however, competition and stereotyped attitudes dominate the discussion about aging and work In the workplaces, what is needed is better cooperation between young and old Healthy and productive work should be based on better combination of different capabilities, knowledge and experience of age and cohort groups

RETIREMENT ATTITUDES IN FINLAND

As the starting point for the "Respects for the Aging" programme in Finland (Ilmarinen, 1991), an interview study was carried out dealing with the attitudes of people toward work and retirement in 1990 It was repeated in 1992 The focus was laid on the first phase of the retirement process, that of anticipation Thoughts of retirement are seen as attitudinal antecedents of retirement, giving a base to better understand, how to address the aging working population (see also e g Beehr, 1986, Tikkanen and Kuusinen, 1992)

The main study problems were How frequent are the thoughts of early retirement? What are the grounds for these thoughts in these groups? What, in general, are the prerequisites for continuing to work up until retirement age? How high is the willingness to take part in the work life once retired? The first study sample in 1990 consisted of 879 persons over 35-years old in the Finnish labour force (51% female, 49% male) The second sample in 1992 comprised 1175 persons with the same age and gender distribution Data was collected by telephone interviews in connection with the Labour Force Survey compiled by the Central Statistical Office of Finland Also a sample of persons on early retirement was interviewed

Thoughts of Early Retirement

57% of respondents in 1992 had at least sometimes thought of retirement before their normal old age pension age (normally 63 or 65 years in Finland). The prevalence of retirement thoughts has decreased during the two economic recession years in Finland. Grounds for thoughts of early retirement were health and work capacity (30% in 1992), attraction of the life sphere outside work (23%), aging as such (21%), content of work and work stress (20%), and other varied reasons (6%). Health and work capacity was more frequent reason for thoughts among women and in manual workers' group. Life sphere outside work was emphasized by upper-level employees (Huuhtanen and Piispa, 1994).

A list of 11 items was given of possible improvements at work, and the respondent was asked to evaluate whether he or she considered the presented issue as important in limiting unnecessary early retirement. In order to reduce unnecessary early retirement, most important issues in 1992 were the security of the work place (85%), rehabilitation (75%), improvements in the amount and haste of work (74%), improvement of the work environment (70%), flexible working hours (69%), and improvements in job content (68%) and in leadership styles (67%). In general, the youngest groups emphasized improvements more than the oldest ones.

The importance of sabbatical year and wage had decreased in two year's interval. Women put more emphasis on improvements at work, especially issues like occupational health care, wages, and sabbatical year. The importance of training and flexible working hours was decreased only in the male sample, the importance of work environment among women.

Attitudes toward Work among the Pensioners

Comparisons of the attitudes of workers and pensioners on early retirement (age 35-64 years) were also made. The participation of already retired persons in work life was determined by the question: "Have you thought of the possibility of somehow still taking part in worklife?" . The following answers were given (persons on early retirement, n=391 in 1992): (1) "I have not thought about it" 74%, (2) "I have thought about it sometimes" 16%, (3) "I am thinking continuously about it" or (4) "I am already working within the limits of regulations (eg. right to pension, taxation, amount of pension)" 10%.

The personal prerequisites of the pensioners for taking part in work life (N=73 in 1992, those who had thought about work sometimes or continuously, 19% of the pensioners) were as follows: (1) good health 85%, (2) suitable job locally available 69%, (3) part-time job 66%, (4) changes in the regulations of pension systems and taxation 63%, (5) autonomous work schedule 59%, (6) possibilities to influence one's work 38%, (7) improving in the work environment 29%, (8) improving work content 27%, (9) better wages 27%, (10) improvement in leadership style 13%, and (11) change in family situation 10%.

DISCUSSION AND DEVELOPMENTAL NEEDS

Retirement Attitudes

Surveys in 1990 and in 1992 revealed that clearly more than half of both men and women over 35 years of age had at least sometimes thought of retirement before their normal old age pension age (normally 63 or 65 years in Finland). Thoughts were relatively common even among quite young persons. Health and work conditions and content of work played an important role as grounds for thoughts of early retirement. The data showed a relationship between work and retirement attitudes and subjective evaluations of mental and physical well-being

It must be emphasized that thoughts of early retirement are not the same as the decision to start the process of retirement and the actual retirement Thoughts represent the first, anticipatory phase in the transition from work to retirement (e g Beehr, 1986, Hanisch and Hulin, 1990) In addition, the five categories of grounds of thoughts were based on the open-end question and are not totally separated, e g health and work can be strongly related to each other

Women put more emphasis on improvements at work This might reflect gender differences both in work conditions and in the sensitivity to problems at work The oldest groups emphasized improvements less than the youngest ones A plausible explanation could be that the older respondents feel that the suggested improvements, however important they might be, are not rapid enough to exert positive effects on work near the retirement age The older may also be sceptical as regards getting real improvements in their work

Interest in taking part in work life after retirement was much higher in the group that was still working than among those already retired Individual resources and capacity influence work and retirement motivation more when people are on early retirement than when they are still at work Aspects other than health or loss of it are more important for those still working For the pensioners, the prerequisites for continuing to work after retirement were very personal issues (Huuhtanen and Piispa, 1992) Those still working evaluated the work as a more important target in order to limit unnecessary early retirement

Those persons who are still working near the old-age pension age are representative of the 'healthy worker effect' Most of those who have already retired at these ages are on the disability pension and have done so because of poor health or low work capacity In Finland, only a fraction of those on early pension have retired voluntarily, without any loss of health

Developmental Needs

In order to enhance the possibilities of elderly people to continue at work up till the 'normal' retirement age, both surveys on attitudes and work-site interventions are needed Reliable and valid psychological research data helps changing attitudes and negative stereotypic beliefs as regards elderly workers Human resource planning and counselling should reflect the attitudinal antecedents of individuals' retirement intentions in addition to their year of birth The standardized attitude measurement scales have also proved to be useful on the company level

In order to ensure a more individualistic choice between work and other life activities for people nearing pension age, working conditions, regulations, and individual capabilities and personal resources must be developed Measures should be adapted better than previously to the type of work and to the increased individual variability among elderly workers

The situation of elderly people at work depends on the changing labour policy of the firms Unfortunately, the integration of older workers into the labour force varies according to the supply and demand of labour, both on societal and company levels The current recession has led governments and companies to reduce benefits and protection for workers and pensioners Pressed by global competition, companies are trying to develop measures to lower monthly pension payments and to raise retirement ages and the years that must be worked to get a full pension Since social and environmental factors vary considerably from country to country and between companies, programmes designed for older workers will need to vary accordingly

Work-site Projects Needed

On the company level, on-going longitudinal studies and developmental projects in the Finn Age program will add to knowledge about the temporal changes of attitudes and behaviour of the aging work force One on-going project is aiming at the development of tasks and cooperation between young and old workers The aims of the project are first, to combine different capabilities, knowledge and experience of age/cohort groups at work, and second, to enhance individual choices of aging workers as regards continuing to work until normal retirement age This is expected to minimize harmful competition between age groups The first step into this direction is the dissemination of unstereotyped knowledge on human aging What is needed are changes of the attitudes of both supervisors and workers themselves This is followed by developing division of tasks between age groups, by developing social support and learning, and by creating developmental methods for personnel administration and for occupational health and safety personnel

Previous studies have shown that the problems of mastering new computer applications increase with age (e g Huuhtanen and Leino, 1992) Based on these research findings and experiences at the workplace level, a specific intervention programme was carried out for elderly (over 50 years of age) sales managers in an insurance company The programme was based on intensive group work Also support at the work site was developed Both baseline and follow-up measurements of self-confidence, organizational commitment, and mastery of applications were carried out Another intervention project was carried out in one department of a multinational steel company The aim of the study was to make the work suitable for aged steelworkers and to maintain the work capacity of aging workers Based on a questionnaire survey and interviews, three developmental groups were formed, with four to five workers in each On work-site level, training for supervisors and workers, discussions between occupational groups, and safety at work were increased Interventions have been planned also among teaching professions (Kinnunen et al , 1992)

Filling the Generation Gap

In order to fill the threatening generation gap, more emphasis should be placed on the evolutionary strategy of using advanced technologies in work life In the development of cooperation between young and old workers, modern technology could be one good tool to combine different capabilities, knowledge and experience of age/cohort groups at work New forms of even international cooperation are possible via information technology tools

The self-selection, training and recruitment of older persons to different occupations have taken place according to criteria based on working life demands during the '50s and '60s The pressure to change one's personal career orientation can be more problematic in these groups under conditions of continuous change It is of special importance for elderly employees that the computer applications can be integrated to each one's own work processes The training should be based on peoples' long work experience and high work motivation Experienced employees have 'tacit' knowledge of the functional ways to organize daily office work in local settings This is too easily missed in technology-led implementation processes

During the technological change process, a course on the basics of the new method and its application is not sufficient Learning takes place in combining people's own long-term experience trying out new tools getting experiences, and forming new concepts, mental models at work By participatory approach, elderly people are giving the possibility to control the change and to learn necessary planning methodology The participation in the planning process may also affect the health risks of the employees The already stressful process of job reorganization can be worse for people when they have less influence over the process or when the changes lead to decreased control.

Relevant Questions

In order to develop psychological theories on adult development and on aging in connection to work and stress, comparative studies and theoretical discussion are needed among researchers in different countries. When thinking of the key psychological issues regarding aging and work, I argue that joint learning and planning at work will increase individual growth, motivation, job satisfaction and organizational effectiveness. Increased challenges at work can also be used as one effective tool in reducing the early exit from the labour force. It is of crucial importance, which kind of questions we are putting and trying to answer, when talking about aging and work. The traditional question has been: "How does aging affect learning?" When changing the sequence of the words vice versa: "How does learning affect aging?", we get a better starting point in developing the mental and social contents and demands of the work of to morrow.

REFERENCES

Ageing at work: consequences for industry and individual, 1993, *Lancet*, 340: Jan 9.

"Aging and Working Capacity, 1993, Report of a study group", 1993, WHO Technical Report Series 835, WHO, Geneva.

Beehr, T. A., 1986, The process of retirement: a review and recommendations for future investigations, *Pers. Psych.*, 9:31-55.

Czaja, S. J., 1988, Microcomputers and the elderly, in: *"Handbook of Human Computer Interaction"*, M. Helander, ed., Elsevier, Amsterdam, pp. 581-598.

Floyd, F.R., Haynes, S.N., Doll, E.R., Winemiller, D., Lemsky, C., Burgy, T.M., Werle, M., and Heilman, N., 1992, Assessing retirement satisfaction and perceptions of retirement experiences, *Psych. Aging* 7: 609-621.

Hanisch, K.A., and Hulin, C.L., 1990, Job attitudes and organizational withdrawal: an examination of retirement and other voluntary withdrawal behaviors, *J. Vocat. Beh.*, 37: 60-78.

Huuhtanen, P., 1993, From Naive End-users to Learning Subjects: the changing focus of psychological field studies in office and administrative work, in: *"Crossroads between Mind, Society and Culture"*, H. Perho, H. Räty and P. Sinisalo, eds., Joensuu University Press, Joensuu.

Huuhtanen, P., and Leino, T., 1992, The impact of new technology by occupation and age on work in financial firms: a 2-year follow-up, *Int. J. Hum. Comp. Int.*, 4: 123-142.

Huuhtanen, P., Nygård, C.-H., Tuomi, K., Eskelinen, L., and Toikkanen, J., 1991, Changes in the content of Finnish municipal occupations over a four-year period, *Sc. J. Work E.*, 17: suppl 1, 48-57.

Huuhtanen, P., and Piispa, M., 1992, Work and retirement attitudes of 50- to 64-year old people at work and on pension, *Sc. J. Work E.*, 2: Suppl 18: 21-23.

Huuhtanen, P., and Piispa, M., 1994, *"Attitudes toward Work and Retirement among Elderly Workers: A Two-Year Follow-Up"*, Proceedings of the 12th Triennial Congress of the International Ergonomics Association. Volume 6. Toronto, Canada, August 15-19, p. 170-172.

Ilmarinen, J., 1991, FinnAge: Action program on health, work ability, and well-being of the aging worker. *Työterveiset* Special issue 2.8.1991.

Kilbom, Å., 1992, Early retirement and social security systems. The situation of elderly workers in Denmark, Finland, Norway and Sweden. *Arbete och Hälsa 29*.

Kinnunen, U., Rasku, A., and Parkatti, T., 1992, "Aging among the Teaching Profession: Work, Well-being and Health among Aging Teachers", Proceedings of International Scientific Symposium on Aging and Work. Organized by Finnish Institute of Occupational Health and ILO, 28-30 May 1992, Haikko. Finland, pp. 157-161.

Kohli, M., and Rein, M., 1991, The Changing Balance of Work and Retirement. in: *"Time for Retirement. Comparative Studies of Early Exit from the Labor Force"*, M. Kohli, M. Rein, A.-M. Guillemard and H. van Gunsteren, eds., Cambridge University Press, Cambridge.

Laczko, F , and Phillipson, C , 1991, *"Changing Work and Retirement Social policy and the Older Workers"*, Open University Press, Philadelphia

Levinson, D J , 1986, A conception of adult development, *Am Psych , 41* 3-13

Lindstrom, K , 1994, Psychosocial criteria for good work organization *Sc J Work E , 20* 123-33

McEvoy, G M , and Cascio, W F , 1989, Cumulative evidence of the relationship between employee age and job performance, *J Appl Psychol , 74* 11-17

Ornstein, S , Cron, W L , and Slocum Jr, J W , 1989, Life stage versus cares stage a comparative test of the theoriens of Levinson and Super, *J Org Behav* 10 117-133

Rhodes, S R , 1983, Age-related differences in work attitudes and behavior a review and conceptual analysis, *Psychol B ,* 93 328-367

Salminen, K , 1993, *"Pension Schemes in the Making A comparative study of the Scandinavian countries"*, The Central Pension Security Institute, Studies 1993 2 Helsinki

Super, D E , 1980, A life-span, life-space approach to career development, *J Voc Behav ,* 16 282-298

Tikkanen, T , and Kuusinen, J , 1992, "Retirement and Retirement Preparation " Proceedings of International Scientific Symposium on Aging and Work Organized by Finnish Institute of Occupational Health and ILO, 28-30 May 1992, Haikko, Finland, pp 188-194

26

A NEW CONCEPT FOR PRODUCTIVE AGING AT WORK

Juhani Ilmarinen

Finnish Institute of Occupational Health
Department of Physiology
Helsinki
Finland

INTRODUCTION

It is a common belief that productivity is age-related younger workers are more productive than older ones Whether this myth is valid or not, depends on the definition of "productivity" It is obvious that the dimensions of quantitative productivity favor younger age, on the other hand, the dimensions of qualitative productivity can depend on skills and experience, which are the strength of older workers Several myths on aging, especial y those related to functional capacities, are not valid for the working-aged population (Ilmarinen, 1991a)

The interaction of the following factors affect human work ability aging, health, work, environment and life-style It should be noted that none of these factors plays a dominating role on work ability between the ages of 45 and 65 years On the contrary, the different interactions between these factors can become decisive in formulating work ability during aging (WHO, 1993)

Is aging and work a serious problem, and is this problem related to productivity or to some other features of work? The rapid changes in the age structure of the working population of several developed and developing countries have drawn attention to the consequences The "graying" world population is a new phenomenon to which even the technically most advanced countries are still trying to adapt The changing age statistics focus our attention on age, and suggest that aging is the reason, and the low participation rate of older age groups in the workforce is the consequence We should note, however, that aging is one side and work is the other side of the same coin We can therefore ask a fair question What are the appropriate working conditions for the "graying work force"? In other words, the main problem can be the work, which does not fit the changed characteristics of the aging workforce One consequence is that the productivity can not be optimal, if the skills and experience of aging workers have not been sufficiently utilized

Preparation for Aging Edited by E Heikkinen *et al*
Plenum Press New York 1995

WORKFORCE PARTICIPATION

Decreased workforce participation rates of older workers can be found both in developed and developing countries. From 1960 to 1989 the rates have decreased from 87 to 76% among men aged 50-59 years and, more dramatically, from 73 to 43% among men aged 60-64 years in 10 industrialized countries (WHO, 1993).

More recent figures from the year 1992 indicate that considerable differences exist between the countries (Moore et al., 1994). While the participation rates of women aged 55-59 are higher than 60% in Sweden, Denmark, Norway and Finland, the corresponding rates for women in the Netherlands, Portugal and Spain are less than 30%. Although the participation rates of women in the age group of 60-64 years drop dramatically in all countries, the difference e.g. between Norway and the Netherlands remains great (Figure 1).

The participation rates among men aged 55-59 years are higher than 80 % in Japan, Sweden and Denmark, while the respective rates are lower than 60 % in Luxembourg and Belgium. The differences between the countries in participation rates are greater among men in the age-group of 60-64 years (Figure 2).

It is hard to believe that, e.g. the differences between the countries as regards aging, health and life-style could explain the differences in workforce participation rates. It is more likely that social and health policy as well as workforce policy modify the participation rates of older workers. On the other hand, the dramatic decrease in participation rates from the age of 55-59 to the age of 60-64 years in nearly all countries is a sign of general difficulties in fitting age with the work.

Because an early exit from work life is very expensive for the society, several changes within the countries can be expected in the near future. At least the early retirement age, but probably also the age entitling to old age pension will increase. Also, the criteria for work disability pension will become more difficult to fulfil. However, it is misleading to believe, that such legislation will as such increase the work ability of the aging workforce. Undoubtedly, the workforce participation rates may increase, but whether the productivity and well-being of older workers will increase, remains unknown.

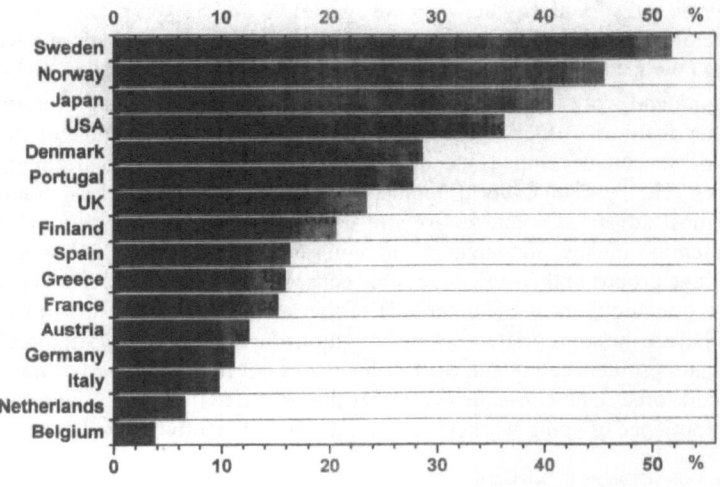

Figure 1. Workforce participation rates for women aged 60-64 years in 1992 in various countries.

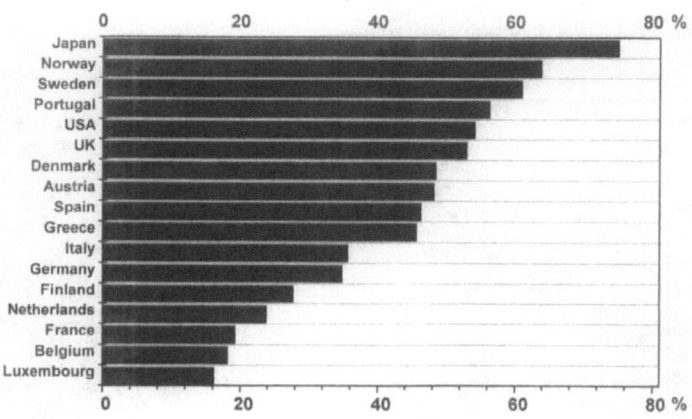

Figure 2. Workforce participation rates for men aged 60-64 years in 1992 in various countries.

The actions most needed relate to the redesign of work, so that the work demands suit the needs and possibilities of aging workers.

SELF-CARE OF FUNCTIONAL CAPACITIES

Although the major problems deal with the changing and increasing work demands in all societies, the role of the functional capacity of older workers should not be underestimated. Physical, mental and social resources create the base for work ability (Figure 3). Such capacities are needed daily at work, the quantity and quality depending on one's occupation. It is well known that the physical capacities decline with age earlier than the mental or social capacities.

Recent longitudinal studies have shown that in the course of 10 years the decline of muscular strength and cardiorespiratory endurance can be much greater than expected, and the decline can be notable already after the age of 45 years. As an example, the decline of the maximal isometric strength of trunk flexion among the men and women is shown in Figure 4 (Nygård et al., unpublished results).

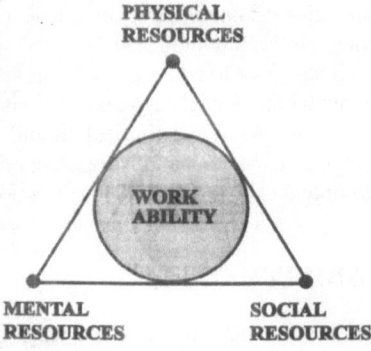

Figure 3. A schematic drawing of functional capacities and work ability

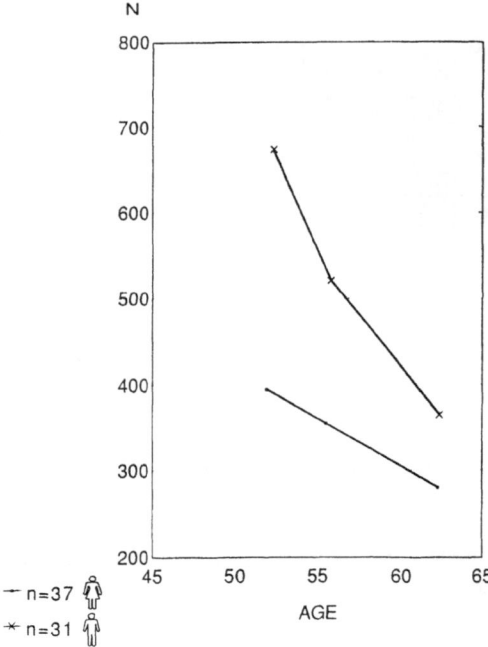

Figure 4. The decline of maximal isometric trunk flexion strength among men and women during 10 years (Nygård et al , unpublished data)

Work ability is based on the physical, mental and social resources of the worker.

With increasing age, the different resources are strongly interrelated. The figure indicates also that more resources should be available than the work demands.

The "extra" resources make the recovery from work strain more efficient and improve the possibilities for learning new skills, etc.

Based on the decline of the physical capacities the question arises whether the lower physical capacity of older workers is still sufficient for work demands? On the other hand, it can also be asked, whether the work demands are suitable for the decreased capacities? It is obvious that the answer to both questions is "no".

Therefore the solution should consist of two actions. First, the premature decline of the physical capacities should be prevented by appropriate physical exercise, and secondly, the physical demands of work should decline with age. As Figure 5 shows, due to the two actions mentioned above, the physical reserves needed can be the same for a 60-year-old worker as for a 30 year-old worker (Figure 5).

Mental and social capacities do not show a similar decline with age as the physical capacities (Berg, 1980; Rabbitt, 1991; Singleton, 1983). However, a concept similar to that presented for physical work can be constructed for mentally or socially demanding work, too. It should be noted that some mental and social capacities can also increase with age. In such cases, higher demands of work (like mentor activities) should be accepted. Generally, the concept of the relationship between capacities and work demands emphasizes, that the work should be designed so that its content changes in the same direction as the worker's capacities.

AGING AND WORK ABILITY

In optimal working conditions, the human work ability should remain at a sufficient level with increasing age. We followed aging workers for 10 years and found that work ability

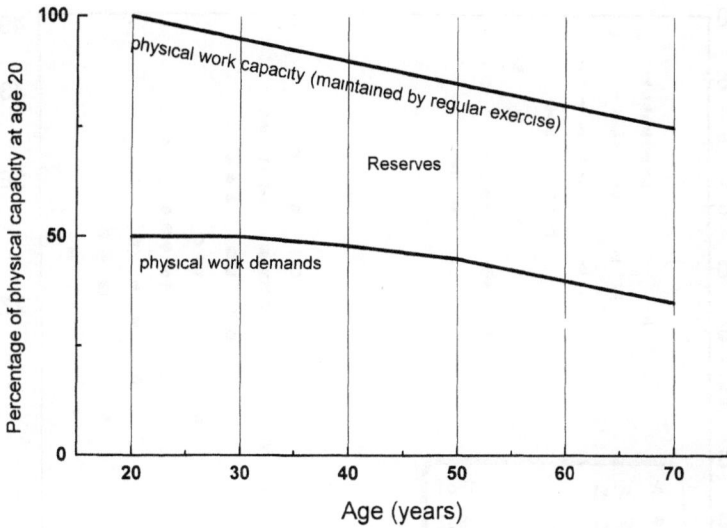

Figure 5. A schematic drawing of physical work demands and physical work capacity related to age. By lowering the physical work demands in line with increasing age, and maintaining physical work capacity by regular exercise, the physical reserves can be kept sufficient with increasing age

decreased with age, independent of the fact that the subjects were able to stay in the same job during the entire follow-up period. Figure 6 illustrates the situation among male installation workers. The work ability index (WAI) (see Ilmarinen and Tuomi, 1992) of installation workers between the ages of 45 and 51 years was fairly good. The majority of the workers had a WAI score higher than 36, which means better scores than on the average. Only two had poor work ability (scores lower than 27). After 10 years, the same group had become more heterogeneous and no one's work ability was good (scores higher than 43). The majority of the installation workers had a lower WAI than the average, and many of them had a poor WAI. It should be emphasized that the decline observed in WAI took place among those men who continued working in the same job; so, working did not prevent the decline of the WAI with aging (see Figure 6). The key question is, whether the decline of WAI with age can be prevented, and what are the measures needed. For prevention purposes, a new concept is needed for maintaining work ability during aging.

A NEW CONCEPT FOR PRODUCTIVE AGING

The basis for productive aging is created already early in working life. The triangle for maintaining work ability includes three groups of actions (Figure 7).

Actions related to work load and the physical work environment can be carried out by processes in ergonomics, hygiene and safety. Actions related to organizational culture include developmental, psychosocial and management issues. Actions related to the individual worker are based on health promotion, where both leisure-time physical activities and other life-style factors that improve health dictate the content of actions (Ilmarinen and Louhevaara, 1994).

As a result of the three-fold actions, sufficient health and work ability can be maintained during aging. The consequences of sufficient work ability are high productivity and quality of work, as well as a good quality of life and well-being.

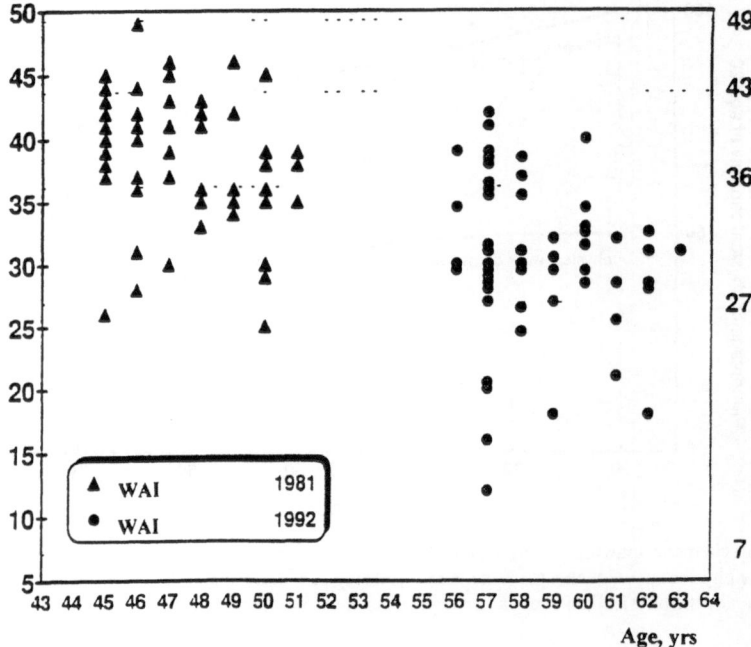

Figure 6. Work ability index of men in installation work The men (n=62) had been followed from 1981 to 1992 In the left panel, when the men were 51 years of younger, only two had poor work ability (score less than 27) The right panel shows that after ten years 10 men had poor work ability and no one had good work ability (score 43 or higher) Working in installation jobs has not prevented the decline of work ability

During the employee's work career, both the employer and the employee benefit from the efforts made.

The entire chain of impacts, however, is not fulfilled during the course of the work life. The main impact, can be realized when the employee enters the "Third Age" upon retirement (Laslett, 1989). A meaningful, successful and productive Third Age is the final goal of actions made in middle age. In most countries the period of the Third Age is increasing in years, meaning 20 to 25 years. It is not meaningless for an individual or for the society, how this long period of life is utilized. The Third Age, as a reference point, is a new concept for evaluating the actions made at "Working Age". It is important that the measures taken to maintain work ability, when we are 45+, are not merely focused on productivity in the short run. The benefits of successful "Working Age" actions can be manifold later in life. Therefore, the pressure to start and carry out ergonomic, psychosocial and health promotion processes are strongly in the interest of the employees.

WORK ABILITY IS THE BASE FOR RETIREMENT ABILITY

The period of "Working Age", lasting 3 to 4 decades, is a predominant period in view of our "Third Age". The prevention of risk factors for work ability is the key issue for meaningful, successful and productive aging. The risk factors that are age-sensitive for work ability have been identified in prospective studies (Ilmarinen, 1991b, Ilmarinen, 1994). Also, a quantity of experience on feasible measures in health promotion is available (WHO, 1993).

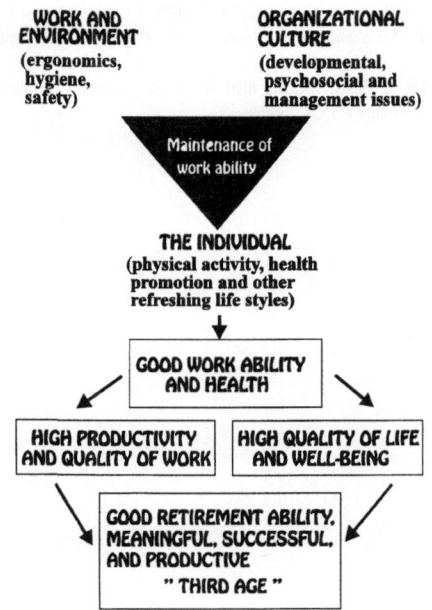

Figure 7. A new concept for productive aging at work (see text)

Practical sets of recommendations concerning the actions promoting the work ability of the aging work force have been published both for international and national purposes (WHO, 1993, HEA ,1994) Thus, the essential knowledge, needed for starting and maintaining work ability during aging, is available It is now time for action

REFERENCES

Berg S 1980, Psychological functioning in 70- and 75-year-old people A study in an industrialized city *Acta Psychiatr Scandi* Suppl 288

Health Education Authority (HFA) 1994 Investing in Older People at Work Contributions Case Studies and Recommendations A Symposium for Employers, Policy Makers and Health Professionals in Europe 11-13 October 1993 Birmingham, United Kingdom, London 1994, 176 p

Ilmarinen, J , 1991a Myths and Facts about the Development of the Capacities of Ageing Individuals in *Towards the 21st Century Work in the 1990s* S Lehtinen et al , eds , International Symposium on Future Trends in the Changing Working Life 13-15 August 1991 Helsinki Finland Proceedings 3 Finnish Institute of Occupational Health, Helsinki 1991, pp 226-236

Ilmarinen, J , ed , 1991b The aging worker, *Scand J Work Environ Health* 17 (Suppl 1) 141

Ilmarinen, J 1994 Promoting the Health and Well-Being of the Older Worker the Finnish Experience, in *Investing in Older People at Work Contributions Case Studies and Recommendations* A Symposium for employers, policy makers and health professionals from Europe 11-13 October 1993 Birmingham, United Kingdom, Health Education Authority, London 1994, pp 90-104

Ilmarinen, J , and Louhevaara V , 1994, Preserving the capacity to work, *Ageing International* June 34-36

Ilmarinen, J , and Tuomi K , 1992, Work Ability Index for Aging Workers, in *Aging and Work* International Scientific Symposium on Aging and Work, 28-30 May 1992, J Ilmarinen ed , Haikko Finland Proceedings 4, Finnish Institute of Occupational Health, Helsinki, pp 142-151

Laslett, P , 1989 *Fresh Map of Life The Emergence of the Third Age '* Weidenfeld and Nicolson, London

Moore, J , Tilson, B , and Whitting, G , 1994, *An International Overview of Employment Policies and Practicies Towards Older Workers '* Employment Department, Research Series No 29 Ecotec Research and Consulting Ltd

Rabbitt, P M A , 1991, Management of the working population, *Ergonomics* 34 775-790

Singleton, W T , 1983, Age, skill and management, *International Journal of Aging and Human Development*
    17(1) 15-23

World Health Organization (WHO), 1993, "Aging and Working Capacity", Reports of a WHO Study Group,
    WHO Technical Report Series 835, Geneva

27

WHAT IF THE DISABILITY PENSION APPLICATION IS DENIED?

Raija Gould

The Central Pension Security Institute
P.O. Box 11
FIN-00521 Helsinki
Finland

INTRODUCTION

Early retirement is very common in Finland. The statutory pensionable age is 65 years, but more than half of the 55-64-year-old population already draws some type of early pension.

In recent years also those elderly people who do not receive a pension have often been without gainful employment. At the end of 1993 the unemployment rate in Finland in age group 55-64 years was 20%. Because of high unemployment the age of exit from work is often considerably lower than the age of retirement—even when it is a question of retiring on an early pension.

It is actually becoming quite exceptional for a person to draw a normal old-age pension immediately after leaving his or her job. As Guillemard and van Gunsteren (1991) point out, the period between exit from work and entry into the old-age pension system amounts to more than just setting the schedule of retirement ahead. The chronological milestones of the life course are being torn up, and functional criteria, such as ability, willingness and possibility to work, are increasingly staking out the later years of life.

Kohli and Rein (1991), call those institutional arrangements that cover the transition process between work and old-age pension 'pathways' of exit. Most of these pathways make use of measures that were originally designed to cover specific risks such as unemployment or disability.

In Finland the most common pathways of early exit from work are disability pension, early disability pension and unemployment benefits, including unemployment pension, which is payable to long-term unemployed persons aged 60-64. The two forms of disability pension are both granted on the grounds of reduced work capacity. However, especially the early disability pension is used not only to cover some individual health hazards but also as a tool for labour market regulation. It is designed for those aging employees who have experienced some reduction of work capacity but who are not sick enough to qualify for a standard disability pension.

Preparation for Aging, Edited by E. Heikkinen *et al.*
Plenum Press, New York, 1995

The applicants of early disability pension, and especially those whose applications are denied are the subjects of this study.

THE PROCESS OF EARLY RETIREMENT

Retirement, both early and old-age retirement, is a process that occurs over time, not a single one-time event. Some personal factors as well as environmental forces may first cause a person to think about retirement. This preference to retire leads to the decision to retire and finally to the act of retirement (Beehr, 1986).

As we see in Figure 1, the ability and willingness to work and working life including the labour market situation, are factors contributing to the decision to retire early. The option to retire is also influenced by the pension scheme; the eligibility criteria and the compensation level of benefits for example affect the attractiveness of retirement.

These factors can be divided into two categories: push and pull factors. On the personal level push factors for example refer to health conditions and on the institutional level to labour-market constraints. Pull factors above all refer to pension policy incentives (Kohli and Rein, 1991; Komiteanmietintö, 1991).

The act of retirement, the entry into the early retirement programs, is not only a matter for the individual to decide. The final decision of eligibility is taken by pension institutions. The same factors - pension rules, individual traits, work and labour market situation - also influence the decision making of the institutions, but the institution may arrive at a conclusion quite different from that of the individual. Thus the pension scheme can act not only as a pull factor for early exit from work but also as a gatekeeper for some of the pathways of early exit.

At the moment nearly half, 45%, of the early disability pension applications are denied. The denial rate has been quite high ever since this pension form was introduced in 1986, but recently, during the economic depression, it has got even higher (Figure 2).

The denial rate is higher for younger applicants than for older ones and also higher for females than for males. The age-bias is partly attributable to the characteristics of the applicants: the older ones are often more severely disabled then the younger ones. But it is also attributable to less stringent eligibility criteria being applied to older applicants, since even with some difficulty the oldest ones are thought to be more disabled and less able to

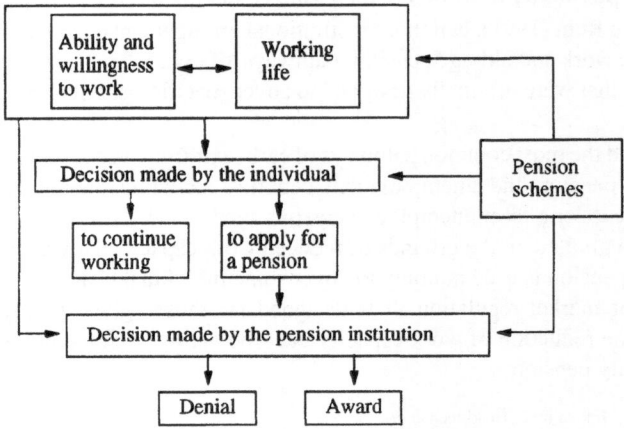

Figure 1. Factors contributing to the retirement decision.

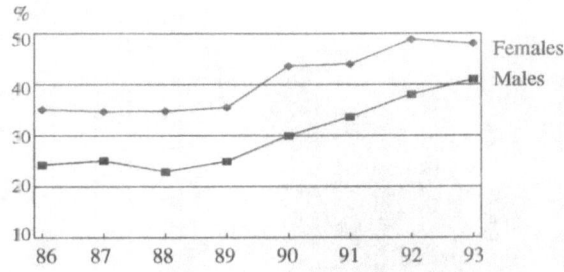

Figure 2. Percentage of denied pension applications in 1986-1993 (early disability pensions)

continue working or to find work than their younger counterparts. The female-male differences in the denial rate originate in the implicit masculinity of the social security system as well as in the physiological, behavioural and cultural disparities of the sexes. Men and women work in different jobs with different occupational prestige - the female jobs are often considered less demanding and thus easier to perform even with some disability. Men and women also have different types of diseases and their sickness behaviour is different, the male model being more in line with the disability pension criteria (Gould and Takala, 1993).

THE DENIAL STUDY

The aim of the study is to find out what happens to people whose early disability pension application is denied. Do they continue working and give up the thought of early retirement, or do they find some other pathway for the early exit from work? The study is interested in the personal plans and feelings of the denied applicants, as well as in the institutional setup of the exit pathways.

The subjects of the study consisted of wage-earners whose early disability pension application was denied in 1988 or in 1992. The 1988 denials, 1094 persons altogether, were 55-60 years of age at the time of the pension refusal. The other 808 persons, denied in 1992, ranged from 55 to 64 years of age at the time of the denial.

The data was collected partly by mail questionnaires (the response rate was 84%), and partly from the pension records. The questionnaires were sent in the spring of 1993, which means that either five years or one year had lapsed from the denial. At the time of the inquiry, all respondents were still below the statutory pensionable age, 65 years, though the oldest ones were just about to reach this age limit.

ALTERNATIVE PATHWAYS OF EARLY EXIT

Figure 3 illustrates the employment situation of the research population after the denial of their pension applications. The post denial careers show a steadily declining trend in gainful employment and a growing trend in early retirement. Unemployment also seems to play an important part in this process.

A few of those denied the pension in 1988 had applied again right away—being more fortunate this time—and were already drawing a pension at the end of 1988. Every year the proportion of early retires was growing, and five years after the first pension denial more

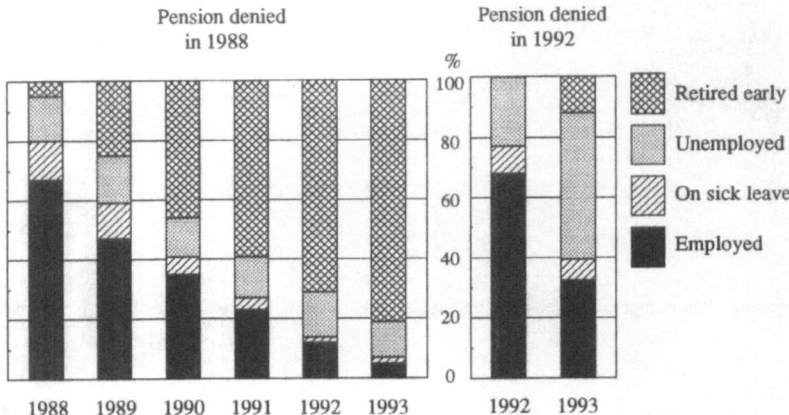

Figure 3. Employment situation in 1988-1993 among persons whose early disability pension applications were denied in 1988 or in 1992

than 80% had found their way into receiving some form of early pension. Only 6% of them were still working in 1993, and 12% were unemployed.

During this five-year-period the number of unemployed has remained the same, but the unemployment rate calculated for those still in the labour force has grown dramatically. In 1993 two thirds of the non-retires were unemployed.

Most of those whose first early disability pension application was denied in 1988 ended up receiving an early pension later on. On an average, it took two years and two months from the first denial for them to finally receive a pension. Since most of them only worked for about one year after the refusal, they had been out of work and not drawing a pension for a period of one year during which most of them lived on unemployment or sickness benefits. Social welfare and family members were also mentioned as sources of income.

The unemployment pathway seems to be a very important substitute for the early retirement programs. This is even more clearly illustrated by the careers of those denied the pension in 1992, at the time of high unemployment. One fourth of them were unemployed at the time of their pension application already, and one year later, in 1993, half of them were unemployed. Most of them worked only for a few months, if at all, after the pension refusal. So for the 1992 denials, the period between exit from work and entering into an early pension is becoming longer than for those who started the process in 1988.

The weight of unemployment in the early retirement process is partly due to the economic depression of the time. But it also demonstrates the links of aging and poor health with unemployment, and the difficulties involved in the efforts of the social policy institutions to keep the health-related and labour-market related programs separated.

MENTAL WELL-BEING

Those people whose pension applications have been denied are usually disappointed and feel that they have been unfairly treated. If they still have a job they often find it difficult or unpleasant to continue working.

To find out about the mental well-being of the research population, the 12-item version of the General Health Questionnaire (GHQ) developed by Goldberg (1978) was

Employment status in 1993

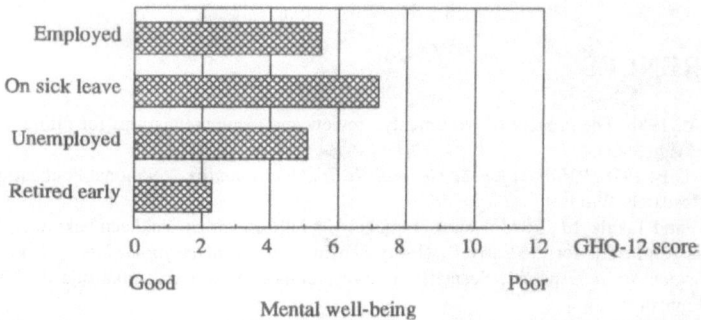

Figure 4. Mental well-being in 1993 of persons whose early disability pension applications were denied in 1992.

employed. This measure includes items like "have you recently lost much sleep over worry", "have you recently been able to enjoy your normal day to day activities", "have you recently felt that you are playing useful part in things", "have you recently been feeling unhappy or depressed", and so on.

Figure 4 shows the GHQ-scores for those whose applications were denied in 1992: the higher the score the more mental difficulties or troubles the people are facing. Those who are on long-term sick leave score highest—their health problems are most likely to affect their mental well-being. Those who are still in the labour force, who are either employed or unemployed, score the second highest. But best off are those who have already succeeded in their option for early retirement.

After the intention to retire early, elderly people do not seem to be satisfied with other alternatives. At this point, having to stay in work or having to stay out as unemployed are both disappointing.

CONCLUSION

When people enter into the process of early retirement, when they first apply for a pension, they are very strongly motivated to retire and not at all motivated to continue working. This means that if we like to keep them in the labour force, if we think that staying on in work is good for them and for the society, something should be done way before their first application, even way before their first serious thoughts about early retirement. After the first application, the pension system's hampering effect, the denial of the application, can only turn them from one pathway of exit to another, usually to unemployment, but its possibilities of keeping these people in gainful employment seem to be very limited.

Improved and more suitable working conditions and restoration to better health, all done in good time, appear to be the key issues involved in persuading aging employees to choose work instead of early retirement. But preferring work to retirement is not enough, there must also be work available. Otherwise unemployment strengthens the willingness to retire, and offers an alternative pathway of exit from work for those who do not meet the eligibility criteria for early pensions.

Early retirement is often seen as a policy for tackling increasing unemployment (Laczko et al., 1988). The findings of this study indicate, however, that the interaction can also be quite the opposite. At the time of depression, unemployment is moulded into a tool for tackling increasing early retirement

REFERENCES

Beehr, T A , 1986, The process of retirement a review and recommendations for future investigation, *Pers Psych* 39 31

Goldberg, D P , 1978, *'Manual for the General Health Questionnaire "*, National Foundation for Educational Research, Windsor

Gould, R , and Takala, M , 1993, Naiset ja miehet yksilollisen varhaiselakkeen hakijoina, [Women and Men as Applicants for the Early Disability Pension], in *' Naisten ja miesten tyokvvvttomyys, erot elakkeiden hylkaamisessa'*, Sosiaali- ja terveysministerio, tasa-arvojulkaisuja A 2/1993, Helsinki, in Finnish

Guillemard, A-M , and van Gunsteren, H , 1991, Pathways and Their Prospects A Comparative Interpretation of the Meaning of Early Exit, in *"Time for Retirement"*, M Kohli, M Rein, A-M Guillemard and H van Gunsteren, eds , Cambridge University Press, Cambridge

Kohli, M and Rein, M , 1991, The Changing Balance of Work and Retirement, in *"Time for Retirement"*, M Kohli, M Rein, A-M Guillemard and H van Gunsteren, eds , Cambridge University Press, Cambridge

Komiteanmietinto 1991 41, *"Elakekomitea 1990 n mietinto'*, [Report of the 1990 Pension Committee], Helsinki, in Finnish

Laczko, F , Dale, A , Arber, S and Gilbert, G N , 1988, Early retirement in a period of high unemployment, *Jnl Soc Pol* 17 313

28

HOW WILL FINNISH PENSIONS BE WORKING IN THE FUTURE?

Simo Forss

The Central Pension Security Institute
P.O. Box 11
FIN-00521 Helsinki
Finland

BASIC FEATURES OF THE FINNISH PENSION SCHEME

Statutory pensions in Finland comprise employment pensions and the national pension. The purpose of the national pension is to guarantee an adequate minimum income for all pensioners. Entitlement to the national pension is based on residence. The purpose of employment pension is to guarantee that the level of consumption attained by employees and the self-employed during their active working career is maintained. The target, or maximum amount, of the pension represents 60% of the wage, and it is accrued over a 40-year-period.

Together, the employment pension and the national pension offer a relatively good income. The income available to pensioner households averages 85% of the income of other households. The staggered implementation of the private-sector employment pension scheme means that new pensions are gradually approaching the target level of 60%. Owing to the composition of the overall pension, the national pension is losing importance.

Employment pensions are financed through contributions that are calculated on the basis of wages. The key role of the labour market organizations in the employment pension scheme means that the rate of the contribution is an essential item on the wage negotiation agenda.

The employment pension is determined on the basis of the duration of employment or self-employment, and the earnings from work or the income from self-employment. Pensions are inflation-proof through the adoption of an employment pension index known as the TEL index.

The earnings which constitute the basis for the TEL pension are determined separately for each contract of employment, and they are calculated on the earnings for the last four years of employment. The earnings are adjusted in accordance with the TEL index to correspond to the level of the award year. As a rule, pension accrues at the rate of 1.5% of pensionable earnings per year of gainful employment.

In Finland, employment pension insurance is financed by an intermediary form of the pay-as-you-go system and the funding system. Until the end of 1992, employees'

Preparation for Aging, Edited by E. Heikkinen *et al*
Plenum Press, New York, 1995

pensions were financed entirely on the basis of contributions made by employers. In 1993, employees started to contribute towards their pensions.

The rate of the contribution is laid down annually. The statutory pension contributions are determined on the basis of uniform actuarial principles. The rates of contribution are confirmed by the Ministry of Social Affairs and Health upon the joint application of the pension institutions (The Central Pension Security Institute, 1993).

FACTORS UNDERLYING THE FUTURE TREND

The future of the Finnish pension scheme is largely determined by the following factors:

1. Growth of the retired population
2. Trend in pension expenditure
3. European integration

1. Growth of the Retired Population

In Finland, people show a high inclination to retire as early as possible. The current pension legislation offering ample scope for early retirement has contributed to this.

The size of the retired population in Finland is exceptionally large in proportion to the whole population and, especially, to the working population. There are already more than one million people receiving some form of pension. The old-age pension beneficiaries alone number approximately 750,000. Early retirement has also been very popular. A good third of people aged 55 - 59 are retired, a good 70% of people aged 60 - 62 and as many as 82% of people aged 63 - 64 (Eläketurvakeskus, Kansaneläkelaitos, 1993). The so-called dependency rate, i.e. the proportion of employed to non-employed, is highly unfavourable. Around 2.9 million people between the ages of 16 and 64 are actively employed. Per every two actively employed persons there are three non-employed persons. This is, however, partly attributable to mass unemployment (Forss, 1993; Korpela, 1994)

There is nothing to indicate that the growth of the retired population is slowing down. On the contrary, the number of old-age beneficiaries is increasing all the time, especially the number of aged people over 80 who are in need of constant care.

2. Trend in Pension Expenditure

The first years of the 1990s have witnessed a surge in pension expenditure. In 1990, pension expenditure amounted to around FIM 56.000 million, representing nearly 11% of GNP. Calculations show that expenditure has increased by this year and now amounts to a little less than 15% of GNP. It is expected to remain at this level until the year 2000 after which the growth will pick up again, and by 2030 pension expenditure will represent 19% of GNP, i.e. around FIM 152.000 million (Forss, 1993; Hänninen, 1994).

Underlying the increase in pension expenditure is also the fact that the private-sector employment pension scheme will become fully operational at the turn of the century. The pension expenditure arising from the employment pension scheme in proportion to the wage bill may even double between the year 2000 and the year 2030. Correspondingly, the pension burden will more or less increase twofold within the following 30 or 40 years (Forss, 1993)

Under such circumstances, the rate at which the national product grows and the way in which returns are distributed between population groups will be decisive for the future of the pension scheme. The decided real growth of the national product took a downturn in the

1990s but an upward trend is anticipated from the middle of this decade onwards In any case, the foreseen increase in pension expenditure raises concern about pension financing (Hanninen, 1994)

3. European Integration

European integration and free competition for labour and markets between countries may put some pressure on the Nordic, universalistic pension model The Finnish model is characterized by a combination of earnings-related income protection and the provision of basic subsistence to everyone The Finnish scheme combines different elements in a way that can very well serve as an example to other countries it is private, it ensures high coverage and it is statutory Indeed, the socio-political dimension of the European Union seems to have taken on more Nordic characteristics The problem is that European insurance companies would gladly enter the Finnish insurance market but Finland lacks means of integrating them with the joint liability mechanism of the Finnish pension insurance companies (Vanamo, 1994)

PROSPECTS FOR THE PENSION SCHEME

The future of the Finnish pension scheme is characterized by an attempt to check pension expenditure and to ensure and increase premium income This is to avoid having to resort to the worst possible alternative, i e cuts in the pension level

One of the measures taken to check pension expenditure was the amendment of the public-sector pension regulations to bring down the maximum level of the future public-sector pensions to 60%, which is the same as within the private sector (Korpela, 1994)

The award criteria applying to disability pensions may have to be adjusted, in other words, more stringent award criteria may have to be introduced Alternatively or concurrently, measures aiming at the development of rehabilitation services, especially early rehabilitation services, will be encouraged Improved working conditions and job satisfaction are also aimed at

It has already been suggested that the retirement age should be raised The matter is still under consideration as it is uncertain what the actual effects of raising the retirement age would be However, the award criteria applying to early retirement pensions are getting more stringent all the time The ages giving entitlement to early retirement pension have already been raised and the same trend is likely to persist, i e still higher age limits may be introduced On the whole, the award criteria for entitlement to early retirement pensions have been tightened, and they are likely to get tighter still This also involves the question of the unemployment pension The abolition of the present unemployment pension plan has been up for discussion and the "fate" of this form of pension is presently considered (Korpela, 1994)

New and positive developments have taken place within the part-time pension plan, in that better terms of pension have been legislated and new propositions pertaining to the provisions governing part-time pension in combination with part time work are likely to be introduced

In order to increase premium income, it was recently decided that those insured start contributing towards their pensions themselves on a permanent basis In line with this, future increases in the contributions payable into pension insurance will be assumed by employees and employers on a 50 50 basis

Cuts in the pension level may have to be resorted to if the national product will not develop as favourably as anticipated and if the above mentioned measures will not bring in enough savings. In fact, amendments to this effect are already under preparation. If materialized, they will not, however, cause any major drop in the pension level. The amendments suggested primarily relate to the pension index and related rules of calculation. So far, cuts in index increments have sufficed, i.e. pensions have been raised less than adoption of the normal index would have required. It has been suggested that the rules of calculating the index be made more flexible to make the index follow economic fluctuations more closely. Another consideration is the possible amendment of the provisions governing the calculation of the future period of service in disability pensions and some other forms of pension (Komiteanmietintö, 1994:9).

An actual reduction in the pension level would, for instance, imply a decrease in the accrual rate of pensions from the present 1.5%. This is, in a way, the same as taking off, say, 5-10% of the present 60% maximum level of pensions. However, we hope and believe that such measures do not have to be resorted to and that the economic trend and the savings measures introduced will be adequate in order to secure the level of pensions in the future, as well.

REFERENCES

Elaketurvakeskus, Kansanelakelaitos, 1993, *"Tilasto Suomen elakkeensaajista", 1992, ["Statistical Yearbook of Pensioners in Finland, 1992"*], The Central Pension Security Institute, The Social Insurance Institution, Helsinki, in Finnish

Forss, S , 1993, Ikaantyvien tyopanosta tarvitaan, ["The Contribution of the Aged Is Needed on the Labour Market] in *"Sosiaalipolitiikan oikeutus", ["The Justification of Social Policy"]*, B Koskiaho, J Lehtinen, H Lehtonen, eds , University of Tampere, Tampere 1993 , Tampereen yliopisto, Tampere, in Finnish

Hanninen, M , 1994, "Antoiko lama valineita tyoelakekeskusteluun?", ["Did the Economic Depression Provide Tools for the Debate on Employment Pensions"] Esitelma Elaketurvakeskuksen laivasemi-naarissa 15 6 , [Talk presented at a seminar arranged by The Central Pension Security Institute on 15 June, 1994], in Finnish

Komiteanmietinto 1994 9 [Committee Report 1994 9], Helsinki

Korpela, T , 1994, Elaketurvan ongelmat, [Problems Facing the Pension Scheme], in *"55+, katsaus ikaanty-vien elinoloihin ", ["55+, Survey of the Living Conditions of the Aged"]*, R Sailas, and S Mikkonen, eds , Helsinki, STAKES, Elinolot 1994 1, [National Research and Development Centre for Welfare and Health, living conditions 1994 1], Helsinki, in Finnish

The Central Pension Security Institute, 1993, The Finnish Employment Pension Scheme, Helsinki

Vanamo, J , 1994, "Elaketurvan kehitysnakymat", ["Future Outlook for Pensions"], [Lecture in a series of lectures arranged by The Central Pension Security Institute], Elaketurvakeskuksen luentosarjan luento, Helsinki, in Finnish

29

LIFE SITUATION AS A FACTOR EXPLAINING RETIREMENT

Jouko Lind

Insurance Institution
Peltolantie 3
FIN-20720 Turku
Finland

BACKGROUND

Withdrawal from labour force has recently become earlier and earlier in most western industrial countries In Finland, this exit has been based on first of all various retirement systems (see Laczko and Phillipson, 1991)

The variation in the number of persons willing to retire depends on many factors, including, e g , health and age structure of the population, changes in working life and labour market, regulations of early retirement and social security, and general values and norms in the society (Hytti, 1993) All these factors are interrelated wherefore it is difficult to separate their effects

The factors affecting a person's labour exit and retirement are often classified as pushing and pulling factors (e g Kohli and Rein, 1991, Hytti, 1993) Poor health, excessive work load, unsatisfactory working conditions, changes in labour market and in work requirements are pushing factors High level of social security and appreciation of leisure are pulling factors Furthermore, the institutional framework of labour market and social security and the relative rates of benefits have their influence on retirement (cf Hytti, 1993) These pushing and pulling factors also emerge in wider context concerning the welfare state (see e g Besseling and Zeeuw, 1993, Olsson et al , 1993)

A number of Finnish studies have shown a strong correlation between age and decreasing health (e g , Aromaa et al , 1989, Ilmarinen et al , 1991, Kalimo et al , 1992, Hytti, 1993) In practice, this means a growing need of health services and also early retirement Variation between individuals is, however, great At macrolevel, these phenomena are manifested in decreasing labour force participation rates

In 1989, the Research Centre of the Social Insurance Institution (SII) in Finland started a study the aim of which was to examine the development of the health status of persons aged 53 years and in working life and to follow up their possible retirement The aim of this paper is to describe the self-perceived health the working capacity and the overall life situation at the beginning of the study of the persons who later applied for ordinary

Preparation for Aging Edited by E Heikkinen *et al*
Plenum Press New York 1995

invalidity pension or for special invalidity pension (early disability pension). The follow-up period ended at the end of March 1994.

The material of this study consisted of 1777 subjects. All persons included in the study underwent a wide health examination between April 1989 and April 1992. The follow-up period thus varies from five to two years, for those examined at the beginning of the study (April 1989) and those examined at the end of the study (April 1992), respectively. The follow-up data is based on the registers of the SII.

RESULTS

At the end of March 1994, 16% (n = 280; 142 men, 138 women) of the subjects in the study had applied for national invalidity pension, either for ordinary invalidity pension or for special invalidity pension. Of all these pension claims, 35% (n = 98) were denied. In this respect, there was no statistical difference between men and women.

Persons who applied for invalidity pension were mostly employees, either salaried employees or workers. They had been in work-life for quite a long time, and they considered their economic situation often moderate or poor (Table 1). Gender or marital status had almost no effect on claims for national invalidity pension. About 7% of all subjects were unemployed at the beginning of the study; there was no difference between men and women in this respect.

Subjects who later claimed for invalidity pension assessed at the beginning of the study their health and working capacity worse than the others (Table 2). They also often estimated that their health and working capacity would deteriorate in the future.

In Finland, as in many other countries, musculoskeletal diseases are an increasing problem for the national health policy. This is clearly seen in the claims for national invalidity

Table 1. Background data for the persons examined at the beginning of the study

	Applied for invalidity pension (280)[a]	Others (1497)	Difference between groups, p[b]	All (1777)
Gender (%)			ns	
Men	51	46		47
Women	49	54		53
Marital status (%)			ns	
Unmarried	8	6		6
Married	80	83		82
Others	12	11		12
Socio-economic status (%)			***	
Self-employed	8	26		23
Salaried employees	43	43		43
Manual workers	49	31		34
Duration of working life (years, mean ± sd)	33 4±5.7	31 9±7 2	**	
Economic situation (%)			***	
Good	6	14		13
Quite good	15	20		19
Moderate	66	56		58
Quite poor	10	7		7
Poor	3	3		3

[a](n) in parentheses
[b]*** p<0 001, ** p<0 01, ns = statistically not significant

Table 2. Self-perceived health and subjectively estimated working capacity at the beginning of the study (%)

	Applied for invalidity pension (280)[a]	Others (1497)	Difference between groups, p[b]	All (1777)
Self-perceived health			***	
Very good or good	10	31		28
Nothing special	38	51		48
Poor of very poor	52	18		24
Health situation in the future			***	
Better	4	4		3
No changes	24	42		40
Worse	72	54		57
Self-perceived working capacity			***	
Good or quite good	15	48		43
Moderate	51	44		45
Quit poor or poor	34	8		12
Working capacity in the future			***	
Good or quite good	10	30		27
Moderate	35	52		49
Quite poor or poor	55	18		24

[a] (n) in parentheses
[b] *** $p < 0.001$

pension (cf. Hytti, 1993). In this study, the persons who applied for invalidity pension during the follow-up time suffered at the beginning of the study more often from musculoskeletal diseases (74%) than those who did not later apply for any national invalidity pension (66 %, $p < 0.05$).

The subjects who applied for invalidity pension estimated more often than the others that during the year before the examination their health had deteriorated ($p < 0.001$) and their health would get worse in the future ($p < 0.001$). Estimated changes in life situation and expectations were corresponding in these two groups.

Subjective estimates of the health status and life situation as well as changes in these at various times (year before the examination, at the time of examination, in the future) correlated strongly with each other (Table 3). Obviously, self-perceived health or expected changes therein are not the only factors affecting a person's working capacity or early retirement (cf. Lind and Mäki, 1994). However, of those subjects who applied for national invalidity pension, 35% estimated at the beginning of the study that they probably are in work-life in the near future while the corresponding figure for the others was significantly higher (71%; $p < 0.001$).

DISCUSSION

Retirement can be discussed at individual or community level (micro- and macrolevel) or from the point of view of social risk or use of pension benefits. Retirement and granting a pension are based on, not only, subjective criteria but also on that the criteria stated by law are met. In studies concerning retirement, a general assumption is that the insured applying for pension tend to maximize their lifetime income with reference to their health situation and leisure preferences (e.g. Aarts and De Jong, 1990).

Table 3. Correlations between status of and changes in perceived health and subjectively estimated life situation

		Applied for invalidity pension (280)[a]	Others (1497)
Perceived health			
Changes in health	– Health at the examination	0 534***	0 495***
Health at the examination	– Health in the future	0 321***	0 339***
Subjectively estimated life situation			
Changes in life situation	Life situation at the examination	0 400***	0 252***
Life situation at the examination	– Life situation in the future	0 291***	0 214***

[a](n) in parentheses
*** p<0 001

When the results of this paper are evaluated it must be pointed out that in studies based on questionnaires or interviews the subjects often proportion their possibilities and choices to the overall structure of the community (Hytti, 1993) In addition it must be taken into consideration that during the follow-up time, a strong economic recession has taken place in Finland and unemployment has rapidly grown At the same time, both health care and social security systems have been exposed to substantial reductions All these changes may have had effects on the decisions of the subjects in this study concerning their applications for invalidity pension

The follow-up time varied among subjects of this study, being quite short for some persons Consequently, it is not possible at this stage to make reliable conclusions on. e g , claims for national invalidity pension according to age Later on, when the number of persons applying for ordinary or special invalidity pension is sufficiently high in this study, we can evaluate how the changes occurring at macrolevel, i e economic recession, unemployment, arrangements in social security system etc , affect the willingness to retire

The duration of one's work-life is an essential factor forming a person's pension When estimating the meaning of the life situation in the retirement process, it must be taken into consideration that reductions of pension security possibly occurring and the high level of unemployment evidently prevailing in the near future may have a retarding effect on the level of living of the early retired persons and thus influence the number of applications for early retirement As to the elderly, this means that more attention must be paid to work conditions in general and especially to those of the elderly This, naturally, has a positive effect on the development of work conditions as a whole (cf Dahlgren and Whitehead, 1992)

The use of unemployment pension, special invalidity pension and invalidity pension depends on the changes in society and legislation During the last years, the number of early retirement pensions has increased and that of invalidity pensions decreased in the older age groups in Finland In this respect, Finland is different from e g Sweden (see Esping-Andersen, 1990, Wadensjo, 1991) It is not possible to estimate the trends of incidence and prevalence of invalidity pensions in isolation of other early pensions Long-term unemployment has rapidly grown during the last years and is estimated to remain at a high level until the end of 20th century These two facts mean that the exit of labour must be seen as a whole, including both retirement and unemployment (cf Kohli and Rein, 1991)

REFERENCES

Aarts, L J M , and De Jong, Ph R , 1990, *Economic Aspects of Disability Behavior* Erasmus Universiteit, Rotterdam

Aromaa, A , Heliovaara, M , Impivaara, O , et al , 1989, *Health Functional Limitations and Need for Care in Finland* , Publications of the Social Insurance Institution, Finland, AL 32 Helsinki and Turku

Besseling, P J , Zeeuw, R F , 1993, *The Financing of Pensions in Europe Challenge and Opportunities* , CEPS Research Report No 14

Dahlgren, G , Whitehead, M , 1992, *Policies and Strategies to Promote Equity in Health* Regional Office for Europe of the World Health Organisation, Copenhagen

Esping-Andersen, G , 1990, *The Three Worlds of Welfare Capitalism* Polity Press Cambridge

Hytti, H , 1993, *Social and Societal Determinants of Disability Pension Incidence* Publications of the Social Insurance Institution, Finland, M 87, Helsinki

Ilmarinen, J , Tuomi, K , Eskelinen, L , et al , 1991, Summary and recommendations of a project involving cross-sectional and follow-up studies on the aging worker in Finnish municipal occupations (1981 - 1985) *Scand J Work Environ Health* 17(Suppl) 13 -141

Kalimo, E , Klaukka, T , Lehtonen, R , Nyman, K , 1992, *Health Security in Finland and Needs for Development* Publications of the Social Insurance Institution, Finland, M 81 Helsinki

Kohli, M , Rein, M , 1991, The Changing Balance of Work and Retirement, in *Time for Retirement Comparative Studies of Early Exit from the Labor Force* M Kohli, M Rein, A M Guillemard, and H van Gunsteren, eds , Cambridge University Press, Cambridge, pp 1 - 35

Laczko, F , and Phillipson, C , 1991, *Changing Work and Retirement Social Policy and the Older Worker* , Open University Press, Philadelphia

Lind, J , and Maki, J , 1994, Aging, self-perceived health and retirement *J Soc Med* 31 75-82

Olsson, S E , Hansen, H , Eriksson, I , 1993, *Social Security in Sweden and Other European Countries—Three Essays'* Rapport till expertgruppen for studier i offentlig ekonomi Finansdepartementet, Ds 1993 51, Stockholm

Wadensjo, E , 1991, Sweden Partial Exit, in *Time for Retirement Comparative Studies of Early Exit from the Labor Force* , M Kohli, M Rein, A M Guillemard, and H van Gunsteren eds , Cambridge University Press, Cambridge, pp 284-323

30

THE PROCESS OF EARLY RETIREMENT AMONG THE 55-YEAR OLD FINNISH URBAN POPULATION

Aira A. Uusimäki, Ulla Rajala, Sirkka Keinänen-Kiukaanniemi, Hannu Virokannas, and Sirkka-Liisa Kivelä

Department of Public Health Science and General Practice
University of Oulu
Aapistie 1
FIN-90220 Oulu
Finland

INTRODUCTION

It is a well-recognized fact that more and more people are exiting from the labour force by retiring before their pensionable age. This growing trend is seen in many industrialized countries (Jokelainen, 1991) and it has been predicted to lead to social and economic problems in the near future because of the aging of the population (Valkonen, 1994).

In Finland the numbers of early retired have grown from 216 000 to 415 000 during the last two decades (Hytti et al., 1992). The growth of early retirement has been most prominent in the oldest age groups of the labour force (55-64 years). One explanation for this trend is the renewals in the Finnish pension system in the last decade: early retirement from the age of 55 years became more attainable because of the changing the pension criteria (Storsjö, 1991). In general, until the age of 60 years the assessment of work disability by a physician is the necessary basis for applying for retirement before the pensionable age. These changes in the pension system offered new pathways for aging people to exit from the labour force and were welcome both to employees and employers (Kohli et. al., 1991).

It has been explicitly recognized that retirement is a process with different phases: 1. Thinking of retirement, 2. Decision to retire and 3. Act of retirement (Beehr, 1986; Atchley 1988; Ekerdt and DeViney, 1993). The progress of this process varies individually and is regulated by the institutions of the society (Kohli et al., 1991; Huuhtanen, 1994). It is a multifactorial life-event where an active, productive role in the labour-market is changing to a less active and less productive role.

In previous studies it has been shown that early retirement is more influenced by subjective factors (self perceptions of health, attitudes toward work and retirement) than the structural sociodemographic factors (Palmore et al., 1982). Poor health, long working history and strenuousity of the work are the most common subjective grounds for seriously thinking

Preparation for Aging, Edited by E Heikkinen *et al*
Plenum Press, New York, 1995

of early retirement (Huuhtanen and Piispa, 1991, 1993; Gould et al., 1992; Gould 1994). In recent years ordinary health risks at work places and physical strain have decreased (Karisto, 1987) and health status of the population is better than before (Klockars, 1994), but early retirement has not diminished. The panorama of diseases is different from that previously; mental ill-being at work has become usual and therefore the assessment of work disability has become indefinite (Karisto, 1987).

To be able to understand and have a positive influence on this process it is important to know its first phase. The purpose of this research is to discover how the process of early retirement had spread in the 55-year old Finnish urban population, what sociodemographic features are typical for those who are in the process of retirement and what are the subjective grounds for the retirement orientation.

SUBJECTS AND METHODS

The population consisted of 1012 persons born in 1935 and living in the city of Oulu 1.10.1990. Postal questionnaires, interviews, clinical examinations and laboratory investigations were used in collecting data in 1990-91. The material was collected by a nurse and two physicians. Altogether 780 (77%) participated. Of the 232 non-participants 53% returned the short version of the mail questionnaire. The proportions of retired and retirement-oriented were similar among participants and non-participants.

The first phase of the process of retirement was measured at the clinical examination by asking" What are you thinking now, are you going to try to retire before your pensionable age?" Those who answered "Yes" were named retirement-oriented and those who answered "No" or "Don't know" were named work-oriented.

The people reported, if they already had retired.

RESULTS

One third of the population had already retired and also one third were retirement-oriented among the study population. Only 36% of this 55-year old population were work-oriented (Table 1).

The profile of marital status was different in these three groups. Among retirement-oriented and retired men there were fewer married than among work-oriented men. Similar differences between the groups were seen also among women.

Among retirement-oriented and retired people the basic educational level was lower than among work-oriented (Figure 1). The level of vocational training was also lower among retirement-oriented and retired (Figure 2).

Table 1. The process of early retirement among 55-year-old
Finnish urban population (%)

The phase of retirement	Men N=345	Women N=435	Both N=780
Work-oriented	30	40	36
Retirement-oriented	30	31	31
Retired	38	23	3
Missing data	2	6	3
Total	100	100	100

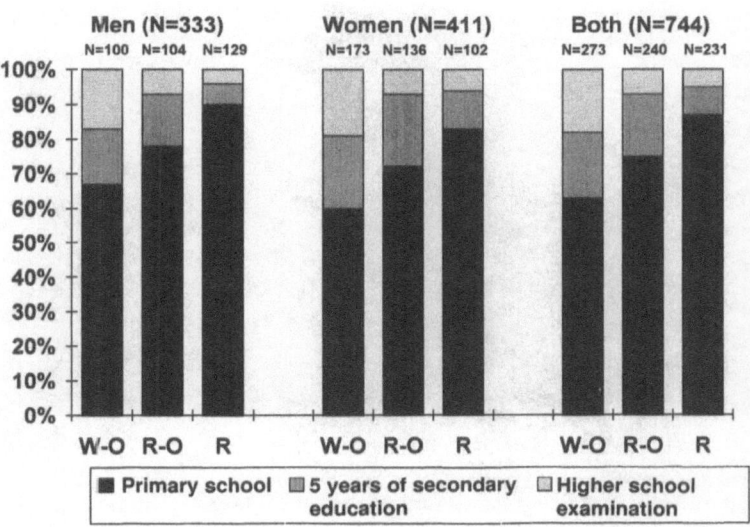

Figure 1. The profile of basic education in the groups of work-oriented (W-O), retirement-oriented (R-O) and retired (R)

The proportion of blue collar workers was higher among retirement-oriented and retired than among work-oriented (Figure 3). Among retirement-oriented there were as many upper white collar workers as among work-oriented.

The main subjective grounds for retirement-orientation were those concerning health status (Table 2).One fifth of men but one fourth of women proposed muscular skeletal symptoms as the main subjective ground for retirement-orientation. 29% of retirement-ori-

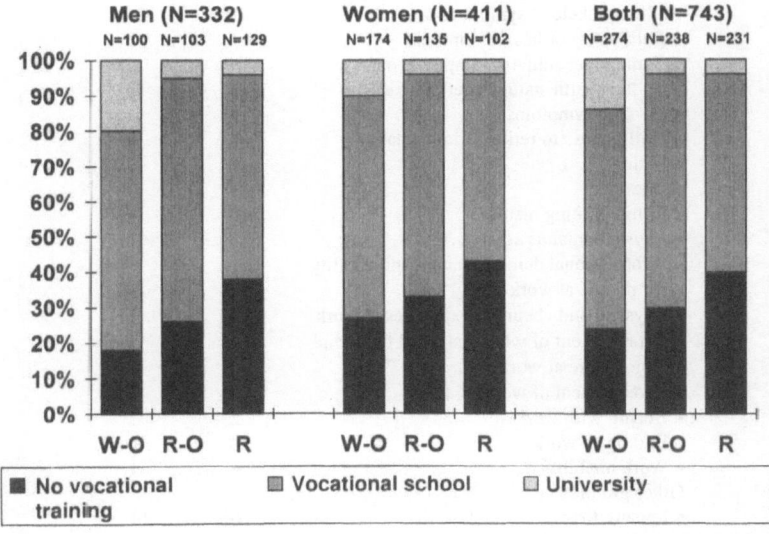

Figure 2. The profile of vocational training in the groups of work-oriented (W-O), retirement-oriented (R-O) and retired (R)

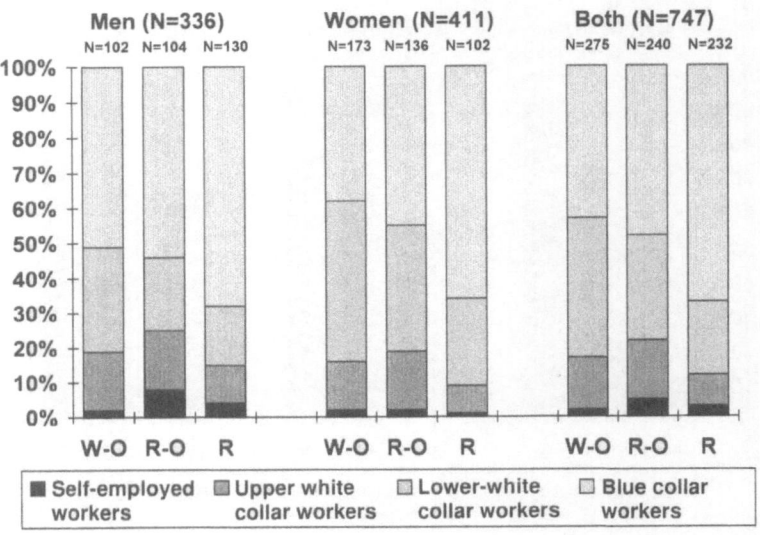

Figure 3. The profile of an occupational status in the groups of work-oriented (W-O), retirement-oriented (R-O) and retired (R).

Table 2. The main subjective grounds for retirement-orientation (%)
(Classified by the authors)

Subjective ground	Men N=105	Women N=134
Health status:		
• Muscle skeletal symptoms	19	24
• Pulmonary or heart symptoms	5	2
• Some other somatic symptoms or "if the health status is getting bad"	14	10
• Psychic symptoms	6	3
• Willingness to retire in good health	6	8
• Fatigue	1	8
Work:		
• Long working history	6	8
• Physic demands at work	6	11
• Informational demands and responsibility for people at work	1	1
• Physical and chemical exposures at work	2	1
• Arrangement of work tasks and the social atmosphere at work	3	2
• Arrangement of working time	0	1
• Boring with work	2	1
• Ending of work	4	3
• Work unability	4	1
Other grounds:		
• Leisure time	20	14
• Feeling too old	2	2
Total	100	100

ented felt that the main ground for retirement-orientation was some difficulty at work. Leisure time activities were also quite a common main subjective ground for retirement-orientation.

DISCUSSION

The process of early retirement seems to be very common among the 55-year-old Finnish urban population. The number of retired was higher among this population than in the whole country (24.2%) (Central Pension Security Institute, 1991). The target population lived in the northern part of Finland. In previous studies it has been shown that incidence and prevalence of most diseases are higher in the northern and eastern part of Finland, where the standard of living is lower (Harni, 1973; Aromaa et al., 1988). Also the development of work disability has been closely dependent on the socioeconomic level of the area (Harni, 1973). Still we can assume that the population who was born in the same year and living in the same country are comparable because of their experiencing the development and changes of the society (Roos, 1987). The target population belongs to a generation who started in working life after the Second World War. Typical for this generation is a low educational level and large changes in working life during their working history.

About one third of the study population was retirement-oriented. In previous studies (Huuhtanen et al., 1991, 1993; Gould et al., 1992) different questions have been used in measuring the thinking of retirement. Still in the age groups of 55-61 years the form of the question about retirement thoughts seems not to have any great impact on results: in both studies about one third had thought seriously or continuously to retire before their pensionable age.

The process of retirement has progressed longer among men than among women, among people with lower educational level or lower occupational status and among unmarried people. These all are features which are associated with poor health (Aromaa et al., 1989). Poor health was also the most common subjective ground for retirement-orientation. According to the previous studies it is known that disabling chronic diseases are common in the adult population (Aromaa et al., 1989) and their number grows during aging.

Poor health is also necessary for applying for retirement before the pensionable age of 60. In spite of the demand of poor health for early retirement almost one half of the retirement-oriented people proposed some other grounds for thinking of early retirement. Difficulties at work and leisure time activities were common subjective grounds for retirement thoughts. Problems at work or the need for more leisure time are not enough to make it possible to retire before the pensionable age - there must be found a medical diagnosis which causes work disability.

The results of this population study support the previous findings concerning early retirement as a process. They fit the previous findings of the studies and complete the picture of the process of early retirement in Finland by describing it among population living in the Northern part of the country.

REFERENCES

Aromaa, A, 1989, Sairaat, vajaakuntoiset ja pitkäaikaisesti hoitoa ja apua tarvitsevat vuosina 1980 ja 2000, in: *"Terveys, toimintakyky ja hoidontarve Suomessa. Mini-Suomi-terveystutkimuksen perustulokset"*, [*"Health, Functional Limitations and Need for Care in Finland. Basic Results from the Mini-Finland Health Survey"*], Publications of the Social Insurance Institution, AL:32, Helsinki and Turku, English summary.

Atchley, R C , 1988, *"Social Forces and Aging"*, 5 edition, Wadsworth Publishing Company, Belmont

Beehr, T , 1986, The process of retirement a review and recommendations for future investigation, *Pers Psych* 39 31-55

Central Pension Security Institute and Social Insurance Institution, 1992, *"Statistical Yearbook of Pensioners in Finland 1991"*, Government Printing Centre, Helsinki

Ekerdt, D J , De Viney, S , 1993, Evidence for a preretirement process among older male workers, *J Gerontol* , 48 S35-S43

Gould, R , Takala,M Lundqvist,B , 1992, *"Varhaiselakkeelle hakeutuminen ja sen vaihtoehdot"*, [*"Applying for an Early Pension and It's Alternatives"*] , Central Pension Security Institute, Studies 1992 1, Helsinki, in Finnish

Gould, R , 1994, *"Tyoelama takanapain?"* [*"Is the Working Life Over?"*], Central Pension Security Institute, Studies 1994 3, Helsinki, in Finnish

Harni, A -L , 1973, *"Sairauksien kehityksen alueittainen vaihtelu Suomessa"*, [*"Regional Variations in the Development of Illnesses in Finland"*], Publications of the Social Insurance Institution A 10, Helsinki, English summary

Huuhtanen, P , Piispa, M , 1991, *"Elake ajatuksissa Tyossa olevien tyo- ja elakeajatukset "*, [*"Thinking of Continuing in Work or Retireing Among Working Population"*], Finnish Institute of Occupational Health and Work Environment Fund, Helsinki, in Finnish

Huuhtanen, P , Piispa, M , 1993, *"Tyo- ja elakeasenteet"* , [*"Attitudes to Work and Retirement"*], Finnish Institute of Occupational Health and Work Environment Fund, Helsinki, in Finnish

Huuhtanen, P , 1994, Tyossa vai elakkeelle?, [To Continue in Working Life or Retire?], in *"Ikaantyminen ja tyo'*, [*"Aging and Work"*], Finnish Institute of Occupational Health, WSOY, Juva, pp 151-166, in Finnish

Hytti, H , Kettunen, S , Lindell, C , Peltonen, H , 1992, *"Varhaiselakkeet Suomessa 1970-90"*, [*"Early Pensions in Finland 1970-90"*], Publications of the Social Insurance Institution T9 44, Helsinki, in Finnish

Jokelainen, M , 1991, *"Elakelaisten osuus 55-64-vuotiaasta vaestosta kymmenessa OECD-maassa vuonna 1989"*, [*"The Proportion of Retired among 55-64 -Year Old Population in the Ten of OECD-Countries in 1989"*], Publications of the Social Insurance Institution T9 42, Helsinki, in Finnish

Karisto, A , 1987, Tyoelaman muutos, [Changes in Working Life], in *"Tyopsvkologia "*, [*"Work Psychology"*], Finnish Institute of Occupational Health, Helsinki, in Finnish

Klockars, M , 1994, Ikaantyminen, tyokyky ja tyokyvyttomyys, [Aging, Work Ability and Work Disability], in *"Ikaantyminen ja tyo"*, [*"Aging and Work"*], Finnish Institute of Occupational Health, WSOY, Juva, pp 232-250, in Finnish

Kohli, M , Rein, M , Guillemard, A -M , van Gunsteren, H , eds , 1991, *' Time for Retirement"*, Cambridge University Press, Cambridge

Palmore, E B, George, L K, Fillenbaum, G G , 1982, Predictors of retirement, *J Gerontol* 37 733-742

Roos, J -P , 1987, *"Suomalainen elama"* , [*"The Finnish Life']*, Karisto Oy, Hameenlinna, in Finnish

Storsjo, R , ed , 1991, *"Tyoelakelait 1991"*, [*"The Legislation Regulating Work Pension"*], Central Pension Security Institute, Helsinki, in Finnish

Valkonen, T , 1994, Tyoikaisen vaeston vanheneminen , [Aging of the Labour Force], in *Ikaantyminen ja tyo"*, [*"Aging and Work"*], Finnish Institute of Occupational Health, WSOY, Juva, pp 16-26, in Finnish

31

EDUCATING HEALTH PROFESSIONALS IN GERONTOLOGY

A Canadian Perspective

A.C. Beckingham

School of Nursing
McMaster University
Hamilton, Ontario
Canada

INTRODUCTION

A new Statistics Canada report, bringing together years of analysis based on 1991 census data offers a comprehensive portrait of a society dominated by an overwhelming trend—aging. And some experts warn that society still hasn't come to grips with the consequences. We're now much more aware of population aging than we were before, but have we come to terms with it?

The traditional pyramid of age groups in a society—with many young people at the base and small numbers of elderly at the top—has been turned upside down over the last 30 years (Kane et al., 1990; Beaujot, 1991; Beckingham and DuGas, 1993). In the 10 years from 1981 to 1991 the number of older adults rose to 3.2 million from 2.4 million. The number is expected to reach 8 million by the year 2031.

Concern about the social effect of aging has largely focused on pension plans and health costs, but the implications spread into every area, from transportation planning to crime prevention. A 50-year-old is less likely to take the subway than a youth. Older people are more likely to engage in white-collar crime than street crime. The aging of society is mainly due to an unprecedented drop in birth rates over the last 30 years. Population growth has continued, largely due to immigration, and the total population increased by 7.9 per cent from 1986 to 1991.

In the past decade scientific and analytic advances have improved our ability to forecast disability rates, and assess the health care services and manpower implications of an aging society. Some forecast that by mid-century the total number of older persons will constitute nearly one fourth of the population. These projections create the need for a radical transformation and restructuring of the health services labour supply including long-term care, and in conceptualizing geriatric/gerontological education and training in aging-related fields.

Preparation for Aging, Edited by E. Heikkinen *et al.*
Plenum Press, New York, 1995

What will be higher education's role in preparing workers and professionals to effectively face the future? Will gerontologists, the health professions, academic health centres, and researchers in aging assume the leadership in envisioning new and different ways to structure and deliver health and social services, and novel approaches to educating those at the centre of the dilemma practitioners, researchers, and future educators?

This paper will focus on one Canadian provincial initiative in the development of a network, and a cadre of health care professionals. A brief review of background information will be followed by major factors, relevant topics clustered together and a few selected individual challenges to educational gerontology. Additional issues impacting on education, challenges and choices for the future with some recommendations will conclude the paper.

BACKGROUND INFORMATION

A one-day Workshop on Educating Health Professionals in Gerontology and Geriatrics organized by the Educational Centre for Aging and Health, McMaster University, Hamilton, Ontario, Canada, was held on July 3, 1993, prior to the XVth International Congress of Gerontology, in Budapest, Hungary, July 4-9, 1993 (Skeet and Beckingham, 1993).

The purpose of the Workshop was to raise awareness of the need for education and to stimulate action at an international level. The programme was designed to consider the subject from a World Health Organization (WHO) Regional (Hermanova, in press) level to a provincial structural level. Discussion ranged from out-dated curricula and perpetrated myths to future international needs and prospects for meeting them.

The organizers put forward Canada as a case example in educational development. In Canada the importance of education about aging and in the care of older adults has been recognized. Referring to academic gerontology Thornton (1992) describes the concepts, the growth of gerontology in Canada, and proposes a schema for study, research, and teaching. Ryan (1993) in an unpublished paper identified five key issues that have emerged over a decade of debate in Canada:

  i. multidisciplinary education for multidisciplinary work in aging and health;
  ii. education of health professionals from diverse ethno-cultural traditions—need for interpreters of culture and customs as well as of language;
  iii. education for health care practice in rural areas;
  iv. education for enabling older adults to maintain independence despite various health problems, including support for informal care-givers; and
  v. integration of research and education.

The Ontario government, in 1986 requested proposals for a centre of excellence in education in aging and health based on:

  i. universal aging education within professional educational programmes;
  ii. development of health professionals who are specialists in aging and may serve as planners, managers, educators or researchers; and
  iii. continuing education for health and social service professionals.

The Educational Centre for Aging and Health (ECAH), (see Figure 1) its five components and supporting activities are described by Macpherson and Blumberg (1992) and Macpherson (in press).

Selected workshop and other papers (Beckingham, in press, Beregi, in press, Greaves, in press, Mold, in press, Watt and Meredith, in press) are included in a special theme

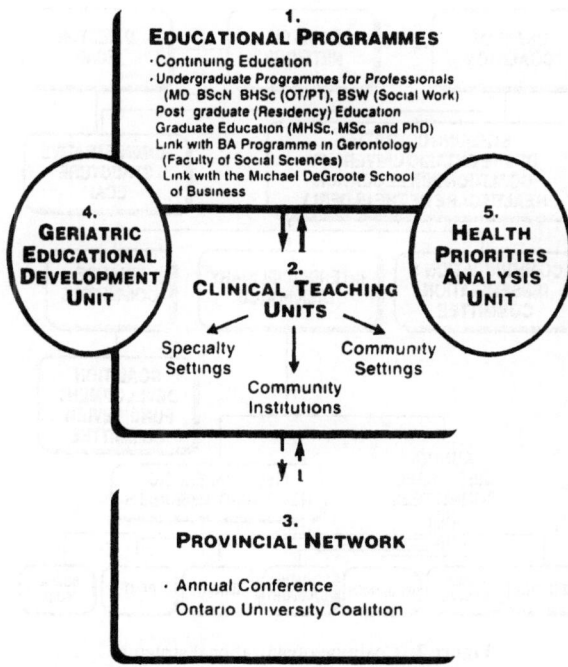

Figure 1. Educational centre for aging and health

issue of Educating health professionals in gerontology and geriatrics, *Educational Geron-tology: An International Journal* (Beckingham et al., 1995).

The ECAH mission statement is as follows:

The Educational Centre for Aging and Health seeks through professional education to enhance the quality of life, self-determination and well-being of older persons living in the community and in institutional settings in Ontario by

- increasing the number and proportion of skilled health professionals who are committed to promoting health and providing excellent care for aging individuals,

and

- developing collaborative interdisciplinary and interprofessional educational approaches and models concerning aging and health and evaluating their effectiveness

Achieving this mission requires a systems approach that includes but is not limited to the educational institutions, stressing the need to build widespread support for educational change within the university, with government and in the community

Because of its province-wide mission and because of expressed need by health professionals in the field of aging to meet regularly to discuss common interests in education, ECAH facilitated the development of the Ontario University Coalition for Education in Health Care of the Elderly (Puxty, in press). The Coalition (see Figure 2) developed the following goals:

i. provide opportunities for mutual support and joint action for improvement of education in health care of the elderly in Ontario;

ii. encourage and support information exchange on educational activities intended to strengthen the aging components in the education of health professionals; and

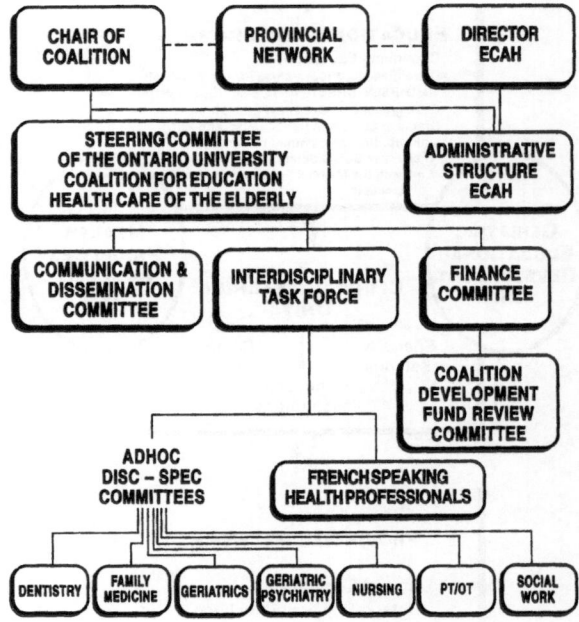

Figure 2. Coalition organizational structure.

iii. ensure the incorporation of appropriate core curricula on aging and health care of
the elderly in all five Ontario health science centres.

CHALLENGES

Population aging does not inevitably pose a "crisis" for Canadian society, it does,
however, present challenges to prevailing conceptions of the life course, to the ways
Canada's public and private institutions deal with older adults and to educational programmes of the future. Some of the emerging issues, selected on the basis of seniors' needs
and rights identified in the following section, suggest that Canada faces and must plan for
some complex transitions in responding to these challenges in the years ahead as we move
into the 21st century. The following are a few factors, clustered with selected sub-headings
and relevant challenges (Health and Welfare, 1986; National Advisory Council on Aging,
1989).

HEALTH AND SOCIAL CARE

Health Services

Canada has been a leader in health care, and all of its citizens have benefited. Seniors
as a group are quite healthy. A relatively small portion of the senior population makes
disproportionate use of the health care system. When seniors develop health-care needs,
however, these needs can be described, in general, as long-term, chronic disorders requiring
long-term caring. Canada's health-care system, on the other hand, may be characterized as

institution-based, high-technology and cure-oriented. It is clear that if Canada continues to deliver health care for seniors that places a heavy emphasis on cure or institutionalization, the costs may become unbearable. The present structures will not work effectively in the first decades of the next century when a much larger proportion of the population will be "old".

Challenges:

- reallocating resources from technologically-sophisticated, acute, institutional care to long-term, chronic, community-based care;
- maintaining high technology while providing humane care;
- moving the system toward a primary health care (PHC) health promotion/disease prevention/disability, postponement /rehabilitation orientation;
- confronting issues of territoriality between levels of government and between health-related disciplines;
- resolving ethical issues such as the rationing of health-care resources, euthanasia, dying with dignity, and respecting individual wishes; and
- ensuring that programs and services meet the changing needs of older adults effectively.

Informal Caregivers

Families provide the bulk (85 to 90%) of the care and support for their senior members in the community. Research reveals that family caregivers are generally women—wives, daughters and daughters-in-law—aged between 30 and 80.

In the future, the pool of available caregivers is likely to change with increasing numbers of women in the labour force, the trend toward smaller families, increases in separation and divorce, and the continuing mobility of the Canadian population.

Challenges:

- finding ways to support family caregivers;
- promoting informal support networks; and
- ensuring funding for services that provide support to caregivers and promoting development of services across the country.

Community Services

There is no doubt that seniors prefer to remain in their own homes in their own communities. Such community-based services as homemakers, meals-on-wheels, congregate meals, and visiting nurses, together with supportive physical environments, can sustain seniors' autonomy and help many to remain in their communities. However, community-based and supportive services for informal caregivers have not been integrated with health-care policies and legislation. In many provinces, they are provided on an inconsistent, piecemeal basis.

Challenges:

- resolving issues concerning the availability and accessibility of community and institutional services for older persons;
- establishing satisfactory levels and standards of care, quality of training and employment conditions for workers; and
- co-ordinating and integrating services, not only geographically, but also among levels of care, among various disciplines and between the formal system and the informal support network.

Institutional Services

As long as senior Canadians suffer from dementias such as Alzheimer's disease or have major disabilities and infirmities that require intensive nursing, Canada will continue to have a need for institutional care.

Challenges:

- resolving problems related to the quality of patient care, patients' rights and patients' security;
- determining the implications of proprietary versus public ownership for patient care;
- meeting the need for more facilities providing a continuum of care with multiple levels of care within the facility; and
- integrating homes for the elderly within the community.

Mental Health

Older adults now constitute the majority of patients in psychiatric hospitals, and a high proportion of residents of nursing homes and other long-term care institutions have mental illnesses. As well, Canada is experiencing increasing rates of later-life alcoholism and suicide for men over 60. Taken together, these indicate the inappropriate and inaccessible care for older adults with mental health problems.

Challenges:

- promoting mental health for older adults by ensuring adequate and appropriate services (health care, income maintenance, housing and transportation);
- encouraging social networks, such as good family and peer relationships;
- encouraging understanding and preparing for aging and the development of positive societal attitudes to aging, and the senior population through educational opportunities, facilitation of information development and communication;
- ensuring accessible assessment and treatment facilities; and
- supporting caregivers of those with cognitive impartment.

Self Care

Self-care - the personal health practices and lifestyle decisions that people make to protect and restore their health - together with health-promotion efforts aimed at fostering people's control over their own health have been important elements of health care in recent years.

As well, people's efforts to plan for other aspects of their later years (financial, residential, social) will increase the likelihood of a fulfilling later life.

Challenges:

- promoting healthy life styles over the life course;
- promoting an active role for older adults in taking responsibility for their own well-being;
- re-educating service providers (including rethinking their role); and
- recognizing the role of informal and support groups in contributing to seniors' self-care (e.g., peer counselling) and integrating them into the health and social services systems.

INCOME

Poverty

Income is a critical determinant of health and life satisfaction for older people. There is considerable literature today on the impact of productive activity across the life course on economic well-being in old age (Kerschner and Nusberg, 1993; Stoller and Gibson, 1994).

We have seen advances in the income of seniors in the last 20 years that reflect such societal changes as recognition of seniors' contributions to society, changing family forms, and the value of women's unpaid domestic work. These advances include increasing and indexing levels of Old Age Security (OAS), Guaranteed Income Supplement (GIS) and Canada/Quebec Pension Plan (C/QPP); splitting C/QPP credits between spouses at retirement or on divorce; requiring employers to provide pensions that are portable, vested, etc.; and ensuring equal opportunity and equal pay for work of equal value in the labour market.

Challenges:

- ensuring that retirement income from the public and private systems and from retirement savings is at a level that maximizes the number of older Canadians, both men and women, who can maintain a reasonable standard of living in their retirement years;
- ensuring that pension funds are protected for the benefit of the participants;
- ensuring that income support programs reduce poverty and meet the needs of at-risk populations; and
- encouraging early planning for retirement.

Intergenerational Equity

Today, older people are not the poorest segment of Canada's population. Single-parent families with children and young adults currently experience higher rates of poverty. Research has shown that children brought up in poverty do not fare as well in areas like health and education. Poverty in childhood can thus have life-long consequences. If Canada continues to sustain its income security programmes through taxes, the tax burden needed to support retired baby boomers may become onerous.

Challenges:

- ensuring the equitable allocation of society's resources among various age groups;
- developing alternatives to avoid onerous tax burdens for future workers; and
- examining the pension system with a view to further reform.

THE ENVIRONMENT OF AGING

Housing

A person's quality of life, which encompasses such factors as self-esteem, a sense of identity, and access to services, is significantly influenced by his or her physical surroundings. Access to housing that is affordable, safe and designed (in terms of size and features) for older adults is therefore of central importance in determining the quality of life. Many of these factors are receiving increasing attention in residential building and community planning.

Most elderly people in Canada (about 87%) live independently in the community; a further 6% live in supported environments. Only 7% are residents of institutions. Two-thirds of older people, in fact, own their own homes and the cost of maintenance and repairs can be high. This may be a particular problem for the large number of older women who will in the future be living alone.

Challenges:

- resolving issues related to jurisdiction in both the funding of affordable housing and the development and co-ordination of community support services;
- designing environments that will help to sustain relationships in old age;
- developing a range of innovative housing options, housing sizes, housing styles and housing locations, with the attendant flexibility in building codes and zoning bylaws;
- reviewing financing alternatives; and
- providing information so that older adults can be aware of and understand the range of possibilities open to them.

Transportation

Transportation is one of the key ingredients of an independent lifestyle for older adults. Although recent improvements in public systems (ramps, shelters, door-to-door service) can assist elderly and disabled people and volunteer driver programmes are more generally available, much remains to be done. Inconsistent or inadequate transportation services in rural and remote communities limit access to essential health and social services, particularly in time of crisis.

A key means of transportation for Canadians - young and old alike - is the car. Although work is being done in the field of traffic safety for older drivers and pedestrians, the traffic planning and control environment is still geared to young adults.

Challenges:

- changing the focus of government and industry from young adults to older adults;
- emphasizing the need for universal, accessible and integrated transportation (public as well as specially adapted services);
- developing a broad range of transportation systems for rural areas; and
- providing more information on the extent, distribution and evolution of needs for mobility assistance so that appropriate planning can take place.

Safety and Security

Older adults are especially vulnerable to accidental injuries, some of which result in long-term disabilities. Vulnerability to accidents stems from many sources, including changes in the older person's physical capacities, such as sensory changes, declining agility, and loss of speed in reacting to stimuli; environmental circumstances such as weather conditions, slippery floors and dim lighting; and the use of medications or alcohol.

As well, seniors' vulnerability to accidents and their justified concerns about having accidents or falling ill and not being able to get help increase their general anxiety. In the future, when even larger numbers of older people live alone, this sense of vulnerability may be a serious impediment to independent living. Victimization and abuse of seniors has received little attention. In the future, when family and institutional caregiving resources may become increasingly strained or over-extended, abuse may be more prevalent.

Challenges include:

- ensuring that all facets of seniors' safety and security concerns are priorities for research and receive responsive service and educational attention.

INFORMATION AND TECHNOLOGY FOR AN AGING POPULATION

Education

Education has traditionally been considered the domain of the young. Yet more and more seniors are recognizing the benefits of lifelong learning and returning to the educational system. Education for older adults is important both as an interest activity and for training or retraining in particular employment skills. In addition, continuing education can be essential in ensuring functional literacy for older adults. Functional literacy is not just the ability to read and write; it is the ability to cope with the demands of the work place and social situations. In the future, older adults will probably demand even more educational opportunities for literacy upgrading, skills maintenance, and retraining occasioned by technological advances or changing labour market requirements.

Education as a leisure activity is also likely to increase; the more educated people are, the more likely they are to seek updating or new learning opportunities. While only one-quarter of today's seniors have graduated from an educational programme beyond the elementary school level, fully two-thirds of the seniors boom generation will have done so.

Challenges include:

- finding the most appropriate methods (including teaching methods and delivery methods such as electronic means) and locations for learning;
- encouraging society to invest in education as a lifelong process;
- finding ways to reduce functional illiteracy among seniors;
- increasing the number of teachers and leaders specially trained in gerontology to plan and guide education about aging;
- creating educational and in-service training and retraining opportunities (age should not be a criterion); and
- finding financing arrangements and other modifications in the service-delivery system so that appropriate education about aging is a prerequisite to working with older adults.

Information and Research

We live in an information-oriented world, but as yet little attention has been given to developing and disseminating useful and reliable information about aging and related issues. Policy analysts, programme planners, product manufacturers and service deliverers need accurate information about the effectiveness and efficiency of their services or products older adults. Older adults, their families and service providers need information about services and products - what they are, what they are for, how to obtain them, and their availability or eligibility criteria. Information must be available when and where it is needed and must be up to date.

The development of information depends on reliable research, which in turn requires trained people, input from older adults, and adequate and secure financing. Challenges:

- promoting better communication between information developers (researchers) and information users;
- establishing standard terminology and definitions across Canada and internationally;
- developing services with continuing responsibility for collecting information, packaging and disseminating it in the most appropriate form to the various information users; and
- providing adequate and secure funding of the infrastructure to support multidisciplinary research and longitudinal studies.

Technology

Technological developments have the potential to assist older adults to avoid, alleviate or overcome potential obstacles to independent living. Technology has already provided potential benefits in such devices as home monitoring devices, emergency response systems, medical aids, apparatuses to assist in walking, remote devices permitting mobility-limited individuals to control a variety of appliances, specially adapted telephones, and improved clothing design, including the use of velcro closings and light-weight, easy-care textiles. Yet in many instances these technologies are not used to the fullest advantage of the older adult. There are many reasons. Attitudes and established practices generate resistance on the part of older adults and service providers. These products can only be useful if people know about them and how to use them. Finally, the products must be accessible and affordable.

As more and more people age, devoting energy and resources to the development and distribution of technologies to promote independent living by seniors will result in benefits for all of society.

Challenges include:

- encouraging the development of new technologies to meet the needs of an aging population;
- establishing and implementing standards to ensure the safety and suitability of technological developments for an aging population; and
- co-ordinating the diverse areas of responsibility.

ADDITIONAL FACTORS

Demographic Changes

The size of the senior population has tripled in the last 55 years. If current assumptions concerning future rates of births, deaths, immigration are correct, the population of older adults can be expected to triple again by 2030.

The senior population itself is aging and will have profound implications for Canadian society. They will be very different than those of today, better educated, probably healthier, and many will have more disposable income. More women will have had a longer history of labour force attachment. Many may not have the same enduring family, religious, and community ties. They may demand a social and physical environment more suited to their requirements as older adults.

Political Impact

The political impact of an aging society is not clear. Older adults today are thought to exercise their right to vote more than younger adults, and are beginning to consider themselves an enduring and coherent political constituency or lobby. The first decades of the next century, when one voter in four will have passed his or her 65th birthday, may evidence an age-based political consciousness. No matter what transpires senior citizens of the future will have a profound influence on Canada's policies and programmes.

Seniors as Consumers

An aging society will have a major effect on the marketplace, with implications for the types of products and services offered in both the private and public sectors. As the baby boom enters its senior years, we can expect to see changes in everything from the kinds of foods demanded to the kinds of leisure activities chosen. Public and private enterprises will ignore the senior market at their peril. The challenges include addressing seniors' rights and needs as consumers, including such issues as print, language, packaging, and protection from fraud and abuse. Also, eliminating ageist or negative attitudes in marketing to or for the senior population.

Multiculturalism and Ethnicity

In Canada the combined effects of aging and ethnicity have been addressed only recently, a conspicuous omission given the multicultural character of Canada's population. Twenty-two per cent of seniors in Canada report an ethnic origin other than British or French.

It is important to ensure that elderly members of the ethnic minorities have access to and are treated responsibly by institutions and health and social service programmes set up largely for the "dominant" English and French cultures. Also ensure the participation of older members of ethnic groups in the planning of health, social services, and access to cultural and language programmes for seniors who are members of ethnic minorities.

Variation in definitions of kin and family obligations can occur within ethnic groups, among generations within families as can family relationships and values (Stoller and Gibson, 1994).

ISSUES IMPACTING ON EDUCATION

You may wish to categorize the following issues or dilemmas which are not in any order of priority under education, practice (clinical) and research, as challenges to educational gerontology:

- explicit education and training needed for professional care providers, and resocialization for current practitioners to change effectively the disempowering approaches fostered by traditional training and the medicalization of aging (Estes and Close 1993);
- three levels of educational intervention; everyday life, professional practice and service delivery, and social structure representing micro, meso and macro system levels (Riley and Riley 1989);
- determinants of population health, and the powerful link between individual health, and social and physical environments (Gerstein et al. 1991)

- biomedical ideology dominating academic health centres making necessary changes difficult to accomplish;
- nature of care provider-care recipient relationship that creates the situation of dependency (Beckingham and Watt, in press);
- centres of academic excellence that develop the flexibility and capacity to respond to the challenge of an aging society;
- use of a life course perspective which merges theoretical orientations from several disciplines as a framework for an inclusive approach to the study of aging. Stoller & Gibson, 1994 outlined four main premises:

  1. The aging process is affected by individuals' personal attributes, their particular life events, and how they adapt to these events
  2. Sociohistorical times shape opportunity structures differently for individuals with specific personal characteristics, such as being in a subordinate position on a social hierarchy. Thus, people's life events, adaptive resources and aging experiences differ
  3. Membership in a specific birth cohort (i e , being born in a particular time period) shapes the aging experience. Within cohorts, however, the experience of aging differs depending on one's position in systems of inequality based on gender, race or ethnicity, and class
  4. Sociohistorical periods shape the aging experiences of cohorts These historical times, however, have different impact on the experiences of disadvantaged and privileged members of the same cohort (p 3)

- disappointed faculty who are searching for a better way to interest students in the field of aging and to strengthen the students capacity of critical thinking about issues with study of aging. Gerontology is a multidisciplinary field drawing on concepts from many disciplines which makes it intellectually exciting. We need a different approach (Moody, 1994).
- the importance of interdisciplinary and team concepts are highlighted by several authors including Clark (1994) Drinka and Ray (in press). Interdisciplinary education is growing in prominence in gerontology and geriatrics, and much descriptive material on a variety of programmes is available. It is now critical to develop conceptual clarity about how interprofessional education programmes should be designed and implemented (Byrne, 1991; Beckingham and Puxty, 1993).
- inequities in aging within and between countries – aging is not a disease and cannot be prevented.
- adult self-directed learning is experienced by all Health Sciences students at McMaster University. The learner is an active participant discovering those things he/she is ready to discover at a particular phase of his/her development. Problem-based learning (PBL) is learning through problems individually or in small group tutorials. PBL encourages the student to define learning goals, select appropriate experiences to achieve these goals, and to be responsible for assessing his/her learning. In PBL the student is exposed to content areas (Norman, 1991) as they relate to specific clients. It is believed that students educated in this system will practice in a holistic manner, establish a broad definition of health that goes beyond illness, and will be a self-directed learner for life.
- using a community development approach—
  i. intervening at all levels of the educational system;
  ii. teaching where the people are; and
  iii. knowing your community. (Macpherson, in press)
- the use of systems theory dealing with encompassing wholes, with the complex patterns of interaction between the wholes and their parts, and with the interrelationships between the wholes and their context or environment. Systems theory is a set of closely interrelated models spanning different domains and providing

explanations of specific phenomena that are different from those that derive from an analytic scientific paradigm. The various sub-theories do overlap and many of the concepts and applications could be introduced under any one of several headings.

The field of gerontology as an evolutionary field needs to develop an emergent unifying paradigm, in order to find and test its scientific foundations. Systems design is the road that makes the difference, but it is the road less travelled by the educational community (Banathy, 1991). Wendt et al. (1993) present a systems model of the dynamic interaction of components of educational programmes in gerontology (p. 3) (see Figure 3).

CHALLENGES AND CHOICES FOR THE FUTURE

Many challenges and choices have been highlighted throughout the paper, and in the future may be framed by four basis principles. Underlying these principles is a conceptual framework that understands a broad array of factors that represent different views of health that impact on the health status of the individual, and the health of the public (Estes and Close, 1993, pp. 14-17).

The principles are.

1 education/research/patient/care· There should be a greater emphasis on meeting the health needs in the community particularly the need for long-term care,
2 specialized training in gerontology/geriatrics should be included in curricula at all levels of health professional education,
3 interdisciplinary approaches· The principles and practices of interdisciplinary training must be integrated into the academic health centre education of health professionals, and
4 patient empowerment Academic health centres should encourage the development of ecucational approaches that empower the elderly in their own care (Lloyd, 1991, Buzzell et al , 1993)

As stated in the workshop report only 10% of the impact on "quality of life" of older adults with chronic disease or disability, was due to health care services. The remainder depended upon "determinants of health" (Skeet and Beckingham, 1993, p. 14).

The need for future manpower supply and demand projections that include assumptions about future government actions as well as including population changes, service utilization, and client decisions (Peterson et al., 1988; Peterson et al., 1994).

RECOMMENDATIONS

At the Budapest workshop it was recognized that educational programmes vary from country to country, and that exchanges of information on what is currently being taught, what education is planned for the future and indeed, collation of successful educational models and materials would be of great interest and value. While a few members of the group had knowledge of small data collections in Europe, it was agreed that an international data base for such information was needed.

The recommendations from the Workshop were:

At national level:
1. Faculty and staff of all educational institutions for health professionals:
   i. should review and evaluate their programmes for actual and potential geron-tology/geriatrics content. The findings should be collated and exchanged.
   ii. should identify the range of target groups for educational initiatives. The groups should include all those whose work brings them into contact with

Larger environmental influences Population characteristics/demographics/imperatives   Political climate   Social and ethical forces
                          Economic environment and resources   Technological environment (manpower demand)

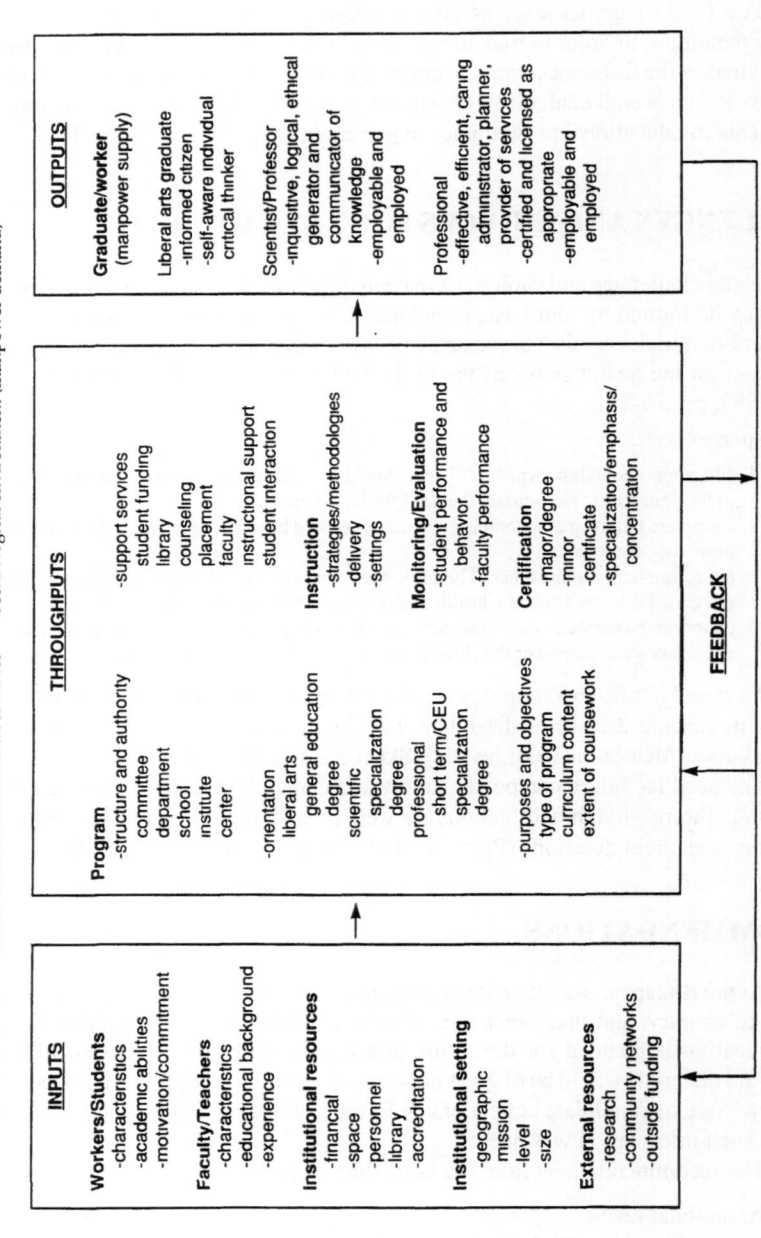

Figure 3. Systems model of the dynamic interaction of components of educational programs

older people in any setting, policy makers and the general public of all ages, including elders themselves.

   iii. should provide programmes for health professionals at graduate, postgraduate and continuing education levels. They should take into account the need for multi-skilled workers as well as specialist gerontologists.

2. Students and practitioners of other human science disciplines should also receive education in appropriate principles and skills in relevant areas of gerontology/geriatrics.

3. A national network should be developed of individuals, faculties, agencies, organizations and institutions interested in the educational aspects of health and aging.

4. National/regional/state/provincial and institutional budgets should include separate and distinct funding for educational endeavours in the gerontological/geriatrics field.

5. Health professionals should work with the media to ensure they have informed perspectives on health and aging and a ready supply of positive role models.

At international level:

6. Collaboration between governments and relevant international and inter- governmental agencies and organizations should be strengthened in order to meet global needs, exchange experiences and achieve educational aims. In the latter context equivalencies in accreditations, and transfer of credits should be established.

7. International networking should be extended to include the exchange of educational resources and materials among individual educators and educational institutions.

8. Consideration should be given to the possibility of "twinning" of faculties of universities on different continents (eg. North America and Europe).

9. An international data bank should be developed to generate, collate and disseminate information on existing educational programmes, and new educational initiatives.

10. Consideration should be given by ECAH to the possibility of organizing an international conference in 1996/97 to enable fuller discussion of the many vital issues raised during this workshop (Skeet and Beckingham, 1993, pp. 16-17).

Although the number and range of post-secondary and continuing education programmes in gerontology have increased in the last decade, there remains a serious lack of education about seniors and aging in occupational training. Because the aging of the Canadian population during the next 40 years will have a significant impact on many services and products, education about aging and seniors will need to be incorporated in the training for several occupations, especially in the health and human service fields.

Education and training are currently a priority on the public agenda. The federal government recently announced its intention to promote post-secondary education, especially in science and engineering, and to encourage training provided by employers to their employees. The development of gerontology education is entirely consistent with the government's stated aim to "reach a national consensus on performance, partnerships and priorities for learning?" The number of persons with gerontological knowledge must be determined, as well as what they will be required to do. Standards of gerontological practice must be established by professional associations. Educational institutions must adapt programmes to provide the required skills and knowledge. Employers must demand gerontology knowledge, provide opportunities and incentives for staff development in gerontology and facilitate the exercise of gerontology knowledge in service delivery. Finally, seniors must insist that the people who serve them know enough about aging to understand and respond appropriately to their needs.

In short, the National Advisory Council on Aging (NACA) recommends that:

1. The Federal and/or provincial/territorial governments promote gerontology education by:

- conducting a national study to determine human resource needs, to compile a complete inventory of current gerontology instruction and to propose content guidelines for existing and new programmes in gerontology;
- providing targeted funding to post-secondary institutions to develop gerontology programmes;
- offering financial assistance to individuals intending to pursue post-secondary education in gerontology, particularly to persons from ethnocultural minorities in health and social service occupations;
- examining requirements for gerontology education in all publicly-funded services and care facilities and providing financial assistance to these agencies and facilities to raise their educational requirements.

2. Post-secondary institutions increase the amount and quality of gerontology instruction by:

- recruiting new faculty with a specialty in aging, providing them with opportunities for development in gerontology and supporting program initiatives in gerontology;
- increasing and diversifying the training experiences with seniors provided to students in health and human service programmes.

3. Educators in gerontology and professional associations develop educational content for pre-employment and continuing education programmes that:

- provide multidisciplinary perspectives on aging and teach service providers to work in multidisciplinary settings;
- address the ethnocultural diversity of seniors and their families;
- train service providers to communicate effectively with informal caregivers and to respond to their needs for information and support;
- are directed to all the employees of service agencies or care facilities whose duties have a significant impact on the lives of seniors;
- can be recognized by the accreditation granted by professional groups for their members.

4. Service agencies and care facilities take the initiative to foster gerontological competence by:

- setting requirements for gerontological competence in the hiring of new employees and assisting existing staff in acquiring the skills and knowledge to meet these requirements;
- identifying and eliminating the barriers that may hinder the exercise of gerontological knowledge and skills by staff.

5. The private sector, both for-profit and non-profit, especially those who provide services for or offer products to seniors, promote gerontology education through grants to post-secondary institutions, professional associations or health and service agencies.

6. Seniors and seniors organizations advocate for and monitor the development of gerontology education in Canada, become involved in defining the service goals

in various occupations and offer their experience of aging to assist instructors and students. (National Advisory Council on Aging, 1991, pp. 5-7).

CONCLUSION

For us as gerontological educators, it is a call for courage. It is a call to prepare ourselves and our students as leaders to challenge the established, dominant, and entrenched institutions. It is a call to revamp the educational system, to empower faculty and our institutions to be agents of social change.

In reviewing goals for the future there is a need to sort out the relative importance of discipline-specific and multidisciplinary experience in education, and there is a need to develop alternate methods of providing education, the open university model, and appropriate use of methods of distance education need to be more widely used. We need to develop alternative models of conceptualizing the relationship between professional and client in order to present students in the professions with new ways of relating to each other and to the client.

As a society, we are entering a new age, one in which old paradigms and old modes of thinking will not serve well to meet the complexities of aging in a world in which technology has surpassed the traditional boundaries of life and death, in which the demographic imperative sets us on a collision course in terms of health care needs and resources availability, and in which social policy cannot decide if it is more interested in cost containment or in commodification.

We need a more global notion of health care in which education and research assure the responsiveness of societal institutions and gerontological practice. There is a strong need for international collaboration among institutions and teachers in the health care professions. As we collaborate in science so we must in our educational approaches, also identifying or developing models of practice appropriate an applicable to the educational concepts referred to in this presentation.

ACKNOWLEDGMENTS

Supported by the Educational Centre for Aging and Health ECAH). ECAH was established in the Faculty of Health Sciences at McMaster University in April 1987 with funding from the Ontario Government through the Ministry of Colleges and Universities.

REFERENCES

Banathy, B.H., 1991, New horizons through systems design. *Educational Horizons*, Winter 83-90

Beaujot, R., 1991, *"Population Change in Canada"*, McClelland and Stewart, Toronto

Beckingham, A.C , 1995, Educational gerontology· Models for continuing nursing education, in· *"Educating health professionals in gerontology and geriatrics"*, A C Beckingham, A S Macpherson, J Puxty, eds , *Educ Geron An International Journal,* 21(1)

Beckingham, A.C., and DuGas, B. , 1993, *"Promoting Healthy Aging A Nursing and Community Perspective"*, Mosby Year Book, Toronto

Beckingham, A C , Macpherson, A S , and Puxty, J , Guest Editors, (1995), Educating health professionals in gerontology and geriatrics, *Educ Geron An International Journal,* 21(1)

Beckingham, A.C , and Puxty, J., 1993, *"Selected Annotated Bibliography on Interdisciplinary Education"*, Educational Centre for Aging and Health, Hamilton, Ontario

Beckingham, A C , and Watt, S , 1995, Daring to grow old Lessons Special theme issue on Education, empowerment and health promotion learning to live at all ages, *in "Educating health professionals in gerontology and geriatrics"*, A C Beckingham, A S Macpherson, J Puxty, eds , *Educ Geron An International Journal,* 21(1)

Beregi, E , 1995, Education to promote healthy aging in Hungary, in *"Educating health professionals in gerontology and geriatrics",* A C Beckingham, A S Macpherson, J Puxty, eds , *Educ Geron An International Journal,* 21(1)

Buzzell, M , Meredith, S , Monna, S , Ritchie, L , and Sergeant, D , 1993, *"Personhood A Teaching Package",* Educational Centre for Aging and Health, Hamilton, Ontario

Byrne, C , 1991, Interdisciplinary education in undergraduate health sciences, *Pedagogue Perspectives on Health Sciences Education* 3(2) 1-8

Clark, P C , 1994, Social, professional, and educational values on the interdisciplinary team Implications for gerontological and geriatric education, *Educ Geron An International Journal* 20 35-51

Drinka, T J K , and Ray, R O , in press, Muddy meanings The presentation of health care teams in the geriatric literature, *J Am Geriatr Soc*

Estes, C L , and Close, L , 1993, *"Long Term Care The Challenge to Education",* Association for gerontology in higher education, Washington, D C

Gerstein, R , et al , 1991, *"Nurturing Health A Framework on the Determinants of Health ",* Premieres Couneil on Health Strategy, Ministry of Health, Toronto

Greaves, G , 1995, Seniors and faculty collaborate in an undergraduate program Paper submitted for publication, in *"Educating health professionals in gerontology and geriatrics",* A C Beckingham, A S Macpherson, J Puxty, eds , *Educ Geron An International Journal* 21(1)

Health and Welfare, Canada, 1986, *"Aging Shifting the emphasis",* Author, Ottawa

Hermanova, H , 1995, Health aging in Europe in the 90's and implications for education and training in care of the elderly, in *"Educating health professionals in gerontology and geriatrics'* , A C Beckingham, A S Macpherson, J Puxty, eds , *Educ Geron An International Journal,* 21(1)

Kane, R L , Macfadyen, D , and Evans, J G , eds , 1990, *"Improving the Health of Older People A World View",* Oxford University Press, New York

Kerschner, H , and Nusberg, C , eds , 1993, *"International Aging A Brief Bibliographv"* Association for Gerontology in Higher Education Washington D C

Lloyd, P C , 1991, The empowerment of elderly people, *J Aging St* 5 125-135

Macpherson, A S , 1995, Educational change exploring the context, in *"Educating health professionals in gerontology and geriatrics"* A C Beckingham, A S Macpherson, J Puxty, eds , *Educ Geron An International Journal* 21(1)

Macpherson, A S , and Blumberg, P , 1992, Building the infrastructure for educational change, *in* Educational Gerontology in Canada, J E Thornton, ed , *Educ Geron An International Journal* 18(5) 529-540

Mold, J , 1995, An alternative conceptualization of health and health care its implications for geriatrics and gerontology, in *"Educating health professionals in gerontology and geriatrics",* A C Beckingham, A S Macpherson, J Puxty, eds , *Educ Geron An International Journal* 21(1)

Moody, H R , 1994, *"Aging Concepts and Controversies",* Pine Forge Press, Thousands Oaks, California

National Advisory Council on Aging, 1991, *"The NACA Position on Gerontology Education'* , Ministry of Supply and Services Canada, Ottawa

National Advisory Council on Aging (NACA), 1989, *"1989 and Beyond Challenges of an Aging Canadian Society'* , Author, Ottawa

Norman, G R , 1991, The fall and rise of the art of teaching, *Pedagogue Perspectives on Health Sciences Education,* 3(2) 1-4

Peterson, D A , Bergstone, D , and Douglass, E B , 1988, *"Employment in the Field of Aging the Supply and Demand in Four Professions",* Association of Gerontology in Higher Education, Washington, D C

Peterson, D A , Wendt, P F , and Douglass, E B , 1994, *"Development of Gerontology, Geriatrics, and Aging Studies Programmes in Institutions of Higher Learning",* Association of Gerontology in Higher Education, Washington, D C

Puxty, J , 1995, Goals, development and strategies of a coalition of Ontario universities, in *"Educating health professionals in gerontology and geriatrics",* A C Beckingham, A S Macpherson, J Puxty, eds , *Educ Geron An International Journal,* 21(1)

Riley, M W , and Riley, S W , 1989, The lives of older people and changing social roles, *Ann Am Poli ,* 503 14-28

Ryan, E , 1993, Education in Aging and Health in Canada Workshop Educating Health Professionals in Gerontology/Geriatrics at the XVth International Congress of Gerontology, Budapest Unpublished document

Skeet, M , and Beckıngham, A , 1993, *"Educatıng Health Professıonals ın Gerontology and Gerıatrıcs"* Workshop Report Budapest, Hungary, McMaster Unıversıty, Educatıonal Centre for Agıng and Health, Hamılton, Ontarıo

Stoller, E P , and Gıbson, R C , 1994, *"Worlds of Dıfference Inequalıty ın the Agıng Experıence"*, Pıne Forge Press, Thousands Oaks, Calıfornıa

Thornton, J E , 1992, Educatıonal gerontology ın Canada, ın *"Educatıonal Gerontology ın Canada"*, J E Thornton, J E , guest ed , *Educ Geron An Internatıonal Journal* 18(5) 415-431

Watt, S , and Meredıth, S , 1995, Integratıng gerontology ınto socıal work educatıon, ın *"Educatıng health professıonals ın gerontology and gerıatrıcs"*, A C Beckıngham, A S Macpherson, J Puxty, eds , *Educ Geron An Internatıonal Journal,* 21(1)

Wendt, P F , and Peterson, D A , 1994, Knowledge of agıng used by full-tıme professıonals ın agıng Implıcatıons for human resource professıonals and gerontology educators, *Educ Geron Ar Internatıonal Journal* 20(4) 365-377

Wendt, P F , Peterson, D A , and Douglass, E B , 1993, *"Care Prıncıples and Outcomes of Gerontology, Gerıatrıcs, and Agıng Studıes Instructıon"*, Assocıatıon of Gerontology ın Hıgher Educatıon, Washıngton, D C

32

PREPARATION FOR AGING—THE ROLE OF THE UTAS—TEN REPORTS*

1. PREPARATION FOR AGING AND RETIREMENT

Jacques Lefèvre, Chantal Declerck, and Francoise Louis
The University for the Aged, Louvain-la-Neuve
Belgium

Ten years ago our "University for the Aged" at Louvain-la-Neuve became conscious of the primary role the work with retirement had to play. The massive retirements and pre-retirements that have been realized since the beginning of the 70ies have modified the structure of our population. A new social group has appeared, the "young retired" group.

Many of our students have witnessed the deep repercussion on the individual and family life that has been caused by involuntary redundancy. This has motivated our UTA to organize a special service, the "Relay-Senior", with the purpose of aiding retired people, and specially the young retired ones, to revise their situation and find a satisfying way to proceed.

Practically, the Relay-Senior works with three seminar groups, combined with lectures on requested subjects:

- Preparation for the retirement;
- Personal development and interpersonal communication;
- Accomplishment of the programmes for retirement.

Preparation for the Retirement

This seminar is arranged annually by the staff of the Catholic University of Louvain, at which our UTA is attached. Primarily the seminar is organized for our UTA students, but it is also open to other interested old students. The seminar attracts mainly near-retired people, but also students already retired, and students still far from retirement. The seminar is composed of five units, and each unit of three hours. Each unit deals with a special theme.

At the first meeting an expert introduces the theoretical background. Next time each participant will have the opportunity to put questions and discuss. At the third meeting dialogues are arranged with reference persons (witnesses) of different background. As reference persons are invited earlier students at UTA. Our experience is that such a dialogue between retired people and the participants has become an important part of the Relay-Senior Programme.

---

*This chapter is a shortened version of the original reports. The reports cover two sessions of the Congress. French text has been translated into English. Editor and translator is Åke Vinterback

The five different themes have so far been:

- Physical and mental health;
- The psychological aspects of retirement;
- The socio-economic aspects (systems of pension, insurances etc.);
- Legal matters and taxes;
- New aspects of retirement (social roles, leisure activities etc.).

Personal Development and Interpersonal Communication

The general theme of the second seminar is: "After the 50ies, to live in harmony with one-self and others".

The first seminar was informative, and ideas among the participants used to be discussed. In the second seminar we try to go deeper with the problems and find a basis for future projects. Thus, we have the same participants as in the first seminar. The seminar lasts for five days, and is more individual. Much of the work has to be done at home. Group meetings with discussions interfoliate. At these meetings a psychologist is present and the group leader acts as an animateur. The participant tries to analyze himself, his experience and intentions. Every corner of life is at stake. The seminar will be an exercise in communication.

Often a period of destabilization cannot be avoided, but normally the problems will be solved in a positive way. The experiment could result in lasting relations, and you will find participants who organize groups of common interest outside the Relay-Senior service.

We have found that experience from the residential seminars lead the "young retired" participants to identify their new objectives, and they have been able to work out individual programmes for their life after retirement. This has caused UTA to develop a new programme that could be a third stage in the preparation for aging, and aiming at the continued aging process.

Accomplishment of Programmes for Retirement

Practically, we are organizing regular meetings in small groups, maximum 15 participants. The intention is to help the participants to work out individual programmes for a better daily life. The challenge is to define in a very strict manner the personal situation, and work out a programme that really corresponds to the participants' needs and desire. With this programme in mind, we have to find out for each participant possible obstacles, and discover existing skills and competence.

If everything goes well, it is wonderful to find that the new phase of one's life is chosen in full freedom, and not pressed upon by professional constraints and other people's expectations.

Lectures

The university is often invited to give external lectures according to agreements with enterprises, that recommend their personnel to attend courses of preparation for aging. The themes are usually leisure time, free time and the transition to retirement. Lectures are also combined with the Relay-Senior Programmes as has been mentioned earlier.

As every UTA our university plays an important role for elderly people. The UTAs contribute to demystify the common picture of old people as sick, poor and isolated, and they continue to gather to their study programmes an increasing number of old people. The University for the Aged at Louvain-la-Neuve counts today 5000 members.

2. A FILM ON AGING AND RETIREMENT

Paulin Duchesne
The University of the Third Age of Namur
Belgium

For some people retirement is the beginning of declination. This idea is a relic from a time when often the end of the active life was the end of life itself.

For others retirement is an opportunity of a new start. For these people the time that follows the retirement is a full period of life. These people know that an extension of life has one single beneficiary, the adult active life. When life expectancy has passed from 60 to 80 years in a relatively short time, it is neither childhood, nor adolescence, nor old age, that gains, but the adult active ages that get a length of 20 -70 years, and sometimes more.

But for both groups, for the optimists and for the pessimists, to enter retirement means changes:

a. Psychological changes: another image, often negative;
b. Physical changes: a decrease in activity, more or less brutal;
c. Economic changes:
   • decrease of income
   • changes of habits of consumption, savings and transfers;
d. Social changes; more or less an end of the role as a social actor;
e. Family changes: family ties could be a support, but also a source of conflict;
f. Cultural changes: very often a greater consumption of news, leisure time and culture.

These are the aspects that we want to inform about in a movie which we are producing in cooperation with the Catholic University of Namur (Notre Dame de la Paix), and with the financial support of the Bank of Savinge and Pensions.

The film will be finished in March 1995. It will be an invitation not to hide the difficulties at the passage from active life to retirement. But above all it will be an invitation specially to observe the period after the retirement. People can admire a piece of art, a forest, the summer night sky, but first when they see it.

The contribution of elderly people to society is considerable in our time. If we do not know it, this is because we have not seen it. And yet, without the benevolent work of old people, would our families be as they are? And our society, would it be what it is without those who are ready to work for others, without expecting anything back?

In our days an old person is not a rarity. Because of that, important sectors in society, such as education, culture and mutual aid could be furnished with resources that otherwise would be scarcity matters. Those who tomorrow investigate the access to labour, will they add all those, who have a profession, all those who are looking for a job, and those who have gone from paid job to voluntary work? Recognizing the old age people for what they do, could finally grant them what they have lost, a status. But no position without recognition.

Our movie at the Namur UTA will try to be an eye, open for the world of retired people in the midst of a society that would be curtailed without the achievements of its old people.

People who we want to invite to look at this movie are first of all the near-retirement staff at our enterprises and in the public sector. To what purpose? Simply, because they will be better prepared to enrich humanity after their active life. But also dominant people in Civil Service and at the medias, too often embarrassed by the public cost of old people but little attentive to economies that they have approved.

To be sure, a society without old people would be a society without pensions, without old people's homes, perhaps without a deficit at social security, but a society without grandparents and without traditions, a society with a life expectancy of not more than 60 years, and a society where nobody will do anything for nothing.

Finally, isn't old age a great period of life, not only for those who live it? 'Paris valait bien une messe', to prepare for an active retirement could be worth while some meters of a movie.

This film has been developed from a thesis for a doctorate, based upon a questionnaire, and introduced in the form of a triptych: to demonstrate changes due to aging, to show the positive aspects in society by the presence of retired people, and finally to inform the different representatives of the society about a world of retirement.

3. LIFE-EXPERIENCE AND LEARNING—OLDER PEOPLE AT THE UNIVERSITY

Sari Poikela
University of Tampere, Institution of Extension Studies, Tampere
Finland

Traditionally learning is considered as a part of human life during the first and second age. Adult learning has been researched, but research has been emphasized mainly on adults in second age. Anyway a recent research shows that preconditions of learning remain relatively unchanged during whole life-course. Later years of one's life can also be active time for studying and self-realization. The idea of lifelong learning forms a basis for learning and studying in the third age. In every age it is important to have a possibility for wide self-development. Learning should be a part of everyday activities in the human life. If learning is understood in a broad sense, for most of the people it goes as long as life itself.

I have studied ideas about learning, life and aging among Third Age University students in Tampere in Finland. Subject of research was a seminar-group, the aim of which was to encourage students to be critical and reflective in their learning. I collected data with interviews, essays and observation. One main question of the research was how did the students feel about their own learning. Others were e.g. what learning actually is and how it affects in different life situations. What is the meaning of learning in life? I will focus on those questions in this presentation.

Learning does not depend only on age. So it is not always relevant to divide learning to pedagogy, androgogy and gerogogy. For this reason the theoretical background of the research was experiental learning instead of gerogogy. The model used has been developed by David Kolb. Experiental learning emphasizes learning as an important part of the whole life cycle. The model states six arguments of learning and development. These arguments formed a basis for my analysis of learning.

The first argument states that learning is a continuing process. It is best conceived as a process, not in terms of outcomes. Learning and knowing are processes, not products. The seminar, which I studied, clearly influenced students' attitudes. Many of them started to change their ideas about learning. Concrete results did not have absolute value any more. In students' opinion clearly defined aims, e.g. examinations, could have been harmful for the learning process itself.

The second argument says that learning is a continuing process grounded on experience. A way to learning is a conflict between previous and new knowledge. What has been learned in one situation becomes an instrument of understanding and dealing effectively in another situation. Students' own definition of learning was for example rethinking, which could mean quite the same. It is important to think openmindedly, because you can always learn something new.

The third argument describes conflicts between opposing ways of dealing with the world. The process of learning requires a resolution of conflicts between different modes of adaptation to world. Students became more analytic when they described their own learning. They also considered creativity as a very essential point in learning.

The fourth argument states that learning is a holistic process of adaptation to the world. Experiental learning is also a concept describing the central process of human adaptation to social and physical environment. When learning is considered as a holistic adaptive process, it provides conceptual bridges across life situations, such as retiring, portraying learning as a continuous, lifelong process. In students' life learning had a very important role when adapting to aging. Studying in the third age university gave a new chance for self development and helped to deal with a new life situation after retiring.

The fifth argument stresses the point that learning involves transactions between the person and the environment. This may seem obvious but learning is often considered mainly as person-centred activity. The transactional relationship between the person and the environment is symbolized in the dual meanings of the term experience. One subjective and personal, referring to the person's internal state, and the other objective and environmental. The interaction between others was a significant aspect of learning for the students. The interaction helped learning and was experienced as very fruitful. One's own learning was reflected with other students' experience and knowledge.

The sixth argument states that learning is a process of creating knowledge. Knowledge is the result of the transaction between social and personal knowledge. Students began to realize learning individually, but also as a result of personal and social interaction. This was considered as a very positive aspect. There was not any specific right or wrong way to learn.

Learning obtained transformative meaning in students' life. E.g.learning was an instrument for evaluating passed and future life. Learning was experienced as an open attitude towards new and different situations during life-course. The ability to reflect one's own learning seemed to increase during the studies. Students' many comments reflected critical evaluation of previous life-experiences. With self-reflection it was possible both to evaluate and create new ideas about life. Student became more aware of their own way of learning and the fact of how much learning can extend thinking.

An essential point of learning in third age is an effort to organize life-experience and in a way refine it to knowledge. Students' ways of analysing and restructuring their own life-experience were individual, so it is not possible to create a general model of how older people learn. However it is important to reflect and evaluate meaning and values which are involved in learning in the third age. This means that a critical approach to deal with the theory and practice of learning in third age is needed. It involves also the fact that moral and ethical issues of education are in a central position. Learning is always individual and personal. For this reason it is questionable to try to influence on students own views and opinions. Dividing learning to pedagogy, androgogy and gerogogy is not a sufficient foundation for lifelong learning. The stage of age forms frames around learning, but despite the age learning is the heart of human development.

4. 21ST CENTURY: MANAGING LIFE CHANGES—A JOINT PRESENTATION BY THE PRE-RETIREMENT ASSOCIATION OF THE GREAT BRITAIN AND THE BRITISH UNIVERSITIES OF THE THIRD AGE

Jean Thompson
The Third Age Trust
Great Britain

The Pre-retirement Association of Great Britain is an independent national charitable organisation, recognised as the national focus for pre-retirement education in the UK. There is a clear link with the University of the Third Age and we work together whenever possible. Unfortunately Dr A. Chiva, The Pre-Retirement Association, was unable to be with us, so I am presenting his views.

The University of the Third Age in Britain was founded eleven years ago on the concepts of Dr Peter Laslett, a keynote speaker at this Congress. Educational conditions vary from country to country and in the UK there were several factors which led to a 'U3A Model' different from the University-based model founded in Toulouse by Professor Vellas.

These factors were: a) British Universities already ran Continuing Education Departments for local people and were reluctant to support this new venture. b) In Britain we already had the very successful Open University, which provided degree-level courses for the public. Our U3A had to be distinct from these organisations and self-sufficient. c) A strong tradition of volunteering to meet perceived needs. Hence came Dr Laslett's concept of self-help groups drawing on their own resources to provide education in the widest sense for Third Agers. U3A in Britain now has over 40000 members in 250 local groups.

Why did we choose this theme of change? Because longer lives, shorter and varied periods of paid work, changes in relationships, all involve change. Static patterns of life are already becoming rarer and in the 21st century dynamic change will be pattern.

Significant changes when individuals move into the Third Age, include retirement from paid work, change of status, financial changes, changing relationships, coping with possibilities of ill-health or immobility, and as Dr Laslett points out, 'Doing the dying for society.' How are the PRA and the U3A contributing to effective management of change?

Effective strategies include: understanding the change, including the feelings involved, letting go of past patterns, and considering options for the future.

This has led Tony Chiva and Allin Coleman to write a book describing their model for managing change (Coleman and Chiva 1991). The model used on some of their Pre-Retirement courses indicates a typical reaction to change. 'The feelings curve' according to the model demonstrates the following statements:

- I am not alone in my change;
- Many people feel the same way as I do;
- I have been here before and will again;
- I have managed other changes and I am therefore a survivor.

The model brings into awareness the way to plan. This externalisation increases the likelihood of resolution and enables greater sharing and mutual support.

Aspects of the individual which will make a positive contribution are:

- Possessing a sense of self;
- Possessing a strong personal philosophy;
- Willingness to learn;
- Viewing change more as a gain than a loss.

The PRA has also written a self-help workbook based on this model, which is being translated into various European languages.

Much work remains to be done in this area, not least in U3A. To quote our International AIUTA President, Jacques Lefèvre (AIUTA news no 1): 'We can no longer be contented merely to open up our cultural and scientific heritage to our students. We also have to prepare them to take an active part in solving problems of society.'

Looking ahead, coping with change is one way in which Third Agers could show the way. We are all rich in experience and have survived many changes in our past. Can we utilise this ability to help others? All our AIUTA students are moving towards shared learning and valuing the experience of themselves and others. Is this an area where we can go even further?

Time does not permit a description of some of the new approaches in the British U3A such as 'Third Age Issue' discussion groups and 'Creative Listening' groups. If AIUTA can assist us to share our ideas and lead the way in cooperating with associations like the PRA, 'Coping with Change' will benefit us all, not only Third Agers but people everywhere.

5. THE ACTIVITIES OF THE UNIVERSITIES OF THE THIRD AGE IN ITALY, REGARDING THE THEME OF PREPARATION FOR AGING AND RETIREMENT

Irma Maria Re
The University of the Third Age UNITRE, Turin
Italy

The University of the Third Age called UNITRE, that is University of the Three Ages open to everybody, was born in Torino in 1975. At present there are about 150 seats of the "National Association UNITRE" in Italy.

Consumer's society tends to set aside elderly people when they leave their working activities. They are left at the mercy of the "retirement shock" and of the "empty nest syndrome",bereft of the wish to project themselves into the future.

Third Age should not be considered a burden but a resource. UNITRE cultural activities help elderly people to turn from working force into cultural force.

UNITRE aims are: to educate, to form, to inform, to prevent, to promote researches mainly concerning aging, to encourage social relations, to make a comparison and a synthesis between former generation cultures and the present one, to privilege "Being" beyond "Knowing"

UNITRE Universities offer their service to all ages to encourage education, which may be permanent, recurrent or renewed, according to the cases. Their projects include some specific objects, such as to contribute to cultural promotion by organizing courses in the scientific, humanistic and literary fields, by opening laboratories (on theatre, painting, photography, etc.) and promoting other activities of social interest.

UNITRE is considered a reservoir of voluntary service, a bridge towards sympathy, positive proposals and participation. It promotes culture in a new way and helps people to learn how to grow old, it opens to solidarity, offers the means to delay physical and spiritual aging, revives lost interests and arises new ones, gives the possibility to communicate and interact with all ages in order to discover again the essential values of "Building together".

For several years the Universities of the Third Age, together with the "150 & PIU" association, have been engaged in researches on the theme: "The Collective Memory, a patrimony of elderly people which must be saved".

The students, led by their teachers, became researchers and gathered the memories of tradition told by the oldest, choosing the most interesting testimonies, objects and documents. A Scientific Committee examined this material, judging it worthy of publication. Then four volumes were printed, making a collection called "Pearls of Memory".

Thus UNITRE students handed down to younger generations the minor history told by the oldest.

UNITRE Universities follow two ways to attain their aims:

- a cultural one, led by the teachers and
- the "Academy of Humanity", where the students are invited to become protagonists.

As they have to co-ordinate the operational secretariats they are no longer passive objects, but find new roles to cover. For instance the students of Turin UNITRE (first seat opened in Italy) co-ordinate 40 secretariats, open also in Summer.

They work in museums, are present in schools and hospitals, animate Old Age Homes, are interested in ecology and environmental problems, in co-operation with Public Institutions.

They are also engaged in a social service of assistance on the phone, very important activity for the oldest trying to ease their loneliness, their anxiety, their depression.

The contacts so established help to restore a friendly network linking elderly people, as far as possible, to the community by which they fear to have been abandoned.

This service is a landmark from which a co-ordinate help starts, supporting the family in case of emergency.

Thus UNITRE teaches its students to conjugate the verbs of life not only using the past tense, but living the present without losing hope in their future.

6. A NEW SOCIAL DIALOGUE

Serge Domí
The University of Free Time of Martinique
Martinique

The University of Free Time of Martinique has been working for five years. It was created in 1988 at the joint initiative by associations and clubs for older people, and a research group (GEREC) at the University of the Antilles and Guyana. The old people's associations felt a need to broaden their programmes to levels outside pure leisure time activities, from the interests of a certain age group to the social field in general.

The GEREC research group had for some years been occupied with investigations on the cultural heritage in Martinique, in particular the Creole language. From these two sources the UTA of Martinique was born. The hope among elderly people today is that they after their active period could change their life and fill it with other matters.

The percentage of retired people in our population increases steadily. Many responsible groups are worried about what will happen when the number of the non-actives will go beyond the number of the active ones. This concern is found specially among people, responsible for the public finance, although the matter is not a purely economic one. It is not the capacity as such that is questioned, but the way of accumulation and partition of means and sacrifices. We have to change our rules a lot, but the problems are manageable. I allow myself this kind of economic deviation, because we have to oppose the shallowness with which the questions of aging is treated. The obvious expectation among the citizens of a better future is a challenge to our contemporary leaders. It is our responsibility to encourage all initiatives to let the growth in number of the old generation be accompanied by a simultaneous broadening of the social care.

Regarded from such a perspective our UTAs will have a strategic importance. At this 17th international congress of AIUTA we would like to put forward three special aspects on the UTAs, and we do it with a growing conviction from the experience of five years UTA work in our country.

We will strongly emphasize that UTA should be

1. a carrier of a new solidarity;
2. a centre for dialogues and fruitful confrontations between the culture of the past and future expectations and
3. an instrument for the creation and growth of a new citizenship.

Carrier of a New Solidarity

With their different and individual attitudes towards life old people could act for a reorientation of modern society towards a better understanding of the importance of human values. By keeping a distance to the dependence of common matter they could cause the active generations to slow down their foolish struggle for more money, more power, more consumption. If we could give old people more opportunities to act and be seen, they could bring to the surface the mechanisms of our society and throw light upon the feelings of isolation and elimination among its citizens.

A Centre for Dialogue

The statement of UTA as a centre for dialogue is based on our experience of intergenerational programmes, initiated by the student association of our university campus. Their local radio station has twice a month organized debates on themes as "Money", "Sex and love", "The parents of today", etc. During these debates members of our UTA and representatives of the young students have expressed their points of view very openly. After a time these emissions attained such an attention that they were systematically repeated during peak hours.

By this type of activities we were seen by the society and could influence the approach to public questions. We could keep our UTA active and its members healthy and attentive. If this will cause a renewed citizenship in our society, we have motivated our existence and work.

7. THE UNIVERSITY OF THE THIRD AGE OF LISBON (ULTI)

Emilia Noronha
Lisbon
Portugal

ULTI in Lisbon, Portugal, was created in 1987 and counts 800 students and 80 theoretical and practical courses. It cooperates with the church of San Domingos de Benfica. ULTI has organized a small research centre, that works with the problems of the third age and cooperates with scientists from the young and active generations. At a school in Lisbon ULTI has activated retired teachers to train young people with insufficient knowledge in their vernacular. This training is followed up by inter-disciplinary classes.

Outside the proper UTA work voluntary members have organized contacts and support for old and sick persons. By its different activities ULTI has contributed to the awareness of the value of old people in the society, and underlined the importance of continued learning to the end of life.

Supported by the political council for the third age in Portugal, ULTI has recently arranged the first national congress of the Portuguese UTAs. A national federation is on way.

8. THE QUALITY OF LIFE OF THE AGED—A QUESTION OF FREEDOM

Josep Bombí i Llopis
The Association of UTAs in Catalonia (AFOPA), Barcelona
Spain

Without freedom you cannot speak of life quality. We do not accept discrimination in any form. What is a Third Age? Does it exist? We are citizens as people in all the other ages, with the same rights, the same duties and responsibilities.

The quality of life has to be evaluated as a function of the cultural tradition of the society, the expectation of a good future, and the chances to succeed.

The length of life has increased, and the old ages with immobility and decrepitude arrive later. The many years you are expected to live after the active period have to be well prepared. The quality is individual, and decisions have to be taken individually. This is not easy. We believe that the great lines have to do with five aspects of quality: health, culture, dignity, economy, and solidarity.

What we regard as life quality in the broadest sense of the word varies with our cultural background, and with interests remaining from our earlier professional life. At the retirement other activities may take over, to which you can devote your time without overworking and in full freedom.

By a well-planned preparation for aging you can reach a physical and mental balance in accordance with the life you are aiming at. Life quality is often only a smile or some words. Various expressions of culture offered by society permit the old citizen to develop his talents at maximum. To practise a sport, to exercise, to show will and discipline is often a necessary behaviour to avoid physical weakness. An ethical change follows retirement: while the active life worries about rivalry and competition, the life of retirement embraces questions of sharing. This perspective will allow us to avoid every kind of discrimination, whether or not it concerns age or social situations.

9. A PROGRAMME FOR SELF-LEARNING AND INSTRUCTOR TRAINING IN HEALTH

Rosita Kornfeld, P.P. Marín, E. Gaete, and V. Orellana
The Catholic University of Santiago de Chile
Chile

The Catholic University of Santiago de Chile has opened a space of education for persons, 50 years or older, who want to return to the university, or who have never studied at a university before. The programme covers three objectives: personal development, general culture and creativity (theatre etc.) The purpose of the programme is to increase the quality of life among old people. Totally 10% of the Chilean population is 60 years or older, the prediction for the year 2025 is 16% or three million citizens. A study programme is composed with a capacity of 700 students a term, three terms a year. Self-training and instructional material has been developed.

Parallel to this, studies have been organized for an academic diploma in gerontology. The training applies to persons in the field of health, members in families with Alzheimer's disease, and members of churches and enterprises. The length of a diploma course is a full year.

Finally a self-study project is developed with practical manuals and self-explaining videos. The material was prepared with the objectives of rendering old people a description of the process of aging and foster self-care abilities. The education will respond to requests from different groups in the society to get information about existing resources. The material is also designed to prepare instructors to work with old people. It can contribute to solve problems in the medical or social field where you are working, and many times to change traditional routines.

Many ideas are very close to the roles of the UTAs. Our universities could engage themselves in parts of this field. We could develop a form of mutual tasks of similar nature in cooperation with related social activities.

10. THE UNIVERSITY OF THE THIRD AGE OF QUÉBEC: SURVEY AND FUTURE PERSPECTIVES

Clermont Simard, D. Drouin, C. Gagné, and J. Laforest
University of Laval, Québec
Canada

Summary

Old people could adapt themselves in many ways to the different functions of a university, its structure, education and service. Generally, little is done to facilitate such an integration. Old people should be offered necessary resources to continue their earlier profession, and also if they wish, to enter new fields of study. The exchange of experiences and ideas between generations should be stimulated. How to engage old people in the various fields of research, educational programmes, and even the internal administration, should be dealt with as an important question by the university. A full integration is recommended.

In the past, two important works at the University of Laval have been published concerning the university of the third age: L.P. Bonneau 1976 ("The Gerontology and the University of Laval") and R. Doyon 1990 ("Mutual Enrichment of the Universities of Québec and their Old Students").

Now we report an investigation using the computer programme "Eric". The outcome was 115 documents. We have consulted several colleagues on the subject. Recently the UTA at the University of Laval has been approved by the government of Québec as a "University Centre for the Third Age" (June 30, 1994).

Introduction

A desire to know and learn characterizes all people, and there is no age limit. Also old citizens should be offered the possibility to activate themselves in whatever field they want, the physical and intellectual, as well as the emotional and social ones. This is the goal of the university of the third age. The universities should engage themselves and put their material and human resources at disposal. The experience and maturity of this new type of students should be regarded as a resource. The purpose is to improve the conditions of life for the old people and open up a role in society.

A university education of old people matches the idea of "non-idleness", now present in all discussions of aging. The pattern of old people as silent and immovable is gone, and replaced by a picture of mobility and alertness. A good aging requires full activity of body and soul. Gerontology is to be redefined as a strict research field, and should enjoy the participation of our "new" students.

According to Bernier the general objective of the UTAs is to develop among the third agers their potentialities and capacity to maintain their autonomy. By autonomy Bernier understands the ability to accept educational achievements and to develop a new identity, that is a new kind of engagement and thinking. (Cf. Bernier, Roger, "Permanent Education and UTA", Sociologie et Sociétés, vol. 2,1984, p. 79-87.)

Four Objectives

The purpose of the UTA:s could be concretized by the following four sub-objectives:

1. To use available teaching resources in order to favour the development of old people:
   a. in the physical field by training programmes and studies of health and food;
   b. in the psychological field by programmes on self-knowledge, and on personal and social identity;
   c. in the intellectual field by stimulation of continued learning in the earlier profession, and new learning in other, for instance cultural fields;
   d. in the social field by participation in community work, also the university administration.

   Bernier has emphasized the role of the university to encourage a social engagement among the students, for instance by the creation of a local association of elderly students, that could give voice to their priorities, or by organizing groups, suitable for social tasks, internally among the students or externally at the community. It is important to create voluntary attitudes and persuade elderly students to accept social tasks according to the personal competence. A society that understands the potential resources of the old age collective will increase its human assets considerably (Mayence, Serge, "The Great Adventure of the UTAs", Marcinelle, Belgium, 1980).

2. To favour the access to libraries, museums, and to other technical resources.

3. To develop research activities in gerontology.

   The university should engage the third age students in its gerontologic courses, and take advantage of their experience and special knowledge. Many examples could be taken from various interdisciplinary fields, as combinations of law and administration or sociology and psychology. Examples already tested are studies of home accidents and programmes of preventions (Geneva), or studies of voluntary functions in different sectors of the community, combined with an inventory of existing competence among the students (Sherbrooke).

4. To intervene among the third age collective, and cooperate with organizations, working with elderly people.

The integration of the elderly students should be total. The university has to put all its resources at disposal, all the things that the society has left into its hands: education, research and administration. This could change the stereotyped attitude towards the elderly, and make education more attractive to all generations. The awareness of the global age situation and its consequences must open our eyes to the need of mutual cooperation.

Mayence has some doubts about the participation of the UTA people as a voluntary resource when it deals with service to the community. Reported successful cases during the last 15 years are few. He agrees with researchers as Michel Loriaux who suggests that the final solution on the questions of aging can be reached only if some problems of political,

Third Agers

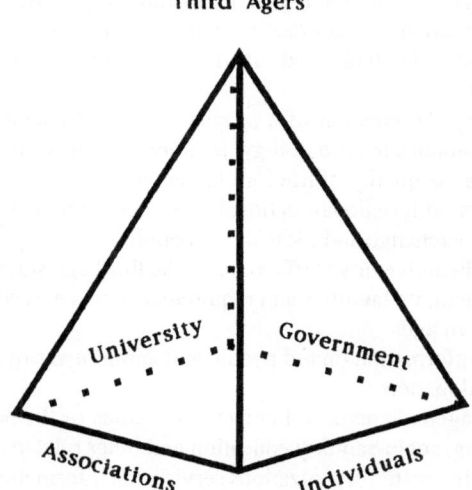

Figure 1. University of the Third Age of Québec *(Centre Universitaire du Troisième Age de Québec).*

social, cultural and economic nature will be left to the elderly people themselves (Mayence, Serge, AIUTA Congress 1990, Univ. of Québec, Hull 1991, p. 108).

The educational programmes of the universities and colleges in Québec are frequented only by 4% of the population older than 65 years, people that to a great extent already have a high education. Due to the greater access to education in our days this percentage will probably increase in the future, but on the other hand our aims must be to reach groups of all social classes and ethnical minorities. This goal surpasses the capacity of higher education. The national council for the third age hints that the important sources of information and knowledge are to be found not only in the educational system, but to a great extent also among the medias and the social and cultural institutions. Managers and trade unions have an important role to play (Wigdor and Hogue-Charlebois, AIUTA Congress 1990, p.93). From this point of view universities and colleges place themselves in a secondary position, but their great importance will be as the educators of the professionals and the heads of the social and cultural institutions: to educate the educators.

Nevertheless, the tasks for the UTA:s have become larger. Today it is accepted that also the third age people need a continuous training for their professional life and their cultural enrichment. The universities have to integrate "the third age dimension" as they have done with the young generation.

The figure illustrates our ideas how we want the University of Laval to function in relation to our centre and the third age people. It emphasizes that our principal attention remains with the old people, and that it should be possible to build bridges between organizations, governments, universities and the individuals.

Goals of the Centre

- To organize and administer the third age centre in order to stimulate contacts between the old people and the university.
- To strengthen the intellectual, psychological, physical, social, cultural, economic and spiritual level of old people facilitating a possible cooperation with the university institutions.

- To favour exchange with the university institutions in order to make available to elderly people existing resources of education, research and culture and other programmes related to their needs of education, cultural development, and professional training.
- To facilitate by the creation of a special research foundation inter-disciplinary research programmes in gerontology, in order to improve the knowledge of aging and strengthen the quality of life for old people.
- To create links with regional, national and international associations in order to favour mutual exchange and ideas of cooperation.
- To stimulate the university staffs to make the third age students familiar with the university system, its facilities and requirements, and extend its service functions to students of all ages.
- To distribute information on old people and aging in various forms, from printed to audio-visual matters.
- To make managers, experts and citizens conscious of the needs of old people as regards training, socio-sanitary education and other related aspects.
- To forward to interested organizations service and information, related to the goals of the third age centre.

REFERENCES

Lefevre, AIUTA News no 1, January 1994, Inviting a New World, p 1

Bernier, R , 1984, Permanent Education and UTA, *Sociologie et Societes,* 2. 79-87

Bonneau, L P, 1976, *"The Gerontology and the University of Laval"*

Doyon, R , 1990, "Mutual Enrichment of the Universities of Quebec and Their Old Students", in. *"Seniors in Higher Education, Development, Opportunities"*, AIUTA,Universite du Quebec a Hull, AIUTA 1990, Ateliere Graphiques Marc Veilleux Inc, pp 335-347, printed in 1991

Mayence, S , 1980, *"The Great Adventure of the UTAs"*, Marcinelle, Belgium

Mayence, S , 1991, Directions and Methods for the Institutetes of Higher Education to Adapt in order to Be Available for Old People", in *"Seniors in Higher Education, Development, Opportunities"*, AIUTA 1990, Universite du Quebec a Hull, Ateliere Graphiques Marc Veilleux Inc, pp 108-117,

Wigdor, B and Hogue-Charlebois, M , 1991, Development of Education Policies at colleges and Universities, in: *"Seniors in Higher Education, Development, Opportunities"*, AIUTA 1990, Universite du Quebec a Hull, Ateliere Graphiques Marc Veilleux Inc, pp 93-101

33

PREPARING FOR RETIREMENT—GOOD HOUSING IN OLD AGE?

A Review of the Provision of Housing and Care for Old People in 7 Systems

John William Murray

University of Westminister
London, England

INTRODUCTION

The concept of housing as shelter responding to a basic human need has been transmuted, for many people in increasing numbers of societies, into the concept of home with all that implies for our attitudes and sense of values (Murray, 1994).

If home is important for all ages, it increases in meaning for the older person. The good home provides security and becomes a treasure trove of memories and adds to the meaning of life. It makes an important contribution to the quality of life enjoyed.

However, housing has 3 characteristics that differentiate it from other goods and services. Housing lasts a long time, it cannot be moved and it is very expensive. As a result both individuals and societies have to make careful provision for homes for the third and fourth generations.

BACKGROUND

For a person to enjoy old age, good quality accommodation is needed. The standards of the society will determine what is regarded as suitable housing and this will also be influenced by individual choice. People can prepare for good housing in old age by making the best of what they have or by moving.

I first became involved in housing elderly people 40 years ago. I worked in the London County Council housing department when it embarked on Britain's great post war clearance programme in the mid 1950s. I was employed in a local housing office in the East End of London where most of the properties demolished were "2 Up and 2 down" terraced cottages. The usual layout of a house consisted of 2 rooms and a kitchen on the ground floor with an external WC and small yard or garden: on the first floor were 2 rooms. The rooms were small and the accommodation was often damp.

Preparation for Aging, Edited by E. Heikkinen *et al.*
Plenum Press, New York, 1995

Thousands of people had to move from these cottages to new styles of building appearing in English towns and cities. It was the old people who suffered most from these compulsory moves. All their lives might have been spent in one neighbourhood: their warmth was provided by a coal fire and few had lived higher than the first floor. Their new life style could involve adjusting to high rise and central heating in accommodation where many pets were forbidden.

This massive uprooting of elderly people was done with the best of intentions. The aim was to clear the mean, insanitary slums and replace them with modern, self-contained accommodation. While all old people in slum areas were confronted with the trauma of forced move, some did of course benefit from the moves.

Britain ended its clearance programme by the end of the 1960s but it is disturbing to read of the situation in Japan whose clearance of inner city areas still proceeds (Tokyo, 1991). In contrast to Britain, much of the redevelopment is carried out by business organisations and where the old people displaced cannot expect, as they could in Britain, to be housed in self-contained, cheaply rented alternative accommodation.

Of course, many people decide to move themselves. It is a matter of their choice. In Britain, many elderly people have chosen to move to the south coast. Usually these have been owner occupiers but there have been examples of local housing authorities building seaside homes to rent for older people wanting to move out of the towns and cities. This concentration of elderly migrants in one area means heavy burdens on local health and social services. One unfortunate aspect of moving by elderly home owners came to light at the end of the 1980s with the collapse of the British housing market. Some owner occupiers had been tempted to home income schemes whereby they hoped not only to buy a new home but also supplement their pension with a small income. The situation became so serious that support schemes were introduced to reduce the losses suffered by the old people who had entered these schemes.

The move southwards is also an important migration pattern for the American elderly who can afford it. One interesting development has been the creation of towns and villages populated by elderly people. Perhaps because it too has an abundance of land this type of elderly resettlement, in villages, has been developed in Australia.

Some elderly people not only move within the country but outside it. In the UK, for many years West Indian old people have been returning to their homelands where the pensions and money they have earned in Britain can buy them a comparatively high standard of living. In contrast, one of the saddest examples of elderly emigration has been that of British old people retiring to Spain. Unlike the West Indians, they are moving to a new culture with a different language and where the cost of living is sometimes higher than in the UK (Age Concern, 1993). Their plight might have been a warning to some Japanese developers who were said to be considering the proposal to develop sites for the Japanese elderly in Australia.

HOUSING OLD PEOPLE IN DIFFERENT SOCIETIES

United Kingdom

The standard of accommodation enjoyed by old people will tend to reflect the standards of the general housing provision although the poorer elderly will usually find themselves in the worst accommodation. Particular problems can emerge because of the tenure pattern of a housing system. Because of its emphasis on home ownership, the UK has many old people who are owner occupiers (Rolfe, 1993) They will usually own a 5 roomed house and, although they will have usually bought the property, some will find the repair and maintenance costs for such a large dwelling too high for their income. This has led to another intervention by a non-interventionist government with its support and funding of voluntary care and repair agencies. (Care and Repair, 1993).

An important response to the housing needs of elderly people is the provision of accommodation specially built and managed for older people. This response has almost a 1,000 years of history. Developing from the Churchs' hospitals for the lepers and its hostels for the wayfarers, the almshouse movement became the first social housing provision. Britain today still has 25,000 units of almshouses accommodating old people (Bailey, 1988). But almshouses are still to be found in parts of Western Europe especially in the Netherlands (with very fine examples in Haarlem and Utrecht).

However, in the 1960s, Britain adopted a type of provision which is known as sheltered housing. Typically, although there are many variations, the sheltered scheme consists of 30 dwellings. with a social room and other facilities, an alarm system with a resident warden (MHLG, 1969). Mostly owned by Councils and Housing Associations, with a few private schemes provided in the 1980s, Britain has the highest number of sheltered units in the world. Its 600,000 sheltered dwellings accommodate 5% of its elderly population. Similar sheltered schemes are to be found in France and West Germany but these do not have the same prominence in housing the elderly as is to be found in the UK (Brink, 1991).

The Netherlands

An interesting contrast with the UK is to be found in the Netherlands. Although in many ways a similar society to the UK, its special housing provision for elderly people developed in another way because of different demographic patterns and attitudes towards institutions. With the highest post war birth rate in Europe, the Netherlands gave priority to housing families. One part of this policy involved persuading older people to move from larger dwellings. And the Dutch provided large, well equipped institutions for their older people. While old people in the UK have tended to resist entering institutions, the Dutch old people accepted institutional accommodation—to such an extent that by the mid 1970s, the Netherlands had the highest proportion (5%) in the world of old people in institutions (Murray, 1993).

The Dutch, however, then changed their policy as they realised the economic and housing consequence of the "greying" of their population and also decided that institution-alisation was not always good for older people (Priemus, 1990). Although they still retain their high numbers of large homes for the elderly, the proportion of elderly in institutions has dropped (Knipscheer et al., 1991). However, rather than follow the British example of sheltered housing, there has been an "explosion" of experiments in housing elderly people with particular emphasis on community care (Houben, 1989).

Vienna

However, still following the Dutch model of large and well equipped homes for elderly people is Vienna. Vienna has a long history of social housing and, with Britain, can claim to be among the first providers of social housing on a large scale (Murray, 1993). Vienna maintains its benevolent tradition of building social housing and regulating all housing. In its special provision of housing for elderly people, there is an emphasis on large scale housing schemes with a scale of provision that, in British eyes, is lavish. Even for such a progressive authority as Vienna, this means that the institutions have to be large—about 300 units—to justify the expense of such a provision (Kuratorum)

Hungary

In Hungary, in the companion city of Budapest of the old empire, the homes for the elderly are well equipped and staffed but, in a poorer society, the provision is not as lavish. However, what is puzzling on an initial survey is that, whereas in all the other systems examined, the special housing for old people reflected the general standard of social housing, in Budapest there appeared to be a discontinuity. While the standard of the old people's homes was good, the standard of the general family housing was very basic reflecting the low standards in amenities and space to be found elsewhere in Eastern Europe. Another surprising factor was that the demand for accommodation for elderly people varied so much from district to district (Murray, 1994).

Hong Kong

The influence of British social housing and specialised housing for old people is to be found in Hong Kong. The Hong Kong Housing Authority with over 600,000 units is one of the largest social housing authorities in the world. The Hong Kong Housing Association, a voluntary housing organisation, has also provided another 33,000 dwellings. Hong Kong has traditionally had a young population but demographic changes and migration of young people have meant the housing of older people has become a significant issue. At the end of the 1980s, the Hong Kong Housing Authority adopted a package of proposals to encourage older people to live either with their relatives or near to them (HKHA 1989, 1993).

It also introduced its version of sheltered housing with wardens, alarm systems and social rooms. However, it has not built for elderly people but has converted the lower floors of selected high rise blocks to use as sheltered housing. To the British observer the standards are very basic with sometimes 3 unrelated single old people occupying the same flat. However, the Hong Kong Housing Association, which has been building for elderly people since the end of the 1970s, has some sheltered schemes of high standards (HKHA, 1992).

Japan

The housing of elderly people in Japan is surprising. For the richest nation in the world, it may seem strange that their housing is of such a low standard. There appears to be three reasons for this. For reasons of policy and land shortage, housing is expensive in Japan and its quality is below that of poorer, Western European nations. Secondly because of its emphasis on free enterprise, there is only a low proportion of social housing in Japan—about 7% of its stock. Thirdly Japan has been a young country for much of this century and the three generation family traditionally provided care and housing for elderly people (Hayakawa, 1990). Now Japan is experiencing the fastest aging process ever experienced by any society. Again social and economic circumstances are changing and increasingly families are reluctant to care for their elderly (Hayakawa). While the Japanese Government appears to have recognised many of the implications of its rapidly aging society, one might wonder whether it is yet making sufficient provision for their housing (MHW, 1990).

CONCLUSION

This short paper illustrates the variety of responses to housing old people. There are no easy answers except that once a birth rate in a society declines then that population will start to age and then very soon that society will be confronted with the difficult and expensive challenge of providing suitable accommodation. The nature of the problem will depend upon the stock of housing, the housing market, the level of expectations and a whole range of political, social, economic and cultural attitudes. While, with economic pressures and demographic changes, there will be a constant temptation to think only in terms of providing shelter, the objective must be to provide home where older people can live in security, dignity and happiness.

REFERENCES

Age Concern, 1993, *"Growing Old in Spain"*, Age Concern, London

Bailey, B , 1988, *'Almshouses"*, Robert Hale, London

Brink, S , 1991, "Housing & Related Policies for the Aged in Canada, Sweden, France & Japan", Tokyo Conference Paper, (unpublished)

Care and Repair, 1993, *"Annual report of Performance 1991/92"*, Nottingham Care & Repair, Nottingham

Hayakawa, K , undated, *"House Poverty in Japan"*, Kobe University,

Hayakawa, K , 1990, Japan, in *"The International Handbook of Housing Policies & Practices"*, W Vliet, ed , Greenwood Press, New York

HKHA, 1992, Annual Report 1991/92, Hong Kong

HKHA, 1993, Annual Report 1992/93, Hong Kong

Houben, P P J , 1989, "Innovation in Housing of the Elderly First Experiences", Hamburg Conference Paper, Hong Kong

Hong Kong Housing Authority (HKHA), 1989, "Report of the Working Party on Housing the Elderly", Hong Kong

Kuratorium Wiener Pensioninstenheime, undated, *"The Vienna Pensioner's Home'*, Vienna City Council, Vienna

Knipscheer, N Y, et al , 1991, "Social & Economic Policies & Older People Netherlands" National Report, (unpublished)

Ministry of Health and Welfare (Japan), 1990, *"Ten Year Strategy to Promote Health Care and Welfare for the Aged"*, Ministry of Health and Welfare, Tokyo

Ministry of Housing & Local Government, 1969, *"Sheltered Housing'*, Circular 82/69, HMSO, London

Murray, B , 1993, *"Housing Elderly People in England & The Netherlands'*, University of Westminster, London

Murray, B , 1993 "Vienna 1900/1939 A City of Hope & Despair", Study Visit Paper, (unpublished)

Murray, B , 1994, "Social Housing in Budapest", Study Visit Paper, (unpublished)

Murray, B , 1994, *"Homes for Dutch Heroes at the Wallace Collection"* University of Westminster, London

Priemus, H , 1990, The Netherlands, in *"The International Handbook of Housing Policies & Practices"*, W Vliet, ed , Steen Wood Press, New York

Rolfe, et al , 1993, *'Age File 93"*, Anchor Housing Trust, London

Tokyo Metropolitan Government, 1991, *'The Tokyo Metropolitan Housing Master Plan"*, Tokyo City Council, Tokyo

34

AGING WELL

European Health Programme for Older People

Sally Greengross

1268, London Road
London SW16 4ER
United Kingdom

Good health in later life is a key factor in promoting individual wellbeing and personal growth. An improvement in the health of the older population is essential to enable older people to continue to contribute socially and economically—maximising that contribution and minimising calls on health services can both help offset the escalating costs associated with an aging population. A major priority now in Europe is to create a better balance between acute and chronic health care provision, self-care and the prevention of disease in later life. The World Health Organisation (WHO) Expert Committee report *Health of the Elderly* (1989), and its Regional Office for Europe adopted Target 6 on 'Healthy Aging' from *The Future of Health in Europe;* both support primary preventive efforts and greater efforts at case-finding for the elderly. Article 129 of the Treaty on European Union provides for new competence on public health. The European Commission's Communication on the framework for action in the field of public health *(Com (93) 55)* identifies the aging of the population as a major challenge for Europe. There is a clear need now for a European initiative promoting practical programmes and the exchange of good practice to encourage preventive measures for the health of older people (A European Community..., 1993)

Within this framework, Age Concern England is implementing a three year health promotion programme aimed at older people, *Aging Well UK,* with targets in line with the Government's 'Health of the Nation' (The Health of the Nation..., 1992) strategy and as a practical response to the Government's commitment to Community Care (Caring for People..., 1989). The programme is a multi-sector initiative, supported by the UK Department of Health and other public and corporate sector partners, and initially involves a network of nine pilot projects which are recruiting and managing 'senior health mentors'—older volunteers professionally trained to encourage and support their peers in health promotion activities. The WHO Regional Office for Europe has expressed interest in the development of the programme as a model which might be adapted in central and eastern European countries, and the *Aging Well Europe* project now forms a key part of Eurolink Age's programme at European Union level, including new national steering groups and 'twinning' and exchange opportunities between projects in different countries.

The programme was launched in 1992 in London during the UK Presidency of the European Union. It aims to achieve measurable improvements in mortality and morbidity rates and thus levels of life expectancy. Aging Well Europe comprises the development of pan-European health programmes, together with development of local, regional and national programmes as part of a European network. Aging Well UK is the model programme of Aging Well Europe and thus the most developed national programme of the network.

AGING WELL UK

Aging Well UK is also a practical response to the Health of the Nation strategy and Community Care changes. It is a partnership between voluntary, statutory and corporate sectors, also supported by the Health Education Authority.

The programme comprises:

Training Older Volunteers is four-day core training with additional modules, drawing from North American models, e.g. Santa Monica Peer Health Counselling project, and focusing on attitudes and expectations for health and old age, communication skills, community resources and record keeping. The training package is being revised in response to volunteers' needs.

Senior Health Mentoring (Drury, 1992) which provides better rapport with peers, offers positive role models, ensures closer contact with and a better understanding of local community and bring health promotion into informal settings (e.g. day centres, social clubs or even people's own homes).

Innovative Projects. Nine pilot projects are being funded to test the concept of Senior Health Mentoring, including participating projects experienced in recruiting and managing volunteers. A variety activities is being tested—discussion groups, health courses, one-to-one counselling, healthy eating and exercise initiatives—in a wide range of settings—clinics, leisure centres, health shops, clubs and outreach into people's homes.

Building Healthy Alliances which reflects the Ottawa Charter (The Ottawa Charter, 1986) definition of health promotion being everyone's responsibility. Healthy alliances are central to the Health of the Nation strategy, and sharing energy and expertise increases effectiveness. Aging Well UK is, most importantly, a partnership with older people themselves.

Encouraging Healthy Lifestyles/Enabling Independent Living. The programme encourages positive health, not just disease prevention, and supports people to get the best out of existing services and cope better with disability and illness.

Evaluating Action is a key part of the programme, measuring the process and outcome. It will focus on one or more of the Health of the Nation Target Areas (coronary heart disease and stroke; cancers; accidents; mental health). Local projects are supported by advisory groups of health professionals, community workers and older people with a national advisory committee overseeing the initiative.

The Projects

The Aging Well projects are spread over Stoke-on-Trent (the Beth Johnson Foundation's 'Senior Health Outreach Project'), Devon ('Fair Exchange Project'), Hereford and Worcester ('Ageing Well'); Liverpool ('Listening Support Scheme'); Macclesfield ('Ageing Well'); Northcumberland ('Coronary Heart Disease and Stroke Project'); South Glamorgan ('Senior Health Support Project'); Wakefield ('Beating Blood Pressure'); Warwickshire ('Healthy Ageing').

Aging Well Liverpool. This is one example of an Aging Well UK project, each one being different. It targets mental health, coronary heart disease and stroke. Liverpool has some of the worst health statistics in the UK and a team of twenty senior health mentors operate in resource centres, homes and health centres, with activities including one-to-one and self-help groups, Look After Yourself courses, relaxation classes and reminiscence and art therapy. It is estimated that 1,000 older people each year will benefit directly from this project. Some referrals to the scheme come from GPs and the project is supported by an advisory group which includes senior managers and practitioners from health and social welfare agencies in the city. A wide range of healthy alliances is being made with primary health care teams and allied professionals. Other links are being forged with providers of education and leisure facilities. Training advice on project development, evaluation and other resources are provided by the national Aging Well UK team. The Liverpool project has proved to be so successful that a second group of volunteers is soon to be trained.

The UK programme puts the slogan "Adding Years to Life - Adding Life to Years" into practical effect.

AGING WELL EUROPE

Since the launch conference other countries have also been working with Eurolink Age to develop national Aging Well programmes. These are at a key stage of development in four other countries—France, Ireland, Italy and The Netherlands—who are all working on identifying and bringing together existing older people's health projects as part of a national network, and stimulating the development of new projects. Their national networks are:

France: 'Bien Vieillir' working group set up and currently examining a new project, an educational board game looking at older people's health issues, created by 'Association Force 3' (Proceedings of Eurolink, 1994).

Ireland: The National Council for the Elderly has older people's health issues as a core area of responsibility and, out of that commitment, is now beginning discussion with potential national partners about the co-ordination of a new Aging Well group (Proceedings of Eurolink, 1994).

Italy: AUSER Nazionale, through their own national network of regional and local organisations, is now planning an ambitious new network of Aging Well projects ; 'Invecchiare Bene Italia' (Proceedings of Eurolink, 1994).

The Netherlands: A new national Aging Well group has just been formed and met for the first time, co-ordinated by the Nederlands Instituut voor Zorg en Welzijn.

Aging Well UK provides a model and a valuable resource for these new programmes but each programme is responsive to national needs and priorities. As national programmes develop they also provide a rich resource for one another and for those countries still planning programmes; this resource is exploited by providing funds to enable exchange and training visits between them (Proceedings of Eurolink, 1994).

Regular European network meetings (twice a year) serve to bring together national programmes and other interested organisations, and to discuss and inform relevant policy at EU level (next meeting is in Manchester 6-8 October). Eurolink Age is also producing a comprehensive Aging Well Europe directory, and a European monitoring group is being formed to establish agreed and measurable targets, and to review progress.

BEYOND EUROPE OF THE TWELVE

Eurolink Age and Age Concern England are also currently working closely with the World Health Organisation regional office for Europe on introducing the Aging Well initiative to four countries in Eastern Europe where there is significant interest - Ukraine,

Bulgaria, Slovenia and Russia. It is hoped to bring a group from these countries to the UK in the near future.

REFERENCES

"A European Community Health Policy for Older People", 1993, Eurolink Age, London

"Caring for People – Community Care in the Next Decade and Beyond", White Paper, Department of Health and Department of Social Security

Com (93) 55, Proposal for a European Parliament and Council Decision adopting a programme of Community Action on health promotion, information, education, and training, Com (94) 202 final

Drury, Michael Sir, 1992, The UK Ageing Well Programme, Paper to UK EU Presidency Conference, Ageing Well – A Call to European Action, Eurolink Age, London

"Health of the Elderly", 1989, World Health Organization, p 84

"Proceedings of Eurolink Age, Ageing Well Europe Third Network Meeting", Eurolink Age, London

"The Health of the Nation A Strategy for Health in England", 1992, Department of of Health, HMSO, London

"The Ottawa Charter", 1986, World Health Organization

35

RETIREMENT PREPARATION IN SUBJECTS OF WORKING AGE

Fiorella Marcellini and Norma Barbini

I.N.R.C.A., Social Economic Ctr.
Via Vanvitelli 18
60100 Ancona
01 Italy

INTRODUCTION

The need to offer the elderly a better inclusion in society after retirement is felt at many different levels: by the old people themselves, in the health care services, in economic programmes and in a new ethical concept of old age (Scortegagna, 1991). Suffering, widely felt experienced by the elderly and expressed as a feeling of marginalization, is an evident sign of poor intervention on their behalf (Marcellini et al., 1989; Anderson and Weber, 1993).

With this in mind, preparation for old age is a necessity, but in order to be really efficacious, it must become an educational process which begins already when young, to then develop in the organization of a productive life and within the local political system (Florenzano, 1991). Retirement represents the passage from a productive life to a state of professional inactivity, which often determines a change of role and prestige, especially in men, often linked with a lower standard of living. Retirement is associated with "entering" old age; this new condition of living, can also be seen as a concomitant cause of illness or disability (Giunta, 1993).

Thus, retirement preparation can play an important role in preventing certain risk factors for physical and mental health and the loss of self-sufficiency (Campione, 1990).

AIMS OF THE PROJECT

In our present society, this phase often coincides with a decrease in social contacts. The "working world" usually provides a large part of an individual's relationships and interests. Our project, carried out by I.N.R.C.A., the Italian National Research Institute on Aging, and 50&Plus-Fenacom, a self-help association, aimed at developing one of the sub-themes of the third network, (Solidarity between Generations through Training and Education) proposed by the Commission of the European Communities within the European Year of Elderly People and Solidarity between Generations (1993) (Commission of the European Communities, 1993).

Preparation for Aging. Edited by E. Heikkinen *et al.*
Plenum Press, New York, 1995

The project is called "Preparation for retirement through older persons".

We found the theme of solidarity among the generations, expressed through education and training, most interesting. Generations must not be conceived of as only young generations: the elderly person has the ability and possibility of being useful even towards adults—in the case of our project towards those approaching retirement (Anderson and Weber, 1993).

The novel approach of this project is therefore characterized by supporting the condition of old age, enabling the elderly to find within themselves and express their numerous capacities. This idea is based more on the aspect of asking than giving, on the discovery of one's abilities rather than being passive. Once retired, many elderly people feel that their "free time" is "wasted time" and don't know what to do or how to spend their day. They often have to discover the value of free time, discover themselves and their potentialities again, and also, as a consequence, discover "a new life style" (Glamser, 1981).

This discovery can be carried out through social commitments, for example voluntary work, and also helping those who are approaching the delicate passage from active life to retirement. In Italy this stage of life is often a traumatic experience, especially for men for whom this means the loss of an active role, and who cannot, at first, visualize the many possibilities that "free time" can offer them. As a consequence, the need of a correct preparation for retirement, through the direct experience of others who have already lived this passage, seems quite obvious (Marcellini et al., 1987; Marcellini et al., 1989).

Two important results are thus obtained:

1. the discovery of an active role in life for the pensioners involved in helping those about to retire; and
2. seeing preparation for retirement as prevention.

Preparation for retirement, meant as educational processes, is well carried out in Italy by the University of the Third Age—UniTre; but in the field of research it is a new theme (Tondi et al., 1992).

For this reason the project on which we are working can be used "experimentally", to lay the foundations for further work in this direction. This is another very important point, another goal to be met. In reality the project, after several meetings by the work group, is now in the operative phase.

The aims are:

1. to find out the problems and needs of those already retired, and the expectations of those about to retire, in order to develop adequate activities which answer their needs;
2. to select a certain number of people who could participate actively in the work involved in retirement preparation. This was carried out with the help of a psychologist who, on the basis of clear and simple biographical notes, could then individually obtain further explanations and motivations. The notes prepared for each participant aimed at identifying those who, due to their past experience, could be recruited as potential group leaders for future activities. In this way future trainers can be selected; and
3. to make contacts with other European Member Countries, in order to collect material, experience, and publications on how other European countries face and deal with retirement preparation (Lansley and Pearson, 1989).

It is extremely important to compare ones own experience with that of others. In this way progress can be made and other peoples' errors avoided.

The material collected could be used as the basis for a publication on this subject.

Letters and messages have been sent to establish contacts with similar projects, already being carried out or planned in other European countries, so as to form an operational and effective network. There is still much work to be done to achieve real European "cohesion", though some solid foundations have been laid. We believe that this project will make an important contribution towards this goal; but in order to be effective, it must be a "common" effort.

QUESTIONNAIRE SURVEY

We have carried out a survey with a questionnaire to achieve the main aim. We analyzed a sample of 582 subjects (M and F) of retired and future pensioners. The considered variables are: sex, age, area of residence, level of education, with whom they live, and work economic sector.

Other important items with multiple-choice answers concerned :

- problems about retirement;
- degree of consent to retirement preparation activities (courses, booklets, meetings etc.), and any ideas the elderly had about them;
- degree of consent about the following issues in retirement preparation activities: care of themselves and their wellbeing, social security- even through local boards and associations, management of their finances and patrimony, interest in developing their own creativity (writing, painting....), discovering the importance of free time, travelling, being with friends and being useful to others.

The questionnaire was for both the retired and future pensioners, asking them what problems they had or think they will have with retirement. Their experience must be used to organize further work, such as, a guide to retirement, seminars or courses on specific themes. The questionnaire, of a sociological kind, made it possible to compare the problems between those already retired and those not yet retired. This can establish a basis for a deeper knowledge of their problems, thus enabling the preparation in advance of retirement preparation programmes by taking advantage of, and using, the life experiences of individuals.

We feel it is important to know the degree of interest in projects which prepare the elderly for retirement, and what themes should be discussed in future courses/up-dating schemes.

REFERENCES

Anderson, C E , and Weber J A , 1993, Pre-retirement planning and perceptions of satisfaction among retirees, *Educ Geron* , 19(5) 397-406

Campione, W A , 1990, Predicting participation in retirement preparation programs, *J Gerontol* , 45·521-531

Commission of the European Communities, 1993, *"European Year of the Elderly and Solidarity Between Generations"*, Europe Sociale, Luxembourg, 1.

Florenzano, F , 1991, Elderly teaching, new and old methodologies between culture motivation and ageing problems, *Geron* , 1·49-57

Giunta,S , 1993, Health advocacy and elderly health advocacy, *Giornale di Gerontologia*, 41·367-369

Glamser, F D , 1981, The impact of pre-retirement programs on the retirement experience, *J Gerontol* , 36·244-250

Lansley, J , and Pearson M , 1989, *"Preparation for Retirement in the Member States of the European Community"*, Commission of the European Communities, Bruxelles-Luxemburg

Marcellini, F , Pavan, R , and Ulisse, M., 1989, *"The Elderly Condition"*, Idelson, Napoli

Marcellini, F , Piscitelli S , and Agostinelli E , 1987, *"The Communicative Processes Between Local Bodies and the Elderly",* Comune di Ancona

Scortegagna, R , 1991, Preparation for Ageing, *Med Geriatr*, XXIII 463-475

Tondi, L , Salvador, L , Zanella, A , and Cavedon, L , 1992, Preparation Course for Retirement, *Med Geriatr*, XXIV 45-49

INDEX

"A reflective community", 29
Absenteeism, 168
Age
  adolescence, 192
  birthday, 12
  biological, 111
  cohort, 206
  elderly, 145, 199
  false consciousness, 15
  golden, 51
  old, 140
  older adults, 87
  social construction of, 10
  third, 220
Aging,
  activity theory of, 113
  ELSA study, 125
  extrinsic factors, 105
  healthy, 105
  intrinsic factors, 105
  normal, 106
  preparation for, 195
  productive, 219
  research of, 110
  theories of, 105
  risk factors of, 106
Apologetic euphemism, 10,

Baroreflex, 169
Biological family as the ideal model, 176
Birthday age calibration, 13
Blood pressure, 168

Care
  collective, 176
  community, 285
  health care professionals, 246
  institutions for older people, 281
  primary care, 126
  primary health care, 145
  secondary care, 126
  view of the patient, 132
Changing values, 79

Chronic condition, 153
Citizen's Commissions, 30
Clinical work, 143
Cognitive performance, 108
Cohort comparison study, 199
Cohort effect, 112
Collective care of old people, 176
Collectivism, 191
Community care, 285
Compression of morbidity, 113, 123
Concept of home, 279
Concept of housing as shelter, 279
Contact frequency, 194
Course attendance, 74
Creativity, 68
Cross-cultural, 109, 116, 191
Cross-national, 116
Cultural changes in retirement, 181
Cultural codes, 175

Decision of eligibility, 224
Demographic patterns and attitudes, 281
Demographic transition, 121
Desire for personal development, 75
Differences between gender, 181
Disability, 122, 151
Disability-free life expectancy, 112, 124
Disablement, 165
Domains of life, 158

Early disability pension, 223
Early disability pension application, 225
Early retirement, 206, 223, 233, 239
Economic trend, 232
Education,
  course attendance, 73
  gerontological, 73
  lifelong, 73, 78, 85
  meaning of, 63
  opportunity for, 66, 77
  participation in, 64, 73
  process of, 289

Educational gerontology, 73
Educational process, 289
Employment pension, 229
Employment situation, 226,
European Longitudinal Study of Aging (ELSA),
    125

False consciousness as to aging, 15
Functional capacity
    activity of daily living, 200
    autonomy, 87
    disability, 110, 122, 151, 158
    disability free life expectancy, 112, 124
    functional ability, 133, 200
    functional disability, 158
    functional status, 110, 126, 140
    generalized, 111
    instrumental activity of daily living, 110
    manage daily life, 196
    muscular strength, 217
    physical activity of daily living, 110
    physical functioning, 148
    physical reservers, 218
    psychological status and well-being, 133
    psychosocial dimensions, 148,
    work ability, 217
    working capacity, 205, 233

Generation
    contact frequency, 194
    family as an ideal model, 176
    generation gap, 212,
    grandchild, 191
    grandparent, 191
    intergenerational norms, 192
    relations between generations, 175
    violence against parents, 177
Gerontology
    gerontological psychology, 195
    The Finnish Centre for Interdisciplinary Geron-
        tology, 105
    educational gerontology, 73
    field of gerontology, 195
Global challenge, 27
Golden Age, 51
Government policy, 156
Grandchild-grandparent relations, 191

Health
    chronic conditions, 153
    compression of morbidity, 113, 123
    health and physical function, 127
    health economics, 132
    health of the nation, 285
    health promotion, 114, 219
    health research, 143
    illness, 151

Health (cont)
    mental health, 126
    mortality, 129
    self-perceived health, 158
    self-rated health, 126, 140, 200
    Senior Health Mentoring, 286
    social definition of, 110
Health and physical function, 127
Health care professionals, 246
Health-related quality of life, 125, 139, 152
Healthy life expectancy, 123
Heart rate, 170

Identity, 63, 67
Illness, 151
Individualism, 191
Initiation, 41
Institution for older people, 281
Interdisciplinary, 247
Intergenerational norms, 192
Intervention strategies, 115

Learning
    learning experiences, 63
    learning city, 23
    life-long learning, 73, 85
    wish to learn, 75
LEIPAD Questionnaire, 145
Leisure, 79
Level of activities, 74
Life-expectancy
    disability-free, 112, 124
    healthy, 123
Life-course
    life-course perspective, 114
    life-history, 63
    life-review, 80
    life-revision, 79
    life situation, 233
    life-story, 63
Life expectancy, 106, 121
Life experiences, 79, 84
Life satisfaction, 80, 113, 126, 200
Life-style
    alcohol consumption, 200
    drinking, 108
    life satisfaction, 80, 113, 126, 200
    living habit, 200
    physical exercise, 200
    smoking, 112, 200
Living standard, 134
Longitudinal Aging Study Amsterdam, 157

Manage daily life, 196
Menopause, 197
Mental health, 126
Midlife, 195
Modernity, 67

"Mosaic" laws, 177
Motive for participation, 75
Moving, 279
Multi-dimensional assessment, 140
Muscular strength, 217

"New" public health approach, 87
New social identity, 197
Non-clinical data, 134

Older volunteers, 286
Older worker, 167
Ottawa Charter, 286
Outcome measure, 129

Participation in courses, 74
Partnership, 286
Pension
    early disability pension, 223
    employment pension, 229
    decision of eligibility, 224
    pension benefits, 235
    pension expenditure, 230
    pension refusal, 225
    pension scheme, 229
    pensioners, 210
Period effect, 203
Physical activity
    and psychological well-being, 133
    mobility, 110
    muscular strength, 217
    walking speed, 108
Physical functioning, 148
Physical reserves, 218
Pleasure of aging, 198
Postliminal phase, 47
Preparation for retirement, 55, 290
Prevention, 114
Preventive strategies, 126
Primary care, 126
Primary health care, 145
Process of retirement, 240
Productive aging, 219
Psychometric characteristics, 125
Psychosocial dimensions, 148

Quality of life,
    assessment of, 133
    definition of, 129
    health-related quality of life, 125, 139, 152
    indicators of, 141
    qualitative research of, 116, 143
    psychometric characteristics of, 125
Quantitative indicators, 143

Rate of aging, 165
Regret, 80
Relations between the generations, 175

Religious ideology, 176
Responsibility of the state, 178
Retirement
    cultural changes in retirement, 181
    early retirement, 206, 223, 233, 239
    preparation for retirement, 55, 290
    preparation programmes, 291
    process of retirement, 240
    retirement ability, 220
    retirement attitudes, 209
    retirement decision, 224
    the rite of passage, 40
    time preceding retirement, 195
    three ritual phases, 41
Rotterdam Symptom Checklist, 146

Screening, 115
Secondary care, 126
Security, 279
Self-confidence, 208
Self-report, 145
Senior Citizen's Universities, 55
Senior health mentoring, 286
Sheltered housing, 281
Significant others, 69
Social construct of old age, 10
Social constructionism, 142
Social contact, 75
Social housing, 281
Social inequality, 124
Social participation, 200
Social risk, 235
Social security, 233
Sociodemographic factors, 239
Socio-economic status, 199
Standard of accommodation, 280
Stigmatization, 10
Stress, 165
Subjective experience, 107
Subjective grounds, 239

Temporal dimensions, 17
Third age
    students, 30
    universities, 9, 27, 195
Ties to tradition, 175
Time preceding retirement, 195
Time structures, 24

University of Third Age
    growth and expansion, 33
    programmes, 32
    Senior Citizen's Universities, 55
    students, 30
    U3A, 9, 27, 30
    values and aims, 33

Vanguard of the Third Age, 9

View of the patient, 132

Welfare state, 179, 233
Whole population epidemiology, 101
Work
    ability, 217
    career development, 207
    demands, 218
    employment situation, 226
    job satisfaction, 206

Work (*cont.*)
    job stress, 205
    unemployment, 225, 236
    work behaviour, 207
    work disability, 239
    work-life, 234
    work motivation, 206
    work-site interventions, 211
    work stress, 165
    working capacity, 205, 233